Systemic and Systematic Project Management

Systemic and Systematic Project Management

Joseph Eli Kasser

CRC Press
Taylor & Francis Group
Boca Raton London New York

CRC Press is an imprint of the
Taylor & Francis Group, an **informa** business

CRC Press
Taylor & Francis Group
6000 Broken Sound Parkway NW, Suite 300
Boca Raton, FL 33487-2742

© 2020 by Taylor & Francis Group, LLC
CRC Press is an imprint of Taylor & Francis Group, an Informa business

No claim to original U.S. Government works

Printed on acid-free paper

International Standard Book Number-13: 978-0-367-07540-8 (Hardback)

This book contains information obtained from authentic and highly regarded sources. Reasonable efforts have been made to publish reliable data and information, but the author and publisher cannot assume responsibility for the validity of all materials or the consequences of their use. The authors and publishers have attempted to trace the copyright holders of all material reproduced in this publication and apologize to copyright holders if permission to publish in this form has not been obtained. If any copyright material has not been acknowledged, please write and let us know so we may rectify in any future reprint.

Except as permitted under U.S. Copyright Law, no part of this book may be reprinted, reproduced, transmitted, or utilized in any form by any electronic, mechanical, or other means, now known or hereafter invented, including photocopying, microfilming, and recording, or in any information storage or retrieval system, without written permission from the publishers.

For permission to photocopy or use material electronically from this work, please access www.copyright.com (http://www.copyright.com/) or contact the Copyright Clearance Center, Inc. (CCC), 222 Rosewood Drive, Danvers, MA 01923, 978-750-8400. CCC is a not-for-profit organization that provides licenses and registration for a variety of users. For organizations that have been granted a photocopy license by the CCC, a separate system of payment has been arranged.

Trademark Notice: Product or corporate names may be trademarks or registered trademarks, and are used only for identification and explanation without intent to infringe.

Library of Congress Cataloging-in-Publication Data

Names: Kasser, Joseph Eli, author.
Title: Systemic and systematic project management / by Joseph Eli Kasser.
Description: Boca Raton: Taylor & Francis, a CRC title, part of the Taylor & Francis imprint, a member of the Taylor & Francis Group, the academic division of T&F Informa, plc, 2019.
Identifiers: LCCN 2019012486 | ISBN 9780367075408 (hardback : acid-free paper) | ISBN 9780429021282 (e-book)
Subjects: LCSH: Project management.
Classification: LCC HD69.P75 K373 2019 | DDC 658.4/04—dc23
LC record available at https://lccn.loc.gov/2019012486

Visit the Taylor & Francis Web site at
http://www.taylorandfrancis.com

and the CRC Press Web site at
http://www.crcpress.com

To my wife Lily, always caring, loving and supportive
To Janey Chng

Contents

List of Figures xix
List of Tables xxiii
Preface xxvii
Acknowledgements xxix
Author xxxi

Chapter 1 Introduction 1
           1.1   The Contents of This Book 2
           1.2   How to Read and Use This Book 7
           References 8

Chapter 2 The Seven Interdependent P's of a Project 9
           2.1   People 10
                   2.1.1   Thinking 10
                          2.1.1.1   Critical Thinking 11
                          2.1.1.2   Systems Thinking 12
                          2.1.1.3   The Two Distinct Types of Systems
                                        Thinking 12
                          2.1.1.4   Beyond Systems Thinking 14
                     2.1.2   Categories of People 18
                          2.1.2.1   The Five Ways of Approaching Problems 18
                          2.1.2.2   The Five Levels of Critical Thinking 18
                   2.1.3   Competence 19
                          2.1.3.1   Using the CMM 21
                 2.1.4   Motivation 22
                          2.1.4.1   Externals and Internals 23
                          2.1.4.2   McGregor's Theory X and Theory Y 23
                          2.1.4.3   Maslow's Hierarchical Classification
                                        of Needs 24
                          2.1.4.4   The Herzberg Motivation-Hygiene Theory ... 25
                          2.1.4.5   Recognition and Rewards 25
                          2.1.4.6   Expectations Affect Performance 26
                          2.1.4.7   Discussion 26
                 2.1.5   Meetings 27
                 2.1.6   Communications 28
                          2.1.6.1   Formal Written Communications 29
                          2.1.6.2   Formal Verbal Communications 29
                 2.1.7   Miscommunications 34
                          2.1.7.1   Cultural Differences in Perception 35
                          2.1.7.2   Emotion 35

		2.1.7.3	Language 35
		2.1.7.4	Signal-to-Noise Ratio 37
	2.1.8	Mitigating the Effect of and Overcoming Miscommunications 37
		2.1.8.1	Questions 37
		2.1.8.2	Answers 41
		2.1.8.3	Active Listening 41
2.2	Politics 42
	2.2.1	Positive Use of Politics 44
	2.2.2	Negative Use of Politics 44
	2.2.3	Seven Steps That a Project Manager Can Take to Become Politically Astute 45
2.3	Prevention 46
2.4	Problems 47
2.5	Processes 47
	2.5.1	The Process Timeline 48
	2.5.2	The Three Streams of Activities 48
2.6	Profit 49
2.7	Products 50
2.8	Summary 50
References 50

Chapter 3 Perceptions of Problem-Solving 55

3.1	Big Picture 55
	3.1.1	Assumptions Underlying Formal Problem-Solving 55
	3.1.2	Selected Myths of Problem-Solving 56
		3.1.2.1	All Problems Can Be Solved 56
		3.1.2.2	All Problems Have a Single Correct Solution 57
		3.1.2.3	The Problem-Solving Process is a Linear Time-Ordered Sequence 58
		3.1.2.4	One Problem-Solving Approach Can Solve All Problems 59
3.2	Quantitative 59
	3.2.1	Components of Problems 59
3.3	Structural 61
	3.3.1	Classifications of Problems 61
	3.3.2	The Level of Difficulty of Problems 61
3.4	Continuum 62
	3.4.1	Problems and Symptoms 62
	3.4.2	The Difference between the Quality of the Decision and the Quality of the Outcome 63
	3.4.3	The Different Decision Outcomes 63
		3.4.3.1	Sources of Unanticipated Consequences or Outcomes of Decisions 65

	3.4.4	Research and Intervention Problems	67
		3.4.4.1 Research Problems	67
		3.4.4.2 Intervention Problems	69
	3.4.5	The Different Categories of Problems	70
		3.4.5.1 Well-Structured Problems	70
		3.4.5.2 Ill-Structured Problems	71
		3.4.5.3 Wicked Problems	71
	3.4.6	The Different Domains of a Problem	71
	3.4.7	The Technological System Implementation Continuum	72
3.5	Functional		72
	3.5.1	Decision-Making	72
		3.5.1.1 Decision-Making Tools	73
3.6	Operational		73
	3.6.1	The Simple Problem-Solving Process	74
	3.6.2	The Extended Problem-Solving Process	74
3.7	Scientific		76
	3.7.1	A Problem Formulation Template	76
3.8	Complexity		78
	3.8.1	Continuum	79
		3.8.1.1 The Complexity Dichotomy	79
		3.8.1.2 Various Definitions of Complexity	79
		3.8.1.3 Partitioning Complexity	80
	3.8.2	Temporal	81
	3.8.3	Scientific	81
		3.8.3.1 Resolving the Complexity Dichotomy	82
		3.8.3.2 The Problem Classification Framework	84
3.9	Remedying Well-Structured Problems		84
	3.9.1	The Two-Part SDP	85
	3.9.2	The Multiple-Iteration Problem-Solving Process	87
		3.9.2.1 The First Problem-Solving Processes	88
		3.9.2.2 The Second Problem-Solving Process	89
	3.9.3	A New Product Development Process	90
3.10	Remedying Ill-Structured Problems		91
3.11	Remedying Complex Problems		91
	3.11.1	Remedying Well-Structured Complex Problems	91
	3.11.2	Remedying Ill-Structured Complex Problems	91
	3.11.3	Remedying Wicked Problems	93
3.12	Summary		94
References			94

Chapter 4 Management: General and Project Management 97

4.1	The Purpose of Management		97
4.2	Perceptions of Management		97
	4.2.1	Big Picture	97

		4.2.2	Operational	98
		4.2.3	Functional	99
		4.2.4	Continuum	99
			4.2.4.1 The Differences between a Manager and a Leader	100
			4.2.4.2 The Difference in Competence	100
			4.2.4.3 The Three Types of Management	100
			4.2.4.4 The Difference between the Requirements for the Project and the Requirements for the Product Produced by the Project	100
		4.2.5	Quantitative	101
		4.2.6	Temporal	101
	4.3	General Management		101
	4.4	Project Management		101
		4.4.1	Projects	102
			4.4.1.1 Purpose	102
			4.4.1.2 Activities	102
			4.4.1.3 Funding	103
			4.4.1.4 Milestones	103
			4.4.1.5 Need	103
			4.4.1.6 Priority	103
			4.4.1.7 Sponsor	103
			4.4.1.8 Stakeholders	104
			4.4.1.9 Customers	104
			4.4.1.10 Timeline	104
			4.4.1.11 The ROI	104
			4.4.1.12 Outcomes	105
			4.4.1.13 Risks	106
		4.4.2	The Triple and Quadruple Constraints of Project Management	106
		4.4.3	Project Organization	107
			4.4.3.1 Project Hierarchy	107
			4.4.3.2 Customer–Project Interfaces	108
	4.5	Taking over a Project		109
		4.5.1	Gain an Understanding of the Situation	109
		4.5.2	Decide If Changes Need to Be Made	110
		4.5.3	Prepare a Change Management Plan	110
		4.5.4	Present the Change Management Plan to the Important Stakeholders	111
		4.5.5	Make the Changes	111
	4.6	Research Projects		111
		4.6.1	Types of Research Projects	111
		4.6.2	Sponsor Management	112
		4.6.3	The Need for a Pollinator	112

Contents

		4.7	Summary	113
		References		113
Chapter 5	Project Planning			115
	5.1	The Project Planning Paradox		115
	5.2	Project Planning and Managing Tools		116
		5.2.1	The PAM Chart	116
			5.2.1.1 Creating a PAM Chart	117
		5.2.2	GANTT Charts	118
			5.2.2.1 Creating a Gantt Chart	120
		5.2.3	PERT Charts	121
			5.2.3.1 Creating a PERT Chart	123
		5.2.4	Timelines	125
			5.2.4.1 Creating a Timeline	125
		5.2.5	The GANTT–PERT Cross-Check	126
		5.2.6	The WP	127
			5.2.6.1 The Benefits of Using WPs	128
	5.3	The Project Lifecycle		130
		5.3.1	The Four-State Project Lifecycle	130
			5.3.1.1 The Project Initiation State	130
			5.3.1.2 The Project Planning State	131
			5.3.1.3 The Project Performance State	131
			5.3.1.4 The Project Closeout State	132
	5.4	The SLC		133
		5.4.1	The States in the SLC	133
		5.4.2	Milestones	135
			5.4.2.1 Informal Reviews	137
			5.4.2.2 Formal Reviews	137
		5.4.3	The Cataract Methodology	138
			5.4.3.1 Build Zero	139
	5.5	The Project Plan		142
	5.6	Generic Planning		142
		5.6.1	Build a Little Test a Little	145
	5.7	Specific Planning		147
		5.7.1	The Process for Creating a Specific PP	147
	5.8	The Planning Process		147
		5.8.1	The Numbering System	148
		5.8.2	Process Architecting	149
			5.8.2.1 The WBS	151
			5.8.2.2 Common Errors in Developing a WBS	152
	5.9	The Systems Approach to Project Planning		152
	5.10	Using 'Prevention' to Lower Project Completion Risk		155
	5.11	The Presentation Exercise		155
		5.11.1	The Grading Criteria for the Presentation Exercise	156

		5.11.2	The Requirements for the Presentation Exercise 156
		5.11.3	The Presentation Exercise Project Planning State ... 158
		5.11.4	The Project Numbering System 158
			5.11.4.1 The Activity Categories............................. 158
			5.11.4.2 The Product Categories 158
			5.11.4.3 The Exercise Activities, Products and WPs.. 158
		5.11.5	The Presentation Planning State (WP 00000)......... 160
		5.11.6	The Presentation Requirements State (WP 10000) 161
		5.11.7	The Presentation Design State (WP 20000) 168
		5.11.8	The Presentation Construction State (WP 30000)... 168
		5.11.9	The Presentation Integration and Testing States (WP 40000)... 176
		5.11.10	The Presentation Delivery and Grading State (WP 5000) 185
		5.11.11	Achievements at this Point in Time......................... 185
	5.12	The Engaporean MCSSRP Exercise 185
		5.12.1	The Requirements for the Exercise.......................... 189
	5.13	Summary 192
	References 193

Chapter 6 Successful Project Staffing.. 195

	6.1	People 195
		6.1.1 Availability 195
		6.1.2 Competencies and Skills 195
		6.1.3 Compatibility... 196
		6.1.4 Permanent or Temporary... 196
		6.1.5 Costs 196
		6.1.6 Teams 197
		6.1.6.1 The Lifecycle of a Team 197
		6.1.6.2 Characteristics of Effective Teams........... 198
		6.1.6.3 Creating and Staffing Effective Teams..... 198
	6.2	The Systems Approach to Staffing a Team 200
		6.2.1 Breaking Out What Needs to Be Done into Specific Activities in the Three Streams of Work...201
		6.2.2 Determining the Necessary Skills and Competencies to Do the Activities............................ 201
		6.2.3 Partitioning the Activities into Jobs 201
		6.2.4 Staffing Each of the Job Positions............................202
	6.3	The Presentation Exercise 203
		6.3.1 Achievements at This Point in Time 209
	6.4	The Engaporean MCSSRP Exercise 209
		6.4.1 Requirements for the Exercise 209
	6.5	Summary 210
	References 210

Chapter 7   Successful Project Scheduling 211

- 7.1 Scheduling 211
- 7.2 Estimating the Correct Amount of Time for Activities 212
- 7.3 Levelling the Workload 213
- 7.4 The Critical Path 214
  - 7.4.1 Slack Time, Early and Late Finishes 216
  - 7.4.2 The Fallacy in the Use of Slack Time in Fixed Resource Situations 217
  - 7.4.3 Accuracies of the Estimated Schedules 218
- 7.5 The Theory of Constraints 219
- 7.6 The Critical Chain 219
  - 7.6.1 Use of the Critical Chain in Project Management ... 220
- 7.7 The Presentation Exercise 221
  - 7.7.1 Achievements at This Point in Time 226
- 7.8 The Engaporean MCSSRP Exercise 226
  - 7.8.1 The Requirements for the Exercise 226
- 7.9 Summary 227
- References 227

Chapter 8   Successful Project Cost Estimating 229

- 8.1 The Three Axes of Cost-Effective Projects 229
- 8.2 The Reason for Estimating Project Costs 231
- 8.3 Methods for Estimating Project Costs 231
  - 8.3.1 Conceptual Design to Cost 233
    - 8.3.1.1 Benefits of Using CDTC 233
    - 8.3.1.2 Identifying and Prioritizing the Needed Capability 234
    - 8.3.1.3 Benchmarking the Capability 234
    - 8.3.1.4 Determining the Target Price for the Product 234
    - 8.3.1.5 Estimating the Cost of the Product 235
    - 8.3.1.6 Adjusting the Cost Estimate 236
  - 8.3.2 Lifecycle Costing 236
  - 8.3.3 Cost as an Independent Variable 238
- 8.4 Accuracies of Cost Estimates 238
- 8.5 Categories of Costs 239
- 8.6 Ways of Controlling Costs 240
  - 8.6.1 Failing to Communicate the Vision 240
  - 8.6.2 Failing to Understand the Customer's Real Requirements 241
  - 8.6.3 Failing to Plan Ahead 241
  - 8.6.4 Failing to Control Changes 242
  - 8.6.5 Failing to Apply Lessons Learned from the Past 242
  - 8.6.6 Failing to Document the Reasons for Decisions 243

		8.6.7	Failing to Maximize Use of COTS	243
8.7	Types of Contracts			244
8.8	The Presentation Exercise			244
		8.8.1	Achievements at This Point in Time	246
8.9	The Engaporean MCSSRP Exercise			247
		8.9.1	The Requirements for the Exercise	247
8.10	Summary			248
References				248

Chapter 9 Successfully Adjusting Project Schedules and Costs 251

9.1	Shortening the Schedule		251
	9.1.1	Overtime	252
	9.1.2	Adding People	253
	9.1.3	Crashing a Project	253
	9.1.4	Improving Productivity	254
9.2	Reducing Costs		255
	9.2.1	Replacing Personnel Assigned to an Activity with Lower-Cost Personnel	255
	9.2.2	Increasing the Schedule	255
	9.2.3	Deleting Some of the Work without Changing the Requirements	256
	9.2.4	Reducing Materiel Costs	256
	9.2.5	Redesigning the Product	256
	9.2.6	Removing the Low-Priority Requirements	256
9.3	Cancelling a Project		256
9.4	The Widget System Project		257
	9.4.1	Changes Due to Delays	258
	9.4.2	Comments	260
	9.4.3	Lessons Learned	261
9.5	The Presentation Exercise		261
	9.5.1	Modifying the Staffing Level	261
	9.5.2	Adjusting the Schedule	263
	9.5.3	Adjusting the Costs	263
9.6	The Engaporean MCSSRP Exercise		265
	9.6.1	The Requirements for the Exercise	265
9.7	Summary		266
References			266

Chapter 10 An Introduction to Managing Risk and Uncertainty over the Project Lifecyle 267

10.1	Definitions of the Terminology		267
10.2	Risks and Opportunities		268
	10.2.1	The Risk Rectangle	268
		10.2.1.1 The Flaw in the Risk Rectangle	269

		10.3	Risk Management 269
			10.3.1 Selected Myths of Risk Management 270
			10.3.2 The Traditional Approach to Risk Management 270
			10.3.3 The Systems Approach to Risk Management 271
		10.4	Risks Based on the Availability of Technology 271
			10.4.1 The Technology Availability Window of Opportunity 273
		10.5	Risk Profiles 276
		10.6	Risk Mitigation or Risk Prevention 277
		10.7	Cascading Risks 277
		10.8	Contingencies and Contingency Plans 277
		10.9	The Engaporean MCSSRP Exercise 278
			10.9.1 The Requirements for the Exercise 278
		10.10	Summary 279
		References 279

Chapter 11 Successful Performance Monitoring and Controlling 281

		11.1	Detecting and Preventing Potential Project Overruns 282
		11.2	Ways to Detect Impending Project Failure 283
			11.2.1 The Top Ten Risk-Indicators 284
			11.2.2 The Six Risk-Indicators Most of the Respondents Disagreed 285
		11.3	Managing Changes in Project Scope 286
			11.3.1 Effect of Change on Project 286
			11.3.2 Change Request Processing 286
		11.4	EVA 288
			11.4.1 The Elements of EVA 288
			11.4.2 The Terminology of EVA 289
			11.4.2.1 Planning or Estimating Terminology 289
			11.4.2.2 Project Monitoring and Controlling Terminology 289
			11.4.2.3 Indices and Summary Terminology 289
			11.4.2.4 EVA Calculations 289
			11.4.3 Requirements for the Use of EVA in a Project 291
			11.4.4 Advantages and Disadvantages of EVA 292
			11.4.4.1 EVA Advantages 292
			11.4.4.2 EVA Disadvantages 292
			11.4.5 Examples of the Use of EVA 292
			11.4.5.1 The Master's Degree Project 292
			11.4.5.2 The Data Centre Upgrade Project 296
			11.4.6 The Systems Approach Perspective on EVA 299
		11.5	CRIP Charts 299
			11.5.1 The Five-Step CRIP Approach 300
			11.5.1.1 Step 1: Identify Categories for the Requirements 301

		11.5.1.2	Step 2: Quantify Each Category into Ranges 301
		11.5.1.3	Step 3: Categorize the Requirements........ 302
		11.5.1.4	Step 4: Place Each Requirement into a Range in Each Category 302
		11.5.1.5	Step 5: States of Implementation 302
	11.5.2	Populating and Using the CRIP Chart 302
	11.5.3	Advantages of the CRIP Approach 305
	11.5.4	Disadvantages of the CRIP Approach....................... 305
	11.5.5	Examples of Using CRIP Charts in Different Types of Projects 305
		11.5.5.1	The Ideal Project....................................... 309
		11.5.5.2	A Project with Requirements Creep 313
		11.5.5.3	The Challenged Project 318
		11.5.5.4	The 'Make Up Your Mind' Project 319
	11.5.6	Comments... 322
11.6	Traffic Light and ETL Charts.. 322
	11.6.1	Traffic Light Charts... 322
	11.6.2	ETL Charts... 323
		11.6.2.1	Creating an ETL Chart for Use in a Presentation 328
		11.6.2.2	Adding Even More Information 328
11.7	MBO 329
	11.7.1	MBO in the Planning State of a Project.................... 330
11.8	MBE 330
	11.8.1	The Key Ingredients in MBE 330
	11.8.2	Advantages and Disadvantages of MBE 331
	11.8.3	Using MBE... 331
11.9	The Engaporean MCSSRP Exercise 332
	11.9.1	The Requirements for the Exercise 332
11.10	Summary 332
References 332

Chapter 12 The Human Element 335

12.1	Time Management... 335
12.2	People Management... 336
	12.2.1	Rewards and Recognition... 336
		12.2.1.1	Performance Evaluations 336
12.3	Conflict 339
	12.3.1	Basis for Conflict... 340
	12.3.2	Response to Conflict.. 340
		12.3.2.1	Gaining an Understanding of the Situation 340
	12.3.3	Perceptions of Conflict 340

12.4	Leadership		341
	12.4.1	Types of Power and Influence	342
	12.4.2	Types of Authority	343
	12.4.3	Leadership Styles	343
		12.4.3.1 Persuading	343
		12.4.3.2 Telling	343
		12.4.3.3 Selling	344
		12.4.3.4 Participating	344
		12.4.3.5 Delegating	344
		12.4.3.6 Situational	345
12.5	Negotiation		346
	12.5.1	Negotiation Positions	346
	12.5.2	Negotiation Outcomes	346
	12.5.3	Negotiating Styles	347
	12.5.4	Negotiating Tips	347
12.6	Managing Stakeholders		348
12.7	The Multi-satellite Communications Switching System Replacement Project Exercise		350
	12.7.1	The Requirements for the Exercise	350
12.8	Summary		351
References			351

Chapter 13 Ethics 353

13.1	Personal Ethics		353
13.2	Organizational Ethics		354
13.3	Professional Ethics		354
13.4	Personal Integrity		354
13.5	The Ethical Dilemma		355
	13.5.1	The Issues	355
		13.5.1.1 The Law	355
		13.5.1.2 Your Motives	356
		13.5.1.3 The Company's Ethics Policy	356
		13.5.1.4 The Consequences of Your Actions	357
	13.5.2	The Approach to Solving the Ethical Problem	357
		13.5.2.1 Analyse the Situation	358
		13.5.2.2 Identify Appropriate Lessons Learned	358
		13.5.2.3 Develop Alternatives	358
		13.5.2.4 Determine the Probable Outcome of Each Alternative Decision	358
		13.5.2.5 Evaluate the Alternatives	359
		13.5.2.6 Taking Action	360
13.6	Lessons Learned		362
13.7	The Multi-satellite Communications Switching System Replacement Project Exercise		363

        13.7.1   The Requirements for the Exercise 363
  13.8   Summary 364
  References 364

Appendix 1: The Engaporian Multi-satellite Operations Control Centre Communications Switching System Replacement Project 367

Appendix 2: Change Management Events 369

Appendix 3: Presentation Guidelines and Requirements 371

Appendix 4: Staffing Information 373

Appendix 5: Resumes for the Session 6 Exercise 375

Author Index 393

Subject Index 397

List of Figures

Figure 2.1 The nine holistic thinking perspectives. 15
Figure 2.2 The problem-solving process as a causal loop. 47
Figure 2.3 The three streams of activities in a project. 48
Figure 3.1 The continuum of solutions. 57
Figure 3.2 The traditional simple problem-solving process. 58
Figure 3.3 Iteration of the problem-solving process across the SDP in an ideal project. 60
Figure 3.4 Problems, causes and effects (symptoms). 63
Figure 3.5 The scientific method. 68
Figure 3.6 The extended problem-solving process. (*Source:* © 2016 IEEE. Reprinted, with permission, from Kasser, Joseph Eli, and Yang-Yang Zhao. 2016c. Wicked problems: Wicked solutions. In *11th International Conference on System of Systems Engineering*, Kongsberg, Norway.) 75
Figure 3.7 Perceptions of objective complexity from the *Temporal* HTP. 81
Figure 3.8 A problem classification framework. 84
Figure 3.9 Activities in the context of the SDP and problem-solving. 85
Figure 3.10 The two-part multiple-iteration problem-solving process. (*Source:* © 2016 IEEE. Reprinted, with permission, from Kasser, Joseph Eli, and Yang-Yang Zhao. 2016b. Simplifying Solving Complex Problems. In *the 11th International Conference on System of Systems Engineering*, Kongsberg, Norway.) 86
Figure 3.11 Modified Hitchins' view of the problem-solving decision-making process. 86
Figure 3.12 A new product development variation of the multiple-iteration problem-solving process. (*Source:* © 2016 IEEE. Reprinted, with permission, from Kasser, Joseph Eli, and Yang-Yang Zhao. 2016c. Wicked Problems: Wicked Solutions. In *the 11th International Conference on System of Systems Engineering*, Kongsberg, Norway.) 90
Figure 4.1 The traditional triple constraints. 106
Figure 4.2 The systemic quadruple constraints. 107

Figure 5.1	The PAM chart.	117
Figure 5.2	A partial PAM network chart for making a cup of instant coffee.	118
Figure 5.3	The initial project schedule shown in a Gantt chart.	119
Figure 5.4	Aggregating activities in a Gantt chart.	121
Figure 5.5	The top-level schedule.	121
Figure 5.6	A part of a typical PERT chart.	122
Figure 5.7	The Gantt and PERT cross-check.	126
Figure 5.8	The project lifecycle.	130
Figure 5.9	The SDP: A planning view.	133
Figure 5.10	The waterfall view—problem-solving perspective (partial).	135
Figure 5.11	The configuration control view of waterfall.	141
Figure 5.12	The cataract methodology.	141
Figure 5.13	A traditional example of a WBS for a cup of instant coffee.	151
Figure 5.14	An alternative example of a WBS.	151
Figure 5.15	Planning—level of detail.	153
Figure 5.16	An example of planned prevention.	155
Figure 5.17	A partial WBS view of the WPs in the presentation requirements state (10000).	167
Figure 5.18	The PERT chart for the presentation design state (WP 20000).	176
Figure 5.19	The PERT chart for the presentation construction state (WP 30000).	185
Figure 5.20	A partial WBS view of the WPs in the presentation integration and test state (WP 40000).	189
Figure 5.21	The WBS view of the 50000 presentation delivery and grading state.	192
Figure 7.1	Corresponding initial schedule based on Tables 6.2 and 6.3.	212
Figure 7.2	Initial schedule estimate.	213
Figure 7.3	The project first estimated staffing profile.	213
Figure 7.4	The project initial and revised estimated staffing profiles for the first four WPs.	214

List of Figures

Figure 7.5	The project initial and revised estimated schedules for the first four WPs.	214
Figure 7.6	A partial PERT chart.	215
Figure 7.7	Highlighting the critical path.	216
Figure 7.8	The corresponding schedule for Table 7.3.	217
Figure 7.9	The critical chain.	219
Figure 7.10	The presentation exercise schedule for WP 00000.	222
Figure 7.11	The presentation exercise schedule for WP 10000.	222
Figure 7.12	The PERT chart for WP 20000.	223
Figure 7.13	The presentation exercise schedule for WP 20000.	223
Figure 7.14	The second revised schedule for WP 20000.	224
Figure 7.15	The PERT chart for WP 30000.	224
Figure 7.16	The presentation exercise schedule for WP 30000.	225
Figure 7.17	The presentation exercise schedule for WP 40000.	225
Figure 7.18	The presentation exercise schedule for WP 50000.	226
Figure 8.1	The generic cost of a process and its product.	230
Figure 8.2	The cost of a process that is exceeding estimated costs.	231
Figure 8.3	The *Generic/Temporal* costing approach (analogy method).	232
Figure 8.4	The effect of failing to communicate the vision.	240
Figure 8.5	The estimated costs for the session 1 presentation exercise (line chart).	246
Figure 8.6	The estimated costs for the session 1 presentation exercise (bar chart).	246
Figure 8.7	Graphical summaries of the estimated costs for the session 1 presentation exercise.	247
Figure 9.1	The widget project: original schedule.	257
Figure 9.2	The widget project: revised schedule.	258
Figure 9.3	The widget project: alternate view of the revised schedule.	258
Figure 9.4	The widget project: extended revised schedule.	259
Figure 9.5	The widget project: modified schedule with delays due to the unanticipated problems.	259
Figure 9.6	The widget project: traditional revised schedule.	260

Figure 9.7	The actual costs of the presentation project without Susan.	265
Figure 10.1	The TRL 1991–2001. (*Source:* © 2016 IEEE. Reprinted, with permission, from Kasser, Joseph Eli. 2016. "Applying Holistic Thinking to the Problem of Determining the Future Availability of Technology." The IEEE Transactions on Systems, Man, and Cybernetics: Systems no. 46 (3):440–444)	274
Figure 10.2	The dTRL. (*Source:* © 2016 IEEE. Reprinted, with permission, from Kasser, Joseph Eli. 2016. "Applying Holistic Thinking to the Problem of Determining the Future Availability of Technology." The IEEE Transactions on Systems, Man, and Cybernetics: Systems no. 46 (3):440–444)	275
Figure 10.3	A project risk profile. (*Source:* © 2016 IEEE. Reprinted, with permission, from Kasser, Joseph Eli. 2016. "Applying Holistic Thinking to the Problem of Determining the Future Availability of Technology." The IEEE Transactions on Systems, Man, and Cybernetics: Systems no. 46 (3):440–444.)	276
Figure 11.1	The systemic and systematic change control process.	287
Figure 11.2	The Master's degree project projected cash flow.	294
Figure 11.3	The Master's degree project financial status after 9 months.	295
Figure 11.4	The data centre upgrade project EVA status at PDR.	296
Figure 11.5	The revised data centre upgrade project EVA status at CDR.	297
Figure 11.6	Measuring the quadruple constraints.	300
Figure 11.7	A typical project status meeting agenda.	324
Figure 11.8	Some of the projects with traffic lights.	324
Figure 11.9	The agenda in tabular form.	325
Figure 11.10	The agenda in tabular form with traffic light blocks	325
Figure 11.11	The ETL chart.	326
Figure 11.12	Project trends using the ETL chart.	326
Figure 11.13	Final version of ETL chart.	329
Figure 12.1	A performance evaluation chart.	339
Figure 12.2	Performance evaluation chart for two time periods.	339
Figure 12.3	Application of different styles of management.	345
Figure 12.4	The relationship between what customers need, want and the structure of the problem.	350

List of Tables

Table 1.1	Acronyms Used in the Book	3
Table 2.1	A CMM for Project Managers	20
Table 3.1	Decision Table for Known Outcomes of Actions	65
Table 3.2	How the Activity in the Problem-Solving Process Relates to the Five Ways of Approaching Problems	77
Table 3.3	Summary of Reasons for the Complexity Dichotomy from Various Perspectives	82
Table 4.1	The Differences between Leaders and Managers	100
Table 5.1	Typical (Planned) WP (Spreadsheet)	129
Table 5.2	The SLC, SDP and Hitchins–Kasser–Massie Framework (HKMF) States	134
Table 5.3	Mapping the SDP into Traditional Simple Problem-Solving Process	134
Table 5.4	The Notional States in the SDP	136
Table 5.5	Type of Project Manager Needed in Each State of the SDP	137
Table 5.6	SDP State and State Numbers	144
Table 5.7	Typical Initial Draft WP for State 200	144
Table 5.8	Typical Initial Draft of WP 210D	145
Table 5.9	To Do Project Control Spreadsheet	146
Table 5.10	The First Digit Represents the State	148
Table 5.11	Activity Numbering	148
Table 5.12	Product Numbering	149
Table 5.13	Numbering Representation	149
Table 5.14	Grading Criteria for the Presentation Exercise	156
Table 5.15	Grading Based on Cognitive skills	157
Table 5.16	Activity Numbering	159
Table 5.17	Versions of Products	159
Table 5.18	The Top-Level Sequence of Events for the Activities in Each State of the Exercise	160

Table 5.19	The Relevant Sections of the WPs for the Presentation Planning State (WP 00000)	162
Table 5.20	The Relevant Sections of the WPs for the Presentation Requirements State (10000)	165
Table 5.21	Typical RTM Template for Presentations	167
Table 5.22	The Relevant Sections of the WPs for the Presentation Design State (WP 20000)	169
Table 5.23	The Relevant Sections of WP 22000 (Discussing Readings)	173
Table 5.24	The Relevant Sections of WP 23400 (Designing Reading 1 Component Templates)	174
Table 5.25	The Relevant Sections of the WPs for the Presentation Construction State (WP 30000)	177
Table 5.26	The Relevant Data Elements in the 34500 Create Briefing 1 WP	183
Table 5.27	The Relevant Sections of the WPs for the System Integration and Test State (WP 40000 Level)	186
Table 5.28	The Relevant Sections of the WPs for the Presentation Delivery and Grading State (WP 50000)	190
Table 6.1	Traditional Competency Levels Based on Years of Experience	196
Table 6.2	Partial Competences, Knowledge and Skills Matrix	202
Table 6.3	Partial Activity Job Position Matrix	203
Table 6.4	Staffing Profile for WP 00000	204
Table 6.5	Staffing Profile for WP 10000	204
Table 6.6	Staffing Profile for WP 20000	205
Table 6.7	Staffing Profile for WP 30000	206
Table 6.8	Staffing Profile for WP 40000	207
Table 6.9	Staffing Profile for WP 50000	207
Table 6.10	Staffing Summary for Session 1 Presentation Exercise	207
Table 6.11	Revised Staffing Profile for WP 20000	208
Table 6.12	Revised Staffing Summary for Session 1 Presentation Exercise	208
Table 7.1	Time for Each Path from Milestones 1 to 8 in Figure 7.6	215
Table 7.2	Original and Adjusted Times for Each Path from Milestones 1 to 8 in Figure 7.6	216
Table 7.3	Slack Times in Figure 7.6	217

List of Tables

Table 7.4	Task Duration Accuracies	218
Table 7.5	Second Revised Staffing Summary for Session 1 Presentation Exercise	224
Table 8.1	Cost Categories	239
Table 8.2	Team Member Salary Rates and Labour Categories	245
Table 8.3	Presentation Project Costs	245
Table 9.1	Staffing Profile for WP 00000 without Susan	261
Table 9.2	Staffing WP Profile for 10000 without Susan	262
Table 9.3	Revised Staffing Summary for Session 1 Presentation Exercise without Susan	262
Table 9.4	Staffing Assignment Changes	263
Table 9.5	Actual Costs of the Presentation Project without Susan	264
Table 10.1	Risk Rectangle	269
Table 10.2	The TAWOO States and Levels	273
Table 11.1	EVA Elements and Equations	290
Table 11.2	EVA - Traffic Light Chart Performance Indicator Definitions	290
Table 11.3	Estimated Budget for Studying for a Master's Degree	293
Table 11.4	Actual Budget for Studying for a Master's Degree after 9 Months	294
Table 11.5	EVA for the First 5 Months of a Master's Degree	295
Table 11.6	Notional ETL Chart Colours for Cost and Schedule Component of PDR	297
Table 11.7	EVA Values at Project Milestones	298
Table 11.8	An Unpopulated CRIP Chart	303
Table 11.9	The CRIP Chart at the Completion of the Proposal	306
Table 11.10	The CRIP Chart at the Start of the Ideal Project	308
Table 11.11	The Ideal Project CRIP Chart at SRR	308
Table 11.12	The Ideal Project CRIP Chart at PDR	309
Table 11.13	The Ideal Project CRIP Chart at CDR	310
Table 11.14	The Ideal Project CRIP Chart at TRR	311
Table 11.15	The Ideal Project CRIP Chart at IRR	312
Table 11.16	The Ideal Project CRIP Chart at DRR	312

Table 11.17	The CRIP Chart for the Project with Requirements Creep at PDR	313
Table 11.18	The CRIP Chart for the Project with Requirements Creep at CDR	314
Table 11.19	The CRIP Chart for the Project with Requirements Creep at TRR	315
Table 11.20	The CRIP Chart for the Project with Requirements Creep at IRR	317
Table 11.21	The CRIP Chart for the Project with Requirements Creep at DRR	318
Table 11.22	The CRIP Chart for the Challenged Project at TRR	319
Table 11.23	The CRIP Chart for the Challenged Project at IRR	320
Table 11.24	The CRIP Chart for the for the 'Makeup Your Mind' Project at PDR	321
Table 11.25	Information Represented by Colours in Different Types of Traffic Light Charts	322
Table 11.26	Values for Traffic Light Colours	323
Table 12.1	Where to Apply Power and Influence	342
Table 13.1	The CRIP Chart for the MCSS Project at DRR	363
Table A4.1	Systems Engineering Salary Ranges	373

Preface

This book was written to update project management for the 21st century. It fills the gaps in the existing literature. For example, when I was teaching project management to master's degree students, I often heard myself saying, 'and you won't find that in the textbooks'. This book:

1. Is written to provide that information as well as the standard project management knowledge.
2. Is based on 35 years of successful project management experience and 10 years of teaching project management.
3. Teaches the systems approach, integrating tools and techniques currently used as well as introducing new tested tools and techniques and explaining how to apply them together with current tools and techniques.
4. Focuses on early detection and prevention of impending project failure.
5. Explains how to apply systems thinking to improve your project management.

While other books focus on:

1. The triple constraints, this book focuses on the quadruple constraints; the fourth constraint being people.
2. Planning projects, this book focuses on managing projects and explains what to do when things don't go according to plan and provides some examples of how to react to unforeseen events.

After reading and inwardly digesting the contents of this book and practising what you learn you will be a better and more successful project manager.

JEK

Acknowledgements

This book would not have been possible without the co-authors of the papers upon which some of these chapters are based, and colleagues and friends who helped review the manuscript:

Eileen Arnold

Dr Amihud Hari

Professor Derek K. Hitchins

Robin Schermerhorn

Dr Xuan-Linh Tran

Victoria R. Williams

Associate Professor Yang Yang Zhao

OTHER BOOKS BY THIS AUTHOR

- *The Systems Thinkers Toolbox: Tools for Managing Complexity*, CRC Press, 2018.
- *Perceptions of Systems Engineering*, Createspace, 2015.
- *Conceptual Laws and Customs of Christmas*, Createspace, 2015.
- *The 87th Company. The Pioneer Corps. A Mobile Military Jewish Community*, Createspace, 2013 (Editor).
- *Holistic Thinking: Creating Innovative Solutions to Complex Problems*, Createspace, 2013.
- *A Framework for Understanding Systems Engineering*, Createspace, Second Edition 2013.
- *Applying Total Quality Management to Systems Engineering*, Artech House, 1995.
- *Basic Packet Radio, Software for Amateur Radio*, First and Second Editions, 1993, 1994.
- *Software for Amateur Radio*, TAB Books, December 1984.
- *Microcomputers in Amateur Radio*, TAB Books, November 1981.

Author

Joseph Eli Kasser has been a practicing systems engineer for almost 50 years, a project manager for more than 35 years and an academic for 20 years. He is a Fellow of the Institution of Engineering and Technology (IET), a Fellow of the Institution of Engineers (Singapore), the author of *The Systems Thinkers Toolbox: Tools for Managing Complexity, Perceptions of Systems Engineering, Holistic Thinking: Creating Innovative Solutions to Complex Problems, A Framework for Understanding Systems Engineering* and *Applying Total Quality Management to Systems Engineering*, as well as two books on amateur radio and many International Council on Systems Engineering (INCOSE) symposia and other conference and journal papers.

He is a recipient of the National Aeronautics and Space Administration's (NASA) Manned Space Flight Awareness Award (Silver Snoopy) for quality and technical excellence for performing and directing systems engineering and other awards. He holds a Doctor of Science in Engineering Management from George Washington University. He is a Certified Manager, a Chartered Engineer in both the United Kingdom and Singapore and holds a Certified Membership of the Association for Learning Technology. He has been a project manager in Israel and Australia, and performed and directed systems engineering in the United States, Israel and Australia. He gave up his positions as a Deputy Director and DSTO Associate Research Professor at the Systems Engineering and Evaluation Centre at the University of South Australia in early 2007 to move to the United Kingdom to develop the world's first immersion course in systems engineering as a Leverhulme Visiting Professor at Cranfield University. He spent 2008–2016 as a Visiting Associate Professor at the National University of Singapore where he taught and researched the nature of systems engineering, systems thinking and how to improve the effectiveness of teaching and learning in postgraduate and continuing education. He is currently based in Adelaide, Australia. His many awards include:

- National University of Singapore, 2008–2009 Division of Engineering and Technology Management, Faculty of Engineering Innovative Teaching Award for use of magic in class to enrich the student experience.
- Best Paper, Systems Engineering Technical Processes track, at the 16th Annual Symposium of the INCOSE, 2006, and at the 17th Annual Symposium of the INCOSE, 2007.
- United States Air Force (USAF) Office of Scientific Research Window on Science programme visitor, 2004.
- Inaugural SEEC 'Bust a Gut' Award, SEEC, 2004.
- Employee of the Year, SEEC, 2000.
- Distance Education Fellow, University System of Maryland, 1998–2000.
- Outstanding Paper Presentation, Systems Engineering Management track, at the 6th Annual Symposium of the INCOSE, 1996.

- Distinguished Service Award, Institute of Certified Professional Managers (ICPM), 1993.
- Manned Space Flight Awareness Award (Silver Snoopy) for quality and technical excellence, for performing and directing systems engineering, NASA, 1991.
- NASSA Goddard Space Flight Center Community Service Award, 1990.
- The E3 award for Excellence, Endurance and Effort, Radio Amateur Satellite Corporation (AMSAT), 1981, and three subsequent awards for outstanding performance.
- Letters of commendation and certificates of appreciation from employers and satisfied customers including the:
  - American Radio Relay League (ARRL).
  - American Society for Quality (ASQ).
  - Association for Quality and Participation (AQP).
  - Communications Satellite Corporation (Comsat).
  - Computer Sciences Corporation (CSC).
  - Defence Materiel Organisation (Australia).
  - Institution of Engineers (Singapore).
  - IET Singapore Network.
  - Loral Corporation.
  - Luz Industries, Israel.
  - Systems Engineering Society of Australia (SESA).
  - University of South Australia.
  - United States Office of Personnel Management (OPM).
  - University System of Maryland.
  - Wireless Institute of Australia.

When not writing and lollygagging he provides consulting services and training.

1 Introduction

The best book on the current process-based management paradigm that I've read is one that I purchased for $0.50 about 20 years ago in the Montgomery County, Maryland, library used book sale. The book was published in 1917 and reprinted in 1920 and it was Volume 6 of a factory management course. The title of the book does not contain the word management; the title is *Executive Statistical Control* (Farnham 1920) and it deals with management from the following perspectives:

1. Financial.
2. Making decisions based on facts instead of hunches or opinions.
3. Treating employees fairly and decently.
4. The efficient use of labour.

About the only major change to management that came after the book was published was the Gantt chart that showed project timelines and how the timelines changed during the reporting period on a single chart (Clark 1922).

The current project management paradigm does not seem to work too well because in spite of all the textbooks[1] and courses, projects still tend to fail. The reasons for these failures are well known, yet continue to manifest themselves; an instance of Cobb's Paradox (VOYAGES 1996). Cobb's Paradox states, 'We know why projects fail, we know how to prevent their failure; so why do they still fail?' Now a paradox is a symptom of a flaw in understanding the underlying paradigm. Perhaps Juran and Deming provided the remedy:

- 80%–85% of all organizational problems (Juran quoted by Harrington (1995)).
- 94% of the problems belong to the system, i.e. were the responsibility of management (Deming 1993).

Peter Drucker wrote that management in the 21st century would be different (Drucker 2011). This book discusses a different project management paradigm; project management as a problem-solving methodology using the systems approach.

> The systems approach is a technique for the application of a scientific approach to complex problems. It concentrates on the analysis and design of the whole, as distinct from the components or the parts. It insists upon looking at a problem in its entirety, taking into account all the facets and all the variables, and relating the social to the technological aspects
>
> **Ramo (1973)**

---

[1] The vast majority of books since 1920 are variations on the same theme to the point where I did not specify a management textbook for my postgraduate classes on project management. Instead of assigning specific readings in a textbook, I assigned readings on specific topics. I gave the students a list of three books and told them they could use any of those books or equivalents they could purchase or find in the library.

This book:

- Provides you with a better understanding of the systems approach to problem-solving and project management, which will enable you to be more successful at managing projects.
- Treats project management as a problem-solving paradigm.
- Shows how to incorporate prevention into planning and show the value of prevention.
- Shows how the tools described in *The Systems Thinker's Toolbox* (Kasser 2018) can be applied to project management.[2]
- Shows how to cope with unanticipated problems that arise during the project implementation state.
- Is based on more than 40 years of research and experience.
- Integrates many published and invented tools and techniques into a practical methodology.
- Applies systems thinking to treat project management in a systemic and systematic manner from a problem-solving perspective.
- Considers a project as a system.
- Divides project management into pure project management, applied project management and domain project management which identifies the reasons for the opinion that a project manager can manage any type of project, and identifies the situational fallacy in that opinion.
- Picks up where the *Systems Thinker's Toolbox* (Kasser 2018) ended for project managers.
- Incorporates new project monitoring tools including categorized requirements in process (CRIP) charts and enhanced traffic light (ETL) charts from the book *The Systems Thinker's Toolbox* (Kasser 2018).
- Provides examples of how the tools are used.
- Provides examples of good and bad resumes in the section on staffing a project to show what to look for when considering candidates for staffing a project.
- Provides a list of acronyms in Table 1.1 to help you read the book.

1.1 THE CONTENTS OF THIS BOOK

This book discusses systemic and systematic project management. The systemic aspects in Chapters 2, 3, 4, 12 and 13 help you conceptualize and understand project management. The systematic aspects in the middle chapters help you to become a better project manager.

Chapter 2 explains the seven P's of project management: people, politics, prevention, problems, processes and products which must be managed (or juggled) to achieve the objectives of a project.

---

[2] The companion book *Systemic and Systematic Systems Engineering* (Kasser 2019) shows how the tools can be applied to systems and software engineering.

TABLE 1.1
Acronyms Used in the Book

AC	Actual Cost
ACWP	Actual Cost of Work Performed
AoA	Analysis of Alternatives
BAC	Budget at Completion
BCWP	Budgeted Cost of Work Performed
BCWS	Budgeted Cost of Work Scheduled
BPR	Business Process Reengineering
CAIV	Cost as an Independent Variable
CCB	Configuration Control Board
CDR	Critical Design Review
CDTC	Conceptual Design to Cost
CM	Configuration Management
CMM	Competency Maturity Model
COBOL	Common Business-Oriented Language
CONOPS	Concept of Operations
COTS	Commercial-Off-The-Shelf
CPI	Cost Performance Index
CRIP	Categorized Requirements in Process
CV	Cost Variance
DERA	Defence Evaluation and Research Agency
DMSMS	Diminishing Manufacturing Sources and Material Shortages
DOD	Department of Defense
DR	Discrepancy Report
DRR	Delivery Readiness Review
DTC	Design to Cost
dTRL	Dynamic TRL
EAC	Estimate at Completion
ETA	Employment and Training Administration
ETL	Enhanced Traffic Light
EV	Earned Value
EVA	Earned Value Analysis
FCFDS	Feasible Conceptual Future Desirable Solution
HKMF	Hitchins–Kasser–Massie Framework
HTP	Holistic Thinking Perspectives
ICDM	Integrated Customer-Driven Conceptual Design Method
INCOSE	International Council on Systems Engineering
IRR	Integration Readiness Review
IV&V	Independent Verification and Validation
MBE	Management by Exception
MBO	Management by Objectives
MBUM	Micromanagement by Upper Management
MBWA	Management by Walking Around
MCSSRP	Multi-satellite Communications Switching System Replacement Project

(*Continued*)

TABLE 1.1 (*Continued*)
Acronyms Used in the Book

NASA	National Aeronautics and Space Administration
OCR	Operations Concept Review
OODA	Observe-Orient-Decide-Act
PAM	Product-Activity-Milestone
PBS	Product Breakdown Structure
PDR	Preliminary Design Review
PDR	Presentation Design Review
PERT	Program Evaluation Review Technique
PID	Project Initiation Document
PIR	Presentation Integration Review
PM	Project Manager
PP	Project Plan
PPMR	Presentation Post-Mortem Review
PRR	Presentation Requirements Review
PV	Planned Value
Q	Quality Leader
RFP	Request for Proposal
ROI	Return on Investment
RTM	Requirements Traceability Matrix
SDP	System Development Process
SLC	System Lifecycle
SPI	Schedule Performance Index
SPPCR	Session Presentation Planning Completed Review
SRR	Systems Requirements Review
SRRS	Systematic Reward and Recognition System
STALL	Stay calm, Think, Ask questions, Listen, Listen
SV	Schedule Variance
TAWOO	The Technology Availability Window of Opportunity
TL	Task Lead
TRIZ	Theory of Inventive Problem-Solving
TRL	Technology Readiness Level
TRR	Subsystem Test Readiness Review
USB	Universal Serial Bus
VAC	Variance at Completion
WBS	Work Breakdown Structure
WP	Work Package

Chapter 3 deals with problems and problem-solving. If a problem didn't need solving there wouldn't be any need for management. The chapter discusses perceptions of problem-solving process from a number of perspectives, explains the structure of problems and the levels of difficulty posed by problems and the need to evolve solutions using an iterative approach. After showing that problem-solving is really an iterative causal loop rather than a linear process, the chapter then explains complexity, and how to use the systems approach to manage complexity. The chapter then shows how to remedy well-structured problems and how to deal with ill-structured, wicked and complex problems using iterations of a sequential two-stage problem-solving process. Reflecting on this chapter, it seems that iteration is a common element in remedying any kind of problem other than easy well-structured problems irrespective of their structure.

The purpose of management is to accomplish a goal by getting other people to do the work. Chapter 4 discusses management, general management and project management, and how to accomplish that goal. After a brief discussion on general management, the chapter focuses on the attributes of projects and project management as a problem-solving activity.

Successful projects are planned. Chapter 5 focuses on product-based planning, the project and system lifecycles and planning methodologies. The chapter discusses plans, the difference between generic and project-specific planning, explains how to incorporate prevention into the planning process to lower the completion risk and shows how to apply work packages (WPs) instead of work breakdown structures (WBS). The chapter also explains why planning should iterate from project start to finish at the conceptual level and project finish back to start at the detailed level.

Because projects are staffed by people, Chapter 6 discusses aspects of staffing projects, teams and distributing assignments to members of the project team. The chapter explains:

1. The need for high-performance teams.
2. That people are not interchangeable; namely, one engineer does not necessarily equal another.
3. How to staff a project.

Chapter 7 explains planning and adhering to schedules, an important function of project management since exceeding schedule is one of the characteristics of failed projects. Consequently, it is important to get the right schedule for the project. Accordingly, Chapter 7 explains:

1. How to create the project network.
2. How to create a project schedule.
3. The critical path and its importance.
4. The fallacy in slack time in fixed resource situations.

Chapter 8 explains that project managers need to estimate costs before a project begins to determine if the project is affordable and/or will deliver the required return on investment (ROI). Consequently, it is important to get the right cost estimate for the project. Accordingly, estimating and adhering to costs is an important function of project management and exceeding cost is another characteristic of failed projects. Project managers also need to re-estimate the costs once the project commences and more detail is known about the seven interdependent P's of a project. However, before trying to estimate costs the project manager needs to understand the nature of costs, where they come from and what they represent. Accordingly, Chapter 8 explains:

1. Factors influencing the quality of estimates.
2. Methods for estimating project costs.
3. The different types of cost contracts.
4. Ways of controlling costs.

A project spends most of its time in the project performance state when predicted and unpredicted events have the potential to increase costs and cause delays. Accordingly, Chapter 9 discusses:

1. The effect of unanticipated events on cost and schedule.
2. Shortening the schedule.
3. Reducing costs.
4. Cancelling a project (the ultimate in cost and schedule reduction).

Chapter 10 explains risk management, another important part of project management. The chapter:

1. Focuses on risk prevention as well as mitigation.
2. Contrasts traditional risk management with the systems approach to risk management.
3. Explains the difference between risk and uncertainty.
4. Explains the difference between the traditional and the systems approach to risk management.
5. Identifies where and how risks and uncertainties arise in the project starting with formulating plans for changing the undesirable situation through the implementation of the plan.
6. Explains ways of estimating risks.
7. Explains the fallacy in the risk rectangle.
8. Explains the application of risk profiles rather than using the risk rectangle to manage risks.

Chapter 11 explains how to monitor, control and communicate the performance of a project. The focus of performance monitoring is on the process and the product in the traditional project management paradigm. As the project proceeds along its timeline, the actual cost and schedule are compared with the estimated cost and schedule for that point in the timeline. The version of the systems approach discussed in Chapter 11 incorporates the remaining interdependent P's of a project allowing project management to be more proactive. The chapter explains:

1. Ways to detect and prevent potential project overruns.
2. Ways to detect impending project failure.

Introduction

3. The 'build a little test a little' development methodology.
4. Managing changes in project scope.
5. Earned value analysis (EVA).
6. Categorized requirements in process (CRIP) charts.
7. Enhanced traffic light (ETL) charts.
8. Management by objectives (MBO).
9. Management by exception (MBE).

Chapter 12 focuses on the human element; the most important attribute of project management because people perform the project, cause, prevent or mitigate problems, and take part in the processes that produce the products. Once the project has begun, project managers spend most of their time dealing with people. The outcomes of these interactions with people affect the cost and the schedule and are reflected in the Gantt and EVA charts presented at the milestone reviews. Chapter 12 introduces the following aspects of the human element:

1. Time management.
2. People management.
3. Conflict.
4. Leadership.
5. Negotiation.
6. Managing stakeholders.

Chapter 13 discusses the ethics of project management because as a project manager you may find yourself in a situation where you think something is wrong. Issues arise when you discover an activity, or are asked to participate in an activity that you feel is wrong, or illegal. Chapter 13:

- Explains personal, organizational and professional ethics.
- Explains the ethical dilemma and the consequences of speaking out according to your conscience.
- Concludes with some personal lessons learned by the author.

The presentation exercise starting in Chapter 5 contains examples of using the techniques described in each chapter and the Engaporian[3] Multi-satellite Communications Switching System Replacement Project (MCSSRP) provides the opportunity to practise the techniques in an individual and group/classroom environment.

1.2 HOW TO READ AND USE THIS BOOK

This is a reference book, so don't read the book sequentially in a linear manner, but prepare for several passes through it. This book is non-fiction. Non-fiction books are different to fiction; stories, novels and thrillers are designed to be read in a linear

---

[3] A fictitious country.

manner from start to finish. This book is designed to help you learn and use the content in the following manner:

1. *Skim the book:* flip through the pages, if anything catches your eye and interests you, stop, glance at it, and then continue flipping through the pages. Notice how the pages have been formatted with dot points (bulleted lists) rather than in paragraphs to make skimming and reading easier.
2. For each chapter:
    - Read the introduction and summary.
    - Skim the contents.
    - Look at the drawings.
    - Go on to the next chapter.
3. *If you don't understand something, skip it on the first and second readings:* don't get bogged down in the details.
4. Work though the book slowly so that you understand the message in each section of each chapter. If you don't understand the details of the example, don't worry about it as long as you understand the point that the example is demonstrating.
5. Refer to the list of acronyms in Table 1.1 as necessary.
6. Successfully manage your project and all subsequent ones.

Step 1 should give you something you can use immediately. Steps 2 and 3 should give you something you can use in the coming months. Step 4 should give you something you can use for the rest of your life. Step 6 is the rest of your life.

REFERENCES

Clark, Wallace. 1922. *The Gantt Chart: A Working Tool of Management.* New York: The Ronald Press Company.
Deming, W. Edwards. 1993. *The New Economics for Industry, Government, Education.* Cambridge, MA: MIT Center for Advanced Engineering Study.
Drucker, Peter F. 2011. *Management Challenges for the 21st Century Classic Drucker Collection.* New York: Routledge.
Farnham, Dwight T. 1920. *Executive Statistical Control.* Vol. 6, *Factory Management Course.* New York: Industrial Extension Institute.
Harrington, H. James. 1995. *Total Improvement Management the Next Generation in Performance Improvement.* New York: McGraw-Hill.
Kasser, Joseph Eli. 2018. *Systems Thinker's Toolbox: Tools for Managing Complexity.* Boca Raton, FL: CRC Press.
Kasser, Joseph Eli. 2019. *Systemic and Systematic Systems Engineering.* Boca Raton, FL: CRC Press.
Ramo, Simon. 1973. The systems approach. In *Systems Concepts*, edited by Ralph F. Miles Jr, pp. 13–32. New York: John Wiley & Son, Inc.
VOYAGES. *Unfinished Voyages, A follow up to the CHAOS Report* 1996 [cited 21 January 21, 2002. Available from http://www.pm2go.com/sample_research/unfinished_voyages_1.asp.

2 The Seven Interdependent P's of a Project

> A project is a temporary endeavor undertaken to create a unique product, service, or result
>
> PMI (2013: p. 3)

Projects are systems that span a range of activities such as cooking a meal, arranging a meal, planning and enjoying a vacation, moving house, developing software, deploying disaster relief teams and fighting a war. From conception to completion, a project begins when someone in an undesirable situation decides to remove the undesirability and faces the problem of turning the undesirable situation into a desirable situation. The project then passes through several states, each of which begins with a problem and is completed when a solution to the problem is developed and the solution becomes a problem for the next state.

The project ends when the solution is provided to the problem posed by the last state and someone verifies that the solution actually remedies the original problem for which the project was commissioned. Traditional project management focuses on managing only three of the perspectives: processes, profit and people. This chapter helps you to understand the systems approach which has a wider focus starting with the following seven interdependent P's[1,2] of a project, which must be managed (or juggled) according to the project plan (PP) (Section 5.5) to achieve the objectives of a project:

1. People.
2. Politics.
3. Prevention.
4. Problems.
5. Processes.
6. Profit.
7. Products.

Consider each of them.

---

[1] In alphabetical order.
[2] The seven P's follow once the PP exists.

2.1 PEOPLE

The literature on 'excellence' focuses on people and ignores process (Peters and Waterman 1982, Peters and Austin 1985, Rodgers, Taylor, and Foreman 1993). People are valuable and not interchangeable, so consider the following factors pertaining to people:

1. Thinking discussed in Section 2.1.1.
2. Categories of people discussed in Section 2.1.2.
3. Competence discussed in Section 2.1.3.
4. Motivation discussed in Section 2.1.4.
5. Meetings discussed in Section 2.1.5.
6. Communications discussed in Section 2.1.6.
7. Miscommunications discussed in Section 2.1.7.
8. Stakeholders discussed in Section 12.6.

2.1.1 THINKING

Thinking is:

- The action that underlies problem-solving and decision-making.
- A cognitive act performed by the brain.

Cognitive activities include accessing, processing and storing information. The most widely used cognitive psychology information processing model of the brain likens the human mind to an information processing computer (Atkinson and Shiffrin 1968) cited by Lutz and Huitt (2003). Both the human mind and the computer ingest information, process it to change its form, store it, retrieve it and generate responses to inputs (Woolfolk 1998). These days we can extend our internal memory using paper notes, books and electronic storage. We use our mental capacity to think about something received from a sense (hearing, sight, smell, taste and touch). As perceived from the *Functional* HTP our mental capacities might be oversimplified in four levels as follows: (Osborn 1963: p. 1):

1. *Absorptive:* the ability to observe and apply attention.
2. *Retentive:* the ability to memorize and recall.
3. *Reasoning:* the ability to analyse and judge.
4. *Creative:* the ability to visualize, foresee and generate ideas.

When we view the world, our brain connects concepts using a process called reasoning or thinking and uses a filter to separate the pertinent sensory input from the non-pertinent. This filter is known as a 'cognitive filter' in the behavioural science literature (Wu and Yoshikawa 1998), and as a 'decision frame' in the management literature (Russo and Schoemaker 1989). Cognitive filters and decision frames:

1. Are filters through which we view the world.
2. Include the political, organizational, cultural and metaphorical.

3. Highlight relevant parts of the system and hide (abstract out) the non-relevant parts.
4. Can also add material that hinders solving the problem.[3] Failure to abstract out the non-relevant issues can make things appear to be more complex and complicated than they are and gives rise to artificial complexity (Kasser 2018: Section 13.3.1.1).

2.1.1.1 Critical Thinking

The literature on creativity and idea generation generally separates thinking up the ideas and applying the ideas. The literature on critical thinking, however, tends to combine the logic of thinking with applying the ideas using the terms 'smart thinking' and 'critical thinking'. The term 'critical thinking' by the way, comes from the word 'criteria' not from 'criticism'. The diverse definitions of the term critical thinking include:

> Disciplined, self-directed thinking displaying a mastery of intellectual skills and abilities-thinking about your thinking while you're thinking to make your thinking better
>
> **Eichhorn (2002)**

> The art of thinking about thinking while thinking in order to make thinking better. It involves three tightly coupled activities: It analyses thinking; it evaluates thinking; it improves thinking
>
> **Paul and Elder (2006: p. xiii)**

> Judicious reasoning about what to believe and therefore what to do
>
> **Tittle (2011: p. 4)**

> The process of purposeful, self-regulatory judgement
>
> **Facione (1990) cited by Facione (2011: p. 6)**

> Purposeful, reflective judgment that manifests itself in giving reasoned and fair-minded consideration to evidence, conceptualizations, methods, contexts, and standards in order to decide what to believe or what to do
>
> **Facione (2011: p. 12)**

Depending on the definition, critical thinking covers:

1. The thinking process.
2. The means to evaluate or judge the ideas.

Chapter 3 of the *Systems Thinker's Toolbox* (Kasser 2018) discusses critical thinking and its application and different methods of assessment. One way of assessing critical thinking can be found in Section 2.1.2.2 because it is used in the competency maturity model (CMM) discussed in Section 2.1.3.

---

[3] For example, the differences between the Catholics and Protestants in Northern Ireland are major to many of the inhabitants of that country, but are hardly noticeable to most of the rest of the world.

2.1.1.2 Systems Thinking

There are books on the history and the philosophy and on what constitutes systems thinking. Similarly, the books on problem-solving tend to describe the problem-solving process. However, they don't generally describe how to do the process. And the few articles that do, describe how to use a specific tool published in domain literature and do not get a wide distribution, so few people hear about it and even fewer people actually use it.[4]

2.1.1.3 The Two Distinct Types of Systems Thinking

One reason for the lack of good ways of teaching systems thinking might be because if you ask different people to define systems thinking, you will get different and sometimes conflicting definitions. However, these definitions can be sorted into two types, namely:

1. *Systemic thinking:* thinking about a system as a whole to gain an understanding.
2. *Systematic thinking:* employing a methodical step-by-step manner to think about something.

Many proponents of systems thinking consider either systemic or systematic thinking to be systems thinking not realizing that each type of thinking seems to be a partial view of a whole, in the manner of the fable of the blind men feeling parts of an elephant and each identifying a single and different animal (Yen 2008). However, both types of systems thinking are needed (Gharajedaghi 1999). Consider each of them:

2.1.1.3.1 Systemic Thinking

Systemic thinking has three steps (Ackoff 1991):

1. A thing to be understood is conceptualized as a part of one or more larger wholes, not as a whole to be taken apart.
2. An understanding of the larger system is sought.
3. The system to be understood is explained in terms of its role or function in the containing system.

---

[4] I found that out the hard way when I started to teach systems thinking at the University of South Australia (UniSA) in 2000. I could describe the benefits and history of systems thinking, but no one at the Systems Engineering and Evaluation Centre (SEEC) could teach how to use systems thinking very well. We could teach causal loops in the manner of '*The Fifth Discipline*' (Senge 1990), but that was all. There weren't any good textbooks that approached systems thinking in a practical manner. So, I ended up moving half way around the world to Cranfield University in the United Kingdom to develop the first version of a practical and pragmatic approach to teaching and applying systems thinking to systems engineering under a grant from the Leverhulme Foundation.

The Seven Interdependent P's of a Project

Proponents of systemic thinking tend to:

- Equate causal loops or feedback loops with systems thinking because they are thinking about relationships within a system, e.g. (Senge 1990, Sherwood 2002).
- Define systems thinking as looking at relationships (rather than unrelated objects), connectedness, process (rather than structure), the whole (rather than its parts), the patterns (rather than the contents) of a system and context (Ackoff, Addison, and Andrew 2010: p. 6).

The benefits of systemic thinking have also been known for a long time. For example:

> When people know a number of things, and one of them understands how the things are systematically categorized and related, that person has an advantage over the others who don't have the same understanding
>
> **Luzatto (circa 1735)**

> People who learn to read situations from different (theoretical) points of view have an advantage over those committed to a fixed position. For they are better able to recognize the limitations of a given perspective. They can see how situations and problems can be framed and reframed in different ways, allowing new kinds of solutions to emerge
>
> **Morgan (1997)**

The number of standardized viewpoints has evolved over the years. For example, C. West Churchman introduced three standardized views in order to first think about the purpose and function of a system and then later think about physical structure (Churchman 1968). These views provided three anchor points or viewpoints for viewing and thinking about a system. Twenty-five years later, seven standardized viewpoints were introduced (Richmond 1993). Richmond's seven streams are:

1. *Dynamic thinking* which frames a problem in terms of a pattern of behaviour over time.
2. *System-as-cause thinking* which places responsibility for a behaviour on internal factors who manage the policies and plumbing of the system.
3. *Forest thinking* which is believing that to know something you must understand the context of relationships.
4. *Operational thinking* which concentrates on getting at causality and understanding how behaviour is actually generated.
5. *Closed-loop thinking* which views causality as an ongoing process, not a one-time event; with the effect of feeding back to influence the causes, and the causes affecting each other.
6. *Quantitative thinking* which accepts that you can always quantify something even though you can't always measure it.
7. *Scientific thinking* which recognizes that all models are working hypotheses that always have limited applicability.

2.1.1.3.2 Systematic Thinking

Systematic thinking:

- Is mostly discussed in the literature on problem-solving, systems thinking and critical thinking. It is often taught as the problem-solving process (Section 3.1.2.3) or the system development process (SDP) (Section 5.4).
- Provides the process for systemic thinking which helps you to understand and remedy the problematic situation. The benefits of systematic thinking have been known for a long time hence the focus on process and controlling the process.
- Is made up of two parts:
  1. *Analysis:* breaking a complicated topic into several smaller topics and thinking about each of the smaller topics. Analysis can be considered as a top-down approach to thinking about something (Descartes 1637, 1965). It has been termed reductionism because it is often used to reduce a complex topic to a number of smaller and simpler topics.
  2. *Synthesis:* combining two or more entities to form a more complex entity. Synthesis can be considered as a bottom-up approach to thinking about something.

2.1.1.4 Beyond Systems Thinking

The 'beyond' part of 'systems thinking and beyond' is:

- Where the problem definitions and solutions come from.
- Sometimes called holistic thinking.
- Emerged from research in 2008, which modified Richmond's seven streams and adapted them into nine systems thinking perspectives (Kasser and Mackley 2008).[5]

The nine systems thinking perspectives introduced nine standardized viewpoints which cover purpose, function, structure and more and were later renamed as the holistic thinking perspectives (HTPs) (Kasser 2018: Section 10.1). The nine HTPs:

- Are summarized in this section.
- Are widely used in this book.
- Are a systemic tool for:
  - Gaining an understanding of a problematic situation.
  - Inferring the cause of the undesirability in the situation.
  - Inferring a probable solution to the problems posed when removing the undesirability from the situation.
- Provide a standard set of nine internal, external, progressive and remaining perspectives (anchor points) with which to view a situation.

---

[5] The number of perspectives was limited to nine in accordance with Miller's Rule (Miller 1956, Kasser 2018: Section 3.2.5).

The Seven Interdependent P's of a Project

- Go beyond systems thinking's internal and external views by adding quantitative and progressive (temporal, generic and continuum) perspectives. This approach:
  - Separate facts from opinion. Facts are perceived from the eight descriptive HTPs; opinion comes from the insights from the *Scientific* HTP.
  - Provides an idea storage template (Kasser 2018: Section 14.2) for organizing information about situations in case studies and reports in a format that facilitates storage and retrieval of information about situations (Kasser 2018: Section 10.1.7).

The nine HTPs shown in Figure 2.1 (Kasser 2018) are organized in four groups as follows:

1. *Two external HTPs:*
    1. *Big Picture:* includes the context for the system, the environment and assumptions.
    2. *Operational:* what the system does as described in scenarios: a black box perspective.
2. *Two internal HTPs:*
    3. *Functional:* what the system does and how it does it: a white box perspective.
    4. *Structural:* how the system is constructed and its elements are organized.
3. *Three progressive HTPs:* where holistic thinking begins to go beyond analysis and systems thinking are the:
    5. *Generic:* perceptions of the system as an instance of a class of similar systems: perceptions of similarity.

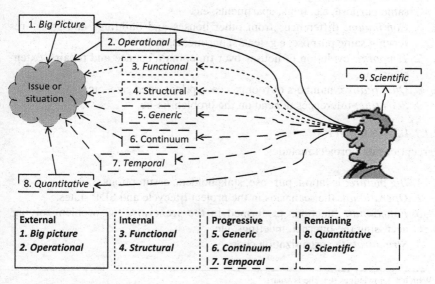

FIGURE 2.1 The nine holistic thinking perspectives.

6. *Continuum:* perceptions of the system as but one of many alternatives: perceptions of differences. For example, when hearing the phrase 'she's not just a pretty face',[6] the thought may pop up from the *Continuum* HTP changing the phrase to 'she's not even a pretty face'[7] which means the reverse.
7. *Temporal:* perceptions of the past, present and future of the system, for example, the system lifecycle (SLC).

4. Two remaining HTPs:
8. *Quantitative:* perceptions of the numeric and other quantitative information associated with the other descriptive HTPs.
9. *Scientific:* insights and inferences from the perceptions from the descriptive HTPs leading to the hypothesis or guess about the issue after using critical thinking (Section 2.1.1.1).

The first eight HTPs are descriptive, while the ninth (*Scientific*) HTP is prescriptive. Consider three examples of using the HTPs

2.1.1.4.1 A House

Perceptions of a house include:

1. *Big picture:* location, purpose and assumptions.
2. *Operational:* scenarios showing what the house is used for each weekday morning, afternoon, evening, as well as weekend and holiday activities.
3. *Functional:* functions performed in the scenarios, e.g. eating, sleeping, reading, talking, accessing the Internet, etc.
4. *Structural:* electrical, plumbing, heating, cooling, etc.
5. *Generic:* similarity with other houses and buildings and structures serving same purpose, e.g. tents, apartments, etc.
6. *Continuum:* differences from other houses and buildings and structures serving same purpose, e.g. tents, apartments, etc.
7. *Temporal:* evolution of houses over time, maintenance and repairs, extensions, etc.
8. *Quantitative:* numbers of rooms, costs, prices, land size, etc.
9. *Scientific:* inferences depend on the problem/issue.

2.1.1.4.2 A Project

Perceptions of a project include:

1. *Big picture:* location, purpose, stakeholders, assumptions.
2. *Operational:* the scenarios in the project lifecycle and SDP states.
3. *Functional:* functions performed in the scenarios, e.g. problem-solving, accessing the Internet, meetings, etc.
4. *Structural:* the organization chart.

---

[6] Which acknowledges that she is smart.
[7] Which means that not only is she not smart, she is also not pretty.

5. *Generic:* similarity with other projects.
6. *Continuum:* differences from other projects.
7. *Temporal:* the situation that gave rise to the project, and what will happen when the project completes.
8. *Quantitative:* all categories of costs, due dates and other quantifiable attributes.
9. *Scientific:* the PP; it is a hypothesis until completed.

2.1.1.4.3 The Ongoing Debate as to Whether a Well-Trained Manager (In Management) Can Manage Any Type of Project in Any Domain

Perceptions of the debate include[8]:

- *Big Picture:* an ongoing debate concerning if a manager does or does not need domain knowledge to manage a project in a domain, one side holds that management competence is enough.
- *Functional:* the functions of project management are planning, organizing, directing and controlling (Fayol 1949).
- *Continuum:* some projects proceed according to the plan while others diverge from the baseline.
- *Scientific:* both parties to the debate are correct but in different circumstances. As long as the project proceeds according to the plan and does not diverge from the baseline, no domain knowledge decisions need to be made and the project manager does not need any domain knowledge (one side of the debate). However, when the project deviates from the plan, the project manager does need domain knowledge to make the decisions to restore the baseline (the other side of the debate).

Use of the HTPs has explained the existence of the different points of view in the debate and removed the need for the debate.

2.1.1.4.4 Further Examples

Further examples of the use of the HTPs to store information in this book include perceptions of:

- Problem-solving in Chapter 3.
- Complexity in Section 3.8.
- Management in Section 4.2.

While the HTPs provide a standard set of perspectives on a perspectives perimeter (Kasser 2018: Chapter 10), perceptions from the *Continuum* HTP point out that there are other perspective perimeters including emotional, cultural, personal, the other party's (in a negotiation) (Section 12.5), etc. These other HTPs should be used as and when appropriate.

---

[8] The other HTPs are not pertinent to the discussion and accordingly are not mentioned.

2.1.2 CATEGORIES OF PEOPLE

Categorization is an example of using perceptions from the *Continuum* HTP to identify the different categories. A good project manager categorizes the members of the team, motivates them and then assigns activities to each according to their ability. While many ways of categorizing people have been developed over the years, two useful ways in project management are:

1. The ways they approach problems.
2. Their levels of critical thinking.

Consider each of them.

2.1.2.1 The Five Ways of Approaching Problems

People can be classified based on the way they approach problems into five different types (Kasser, Hitchins, and Huynh 2009):

- *Type I - apprentices:* have to be told *what* to do and *how* to do it.
- *Type II - doers:* once told *what* to do (i.e. provided with the PP), Type IIs have the ability to follow the process (the *how* to do it).
- *Type III - problem solvers:* once given a statement of the problem, Type IIIs have the expertise to conceptualize the solution and to determine *how* to do it (create the PP for realizing the solution).
- *Type IV - problem formulators:* have the ability to examine the situation and determine *what* needs to be done about it (define the problem), but cannot conceptualize *how* to do it (the solution).
- *Type V:* combine the abilities of the Types III and IV, namely have the ability to examine the situation, define the problem, conceptualize the solution system and plan and manage the process.

2.1.2.2 The Five Levels of Critical Thinking

Critical thinking (Section 2.1.1.1) can be measured in various ways (Kasser 2018: Chapter 3). An appropriate method of measuring critical thinking for project management is the following five levels of critical thinking (Wolcott and Gray 2003), where the five levels (from lowest to highest) are:

1. *Confused fact finder:* a person who is characterized by the following:
    - Looks for the 'only' answer.
    - Doesn't seem to 'get it'.
    - Quotes inappropriately from textbooks.
    - Provides illogical/contradictory arguments.
    - Insists professor, the textbook or other experts provide 'correct' answers even to open-ended problems.
2. *Biased jumper:* a person for whom facts don't influence options. This person is characterized by the following:
    - Jumps to conclusions.

- Does not recognize own biases; accuses others of being biased.
- Stacks up evidence for own position; ignores contradictory evidence.
- Uses arguments for own position.
- Uses arguments against others.
- Equates unsupported personal opinion with other forms of evidence.
- Acknowledges multiple viewpoints but cannot adequately address a problem from viewpoint other than own.

3. *Perpetual analyser:* a person who can easily end up in 'analysis paralysis'. This person is characterized by the following:
   - Does not reach or adequately defend a solution.
   - Exhibits strong analysis skill, but appears to be 'wishy-washy'.
   - Write papers that are too long and seem to ramble.
   - Doesn't want to stop analysing.

4. *Pragmatic performer:* a person who is characterized by the following:
   - Objectively considers alternatives before reaching conclusions.
   - Focuses on pragmatic solutions.
   - Incorporates others in the decision process and/or implementation.
   - Views task as finished when a solution/decision is reached.
   - Gives insufficient attention to limitations, changing conditions and strategic issues.
   - Sometimes comes across as a 'biased jumper', but reveals more complex thinking when prompted.

5. *Strategic revisioner:* a person who is characterized by the following:
   - Seeks continuous improvement/lifelong learning.
   - More likely than others to think 'out of the box'.
   - Anticipates change.
   - Works towards construction knowledge over time.

2.1.3 COMPETENCE

Perceived from the *Continuum* HTP, the literature on:

- Project management focuses on process.
- 'Excellence' focuses on people and ignores process (Peters and Waterman 1982, Peters and Austin 1985, Rodgers, Taylor, and Foreman 1993).

Yet people are generally ignored in mainstream project management, although that is beginning to change. Even so, many organizations measure competence based on years of experience, something that is easy to measure. However, it only measure years not competence. This section summarizes a two-dimensional CMM (Kasser 2015b: pp. 226–223) modified for project management shown in Table 2.1. This maturity model follows Maslow's approach of aggregating lists into broad areas of generic characteristics (Maslow 1966, 1968, 1970) and groups the knowledge, traits, abilities and other characteristics of successful systems engineers into a two-dimensional maturity model.[9]

---
[9] Due to space limitations, where prior work covers a topic in detail, the work is cited and summarized.

TABLE 2.1
A CMM for Project Managers

	Type I	Type II	Type III	Type IV	Type V
Category 1: Knowledge Areas					
Systems engineering	Declarative	Procedural	Conditional	Conditional	Conditional
Problem domain	Declarative	Declarative	Conditional	Conditional	Conditional
Implementation domain	Declarative	Declarative	Conditional	Conditional	Conditional
Solution domain	Declarative	Declarative	Conditional	Conditional	Conditional
Category 2: Cognitive Characteristics					
Systems Thinking					
Descriptive (8)	Declarative	Procedural	Conditional	Conditional	Conditional
Prescriptive (1)	No	No	Procedural	No	Conditional
Critical Thinking	Confused fact finder	Perpetual analyser	Pragmatic performer	Pragmatic performer	Strategic revisioner
Category 3: Individual Traits (Sample)					
Communications	Needed	Needed	Needed	Needed	Needed
Management	Not needed	Needed	Needed	Needed	Needed
Leadership	Not needed	Not needed	Needed	Needed	Needed
Others (specific to situation)	Organization specific	Organization specific	Organization specific	Organization specific	Organization specific

The vertical dimension is based on three categories:

1. *Propositional knowledge:* of project management. Knowledge has been categorized in several different ways usually by content. However, one content-free classification of propositional knowledge described the following three types (Woolfolk 1998, Schunk 1996: p. 166):
   1. *Declarative knowledge:* knowledge that can be declared in some manner, e.g. facts, subjective beliefs and organized passages. It is 'knowing that' something is the case. Describing a process is declarative knowledge.
   2. *Procedural knowledge:* knowing how to do something. It consists of rules and algorithms and must be demonstrated; performing the process demonstrates procedural knowledge.
   3. *Conditional knowledge:* knowing when and why to tailor and apply the declarative and procedural knowledge and why it is beneficial to do so.
2. *Cognitive characteristics:* systems thinking and critical thinking provide the pure management (Section 4.2.4.3) problem identification and solving skills[10] to think, identify and tackle problems by solving, resolving, dissolving or absolving problems (Section 3.1.2.1).

---

[10] Problem-solving and identification skills have been listed separately to map into Types IV and V as discussed below.

3. *Individual traits:* the traits providing the skills to communicate with, work with, lead and influence other people, ethics, integrity, etc. These traits include communications, personal relationships, team playing, influencing, negotiating, self-learning, establishing trust, managing, leading, emotional intelligence (Goleman 1995), and more (Covey 1989, Frank 2010, ETA 2010).

The horizontal dimension provides a way to assess the competence of a person in each broad area of the vertical dimension against the five ways of approaching problems.

2.1.3.1 Using the CMM

In order for an organization to use the CMM, the contents of each of the three categories must be determined and the CMM populated. If the organization already has a competency model then the competencies need to be transferred from the organizations' competency model into the appropriate areas in the CMM. If the organization does not have a competency model and wishes to develop one, then the CMM allows standardization of groupings, which helps to identify both errors of commission and errors of omission (Section 3.4.3.1). However, before developing a competency model, a cost-benefit trade-off should be performed since the amount of effort will depend on the level of detail required. The development effort should be for a model that will be useful, not something that will keep the human resource department busy.

Candidates must qualify at the appropriate proficiency level in all three categories to be recognized as being competent at that competency level. Assessment of knowledge, cognitive skills and individual traits is made in ways already practised in the psychology domain and do not need to be reinvented by systems engineers. Where knowledge is required at the conditional level, it includes procedural and declarative. Similarly, where knowledge is required at the procedural level, it includes declarative knowledge. While examination questions can require the respondent to use conditional knowledge, the successful application of conditional knowledge in the real world must be directly demonstrated by results documented in the form of knowledge, skills and abilities[11] supported by awards, letters and certificates of appreciation from third parties (e.g. employers, customers, etc.). The requirement for supporting documentation overcomes the current deficiency in the use of KSAs. The assessment could be in two parts, one part by examination for declarative knowledge and the second by a portfolio demonstrating successful experience for the procedural and conditional knowledge. Perceptions from the *Generic* HTP show that the portfolio model is already used for qualifications by professional societies including recognition as a:

- Certified Member of the Association for Learning Technology (ALT) (CMALT), a qualification awarded by the ALT.

---

[11] A series of narrative statements that is required when applying to U.S. Federal government job openings.

- Fellow of the Institution of Engineering and Technology (IET) (FIET) a qualification awarded by the IET in the United Kingdom.

Assessment of a candidate is simple in concept as follows:

1. *The project management domain knowledge:* the knowledge in the three domains of a problem (Section 3.4.6) is assessed as being declarative, procedural and conditional. While knowledge of project management applies to all projects, the problem, implementation and solution domain knowledge is situational. The question then arises as what is the domain knowledge is to be? That the domain knowledge competency is situational rather than generic does not stop the CMM being populated by organizations needing competency assessments for *their* personnel working in their situation on *their* projects in their problem and solution domains.
2. *The cognitive skills:* a number of ways of assessing the degree of critical thinking as declarative, procedural and conditional have been described in the literature (Kasser 2018: Section 14.2.1.2). The method used by Wolcott and Gray (Section 2.1.2.2) is used in the CMM.
3. *The individual traits:* the appropriate individual traits are assessed as being 'needed' or 'not needed' at a specific level of ability to suit the role of the project manager in the organization and assessed in the way that the employment and training administration (ETA) industry standard competency models assess those traits (ETA 2010). There is no need to reinvent an assessment approach.

2.1.4 Motivation

One of the problems faced by a PM is how to keep the members of the project team who are externals (Section 2.1.4.1) motivated to complete the project within the constraints of cost and schedule. Each person working in a project has his/her own goals. The goals of project will be met in the optimal manner if the individual and project goals can be aligned. Even self-starters may need to be motivated to do what the project wants them to do. Finding out why people have goals, means that we have to understand why people behave in the way they do. The study of motivation has to do with:

- The analysis of the various factors, which incite and direct an individual's actions (Atkinson and Shiffrin 1968).
- How behaviour gets started, is energized, is sustained, is directed, is stopped and what kind of subjective reaction is present in the project while all this is going on (Jones 1955).

The literature on behavioural psychology contains a number of different theories of behaviour and motivation. Consider the following brief summaries of six of the better-known ones.

1. Gallagher et al.'s externals and internals in Section 2.1.4.1
2. McGregor's Theory X and Theory Y in Section 2.1.4.2.
3. Maslow's hierarchical classification of needs in Section 2.1.4.3.
4. The Herzberg Motivation-Hygiene Theory in Section 2.1.4.4.
5. Recognition and rewards in Section 2.1.4.5.
6. Expectations affect performance in Section 2.1.4.6.

2.1.4.1 Externals and Internals

Gallagher et al. discuss the following types of personality: (Gallagher, Wilson, and Levinson 1992):

1. *Externals:* who work based on inputs, statistics and testimonials. These types of people tend to be motivated by regular praise, recognition and feedback.
2. *Internals:* self-starters and self-directed who work based on their own opinions, feelings and values. These types of people let actions speak for themselves, are not interested in recognition and awards, but work to meet their internal standards and goals.

2.1.4.2 McGregor's Theory X and Theory Y

McGregor described two opposing sets of assumptions, which summarize the different theories regarding the nature of man (McGregor 1960). He called these Theory X and Theory Y. The assumptions underlying Theory X are as follows:

1. The average human being has an inherent dislike of work and will avoid it if he can.
2. Because most people dislike work, they must be coerced, controlled, directed or threatened with punishment to get them to put forth adequate effort towards the achievement of organizational objectives.
3. The average human being prefers to be directed, wishes to avoid responsibility, has relatively little ambition and wants security above all.

The Theory Y assumptions are as follows:

1. The expenditure of physical and mental effort in work is as natural as play or rest.
2. Man will exercise self-direction and self-control in the service of objectives to which he is committed.
3. Commitment to objectives is a function of the rewards associated with their achievement.
4. The average human being learns, under proper conditions, not only to accept but to seek responsibility.
5. The capacity to exercise a relatively high degree of imagination, ingenuity and creativity in the solution of organizational problems is widely, not narrowly, distributed in the population.

Inferences from the *Scientific* HTP include:

- The opposing assumptions are at opposing ends of a continuum of motivation.
- An individual's position on the continuum of motivation may change depending on circumstances.[12]
- The management style should match the individual's position on the continuum of motivation. So, a person demonstrating Theory X characteristics should be managed via Theory X techniques, and a person demonstrating Theory Y characteristics should be managed via Theory Y techniques (Section 12.4.3.6).
- Theory Z (Ouchi 1982) lies somewhere in the middle of the continuum of motivation.

2.1.4.3 Maslow's Hierarchical Classification of Needs

Maslow's hierarchical classification of needs has been by far the most widely used classification system in the study of motivation. Maslow developed his theory in three books (Maslow 1954, 1970, 1968). Maslow's list of needs is:

- *Arranged in a hierarchy:* commonly drawn as a pyramid, and contains a set of hypotheses about the satisfaction of these needs.
- *Short:* the five need categories are (from lowest to highest) as follows:
  1. *Physiological:* i.e. the need for food, water, air and so on.
  2. *Safety:* i.e. the need for security, stability and the absence from pain, threat or illness.
  3. *Social:* i.e. the need for affection, belonging, love, etc.
  4. *Esteem:* i.e. the need for the personal feelings of achievement or self-esteem and the need for recognition or respect from others.
  5. *Ego:* The need for actualization, i.e. the feeling of self-fulfilment or the realization of one's potential.

According to Maslow:

- The categories of needs are arranged in a hierarchy, so that in general, the lower level needs must be satisfied before the higher-level needs act as motivators.
- The behaviour of an adult is determined by a number of motivating factors at the same time.
- Once a need was satisfied, it stopped being a motivating factor and another higher need then became the motivating factor.
- The higher-level needs can be satisfied in more ways than the lower level needs.

---

[12] My children provided a perfect example of this continuum when they were young. They needed no motivating for some tasks (e.g. eating ice cream), while other tasks required an enormous amount of parental energy expenditure to get the children to perform them (e.g. cleaning up their rooms).

- For ego or self-actualization needs, increased gratification led to an increase in the need.
- In addition, the hierarchy is not a rigid fixed order, which is the same for all individuals the order varies somewhat from person to person especially near the middle of the hierarchy (Maslow 1968: p. 30). Maslow described the situation in terms of a dynamic hierarchy of needs.

2.1.4.4 The Herzberg Motivation-Hygiene Theory

The Herzberg motivation-hygiene theory proposes that the primary determinants of employee satisfaction and dissatisfaction are factors intrinsic to the work that employees do (Herzberg 1966). These factors are called:

1. *Motivators:* employees are presumed to be motivated to obtain more of them. Examples of motivators are:
   - Achievement.
   - Recognition.
   - Responsibility.
   - Advancement.
2. *Hygiene factors:* determine the degree of dissatisfaction. Examples of hygiene factors are:
   - Poor working conditions.
   - Company policy and administration.
   - Lack of tools.
   - Low salaries.
   - Poor benefit package.

Herzberg's theory indicates the need for motivators and the need to reduce the hygiene factors. There seems to be a close relationship between Maslow's hierarchy of needs theory and Herzberg's motivation-hygiene theory. Herzberg's hygiene factors map into the Maslow's lowest two needs, while the motivators map into Maslow's three highest needs.

2.1.4.5 Recognition and Rewards

Skinner proposed that people's behaviour could be influenced by rewarding desired behaviour (Skinner 1954). This approach treats the person as a black box and ignores the internal processes within the person (feelings and thoughts) (Kast and Rosenzweig 1979: p. 245). The output (behaviour) is a function of the input (rewards). Behaviour reinforced by recognition of accomplishments will continue, as long as the rewards are linked to performance objectives. Strokes or immediate recognition of achievements is a major motivator. The more specific a stroke, the more positive it is. Recognition can take many forms, including:

- Awards.
- Badges.
- Certificates of appreciation.
- Choice of future assignments.

- Financial rewards (raises and bonuses).
- Offices in desirable locations.
- Parking spots of the month, etc.
- Publication of names in periodicals.
- The 'Thank You' (Kasser 2018: Section 8.13).
- A mixture of the above.

2.1.4.6 Expectations Affect Performance

Studies have also shown people:

- Behave in the way they are expected to behave.
- Behaviour can best be influenced by the following three aspects:
  1. Specify the desired goals in a quantifiable manner.
  2. Commitment towards accomplishing the goals.
  3. Feedback about performance. Stroke or recognize performance in an appropriate manner. Different people need different types of strokes. Sometimes all it takes is a 'thank you' or a 'well done' (Kasser 2018: Section 8.13).
- Achieve the results they wish to achieve. This phenomenon is sometimes termed the 'self-fulfilling prophecy' (Merton 1948).
- Observe what they expect to observe. A typical example is in proofreading. People read the words they expect to read, and miss the mistakes.
- Do actions they expect will provide them with the greatest benefit (Vroom 1964).

2.1.4.7 Discussion

Many theorists have suggested that the following equations express the relationship between ability, motivation and performance (Lawler III 1973: pp. 8–9):

$$\text{Performance} = (\text{Ability} * \text{Motivation})$$

$$\text{Ability} = \big(\text{Aptitude} * (\text{Training} + \text{Experience})\big)$$

where,

- *Motivation:* is a function of attitude, experience, recognition and expectations.
- *Ability:* refers to how well a person can perform at the present time. If a person lacks the ability to accomplish a task, no amount of motivation or effort will lead to better performance (Kast and Rosenzweig 1979: p. 246).
- *Aptitude:* refers to whether an individual can be brought through training and experience to a specified level of ability.

Although these theories are complex, there are enough common threads to make some practical use of them (Kast and Rosenzweig 1979: p. 250). The key point in motivation is that people motivate themselves based on an attempt to satisfy the

The Seven Interdependent P's of a Project

need that is *most important* to them at a point in time. For example, when faced with an unexpected bill for a large amount of money, and a low bank balance, a person will experience a dramatic change in the relative order of needs. Consider the category level in Maslow's hierarchy as a weighting factor defining the urgency of a need. Then, if, instead of a hierarchy or pyramid, think of the situation at a point in time, as a pie chart where the size of the slices reflects the weighting factor, the answer to the motivation question 'what's in it for me?' is 'a slice of the pie'. The manager's problem then becomes to apply the theories summarized above to figure out which slice, and its size (the incentive) to use to motivate a person (Kasser 1995).

2.1.5 MEETINGS

People take part in meetings most of which tend to be a waste of time (Augustine 1986). However, if organized systemically and systematically, meetings can be effective tool for achieving desired results (Kasser 1995), including:

- Obtaining consensus for a course of action, e.g. milestone reviews (Section 5.4.2).
- Transferring information, e.g. staff meetings.

Effective meetings have a (Kasser 1995):

1. *Purpose:* which provides focus, determines who is to attend and why.
2. *Agenda:* which serves several purposes:
    - Published ahead of time, it allows participants to think ahead and come to the meeting having thought about the issues.
    - Helps keep the meeting on track. People tend to respect and abide by written guidelines more than they do for verbal guidelines.
3. *Restricted attendance:* each person attending a meeting is a cost to the project. Minimize the cost by restricting attendance to those who have a need to be there as:
    - *Contributors:* people who will make a contribution.
    - *Recipients:* people who need to receive information and their attendance at the meeting is the optimal way for them to receive it.
4. *Time limit:* people have limited attention spans. They tend to be more active at the start of a meeting, so the effectiveness of the meeting decreases over time (Mills 1953). After an hour or so, terminate the meeting or take a break. If you do not, after an hour and half or so, there is a good probability that at least one person will need to answer the call of nature. If they are counting down the seconds until the break, because they do not wish to disturb the meeting, they are not participating in the meeting.
5. *Prompt beginning:* there is a cost associated with each person attending. If people wait around for a latecomer, they are being paid to waste time. Unless you have a line item in the budget for wasted time, there is going to be a cost overrun or some other activity will not be performed as well

it should be. People learn by observing. If they see meetings starting late, they will estimate the real start time and arrive accordingly. If they see the meetings start on time, they will arrive on time.
6. *Leader:* facilitates the meeting by:
   - Guiding it through the agenda.
   - Encouraging discussion without deviation in a tactful manner.
   - Ensuring the meeting does not digress or adjourn without reaching a conclusion or assigning an action item to achieve a conclusion.
   - Managing disruptors.
   - Merging the official agenda with each person's hidden agenda.
7. *Action items:* make sure the action item:
   - Is relevant.
   - Can be concluded without cost and schedule impact to the project.
   - Is assigned to one person to simplify accountability. If the action item must be carried out by more than one person, then assign it to the person who will lead the group performing the action.
   - Is completed on time. Use the just in time approach to assigning completion dates.
   - Is followed up on, to ensure timely completion. This may be done by briefly reviewing its status at a project progress meeting, or sending out periodic progress reports as appropriate.
   - Is not arbitrarily assigned a completion date, people soon learn which managers 'cry wolf' and act accordingly.
8. *Summary:* review of what was achieved or agreed to at the meeting. This can be done effectively when reviewing the action items assigned during the meeting.
9. *Timely termination:* people have other things to do and want to do them; consequently, they lose interest in the proceedings as the meeting stretches out.
10. *Metric to determine the degree of success of the meeting:* define the criteria for success at the same time as you set the objective of the meeting. Make the measurement and determine the effectiveness of the meeting.

2.1.6 COMMUNICATIONS

According to Talleyrand, writing some 200 years ago, language was given to man to conceal thought. This would certainly appear to be the case in the 20th century. There are, for example, engineers working in the Information Systems [sic] field who can speak entire paragraphs without using a single word.

Augustine (1986)

Project managers need to communicate effectively to stakeholders and other parties. Clear and concise communications are a key element in successful projects (Section 4.4.1.12.1). Communications may be formal and informal, written and verbal (Kasser 2015a).

2.1.6.1 Formal Written Communications

Formal written communications tend to be in the form of documents, notes and emails. This section focuses on documents. Thinking before sending a written communication is crucial, because once sent it cannot be taken back. Letters can be kept for years, these days they can be scanned and copies kept in an electronic archive. When an email is received at the recipient's computer host, the email is archived and can be accessed for up to several years if not longer. It is often useful, to write a letter, an email or a document and then wait 24 h before editing it and sending it. If you make major changes to the letter or email, then another waiting period would be useful.

When you sign off on a document you have written, reviewed or approved, you are stating that the document meets your personal quality standards. Your signature on that document shows your level of competency to everyone who subsequently reads that document (Kasser 1995).

An effective process for creating documents is discussed in (Kasser 2018: Section 11.4).

2.1.6.2 Formal Verbal Communications

A prepared statement is a formal verbal way of communicating ideas between the speaker and the audience. A presentation is a speech reinforced with text and graphics used in the presentation slides. Presentations can be formal and informal. In the same away as the abstract or executive summary is a hook to entire the reader to read the full document, in many instances, the presentation is the advertisement for the document/paper it is summarizing. That means you must use the presentation to excite the audience into reading your document/paper and making use of the content. Generally, you have two to three slides (minutes) to catch their attention. This section discusses:

1. Factors that contribute to effective presentations in Section 2.1.6.2.1.
2. Effective use of text and graphics in Section 2.1.6.2.2.
3. Using backup slides in Section 2.1.6.2.3.
4. Rehearsing the presentation in Section 2.1.6.2.4.
5. Making the presentation in Section 2.1.6.2.5.
6. What happens after the presentation in Section 2.1.6.2.6.
7. Learning from other people's presentations in Section 2.1.6.2.7.

2.1.6.2.1 Effective Presentations

Presentations need to be effective. Some of the factors that contribute to effective presentations are:

- Starting the presentation with a summary of the benefits to the people attending the presentation. This could be in the form of:
    - The objectives of the presentation.
    - An agenda.
    - A list of topics.

- Organizing the presentation about a key idea and using examples and graphics.
- Not speaking for more than 2 min on the same slide. Boredom sets in quickly and attention span wanders (Mills 1953).
- Using redundant communications channels; written and verbal. People receive concepts in various ways, including listening and seeing. Redundant channels can be good when presenting to people from other countries who may not understand the spoken language as well as the written language.
- Not reading the slides word for word.[13] Many members of the audience can read faster than you can speak and they will get bored quickly. At that point, they will tend to tune you out and not receive any further thoughts.
- Periodic progress slides which act as a road map to allow the audience to keep track of where you are in the flow of ideas.
- Numbered slides to allow for quick and easy reference to specific sides during a discussion on the presentation.
- One or more concluding slides to allow you to summarize the presentation and let the audience know that the presentation is over, namely to provide closure. The concluding summary slide can also act as a reminder to give you a chance to mention something you forgot to mention during the presentation.
- A metric to determine the degree of success of the presentation. The metric will depend on the purpose of the presentation. It may be as simple as the:
  - Number of people clapping at a conference.
  - Number of units sold following a sales presentation.
  - Grade received by the student making the presentation in class.

2.1.6.2.2 Effective Text and Graphics

The text and graphics in each slide should be effective, namely they should communicate in an easy way. There is no reason to have the audience try to work out what you are trying to communicate when you can easily make it simple. Factors that contribute to the effectiveness of text and graphics include:

- *Simplicity:* using slides that conform to Miller's rule of no more than 7 ± 2 items (Miller 1956) to assist in comprehension of the contents of the slide.
- *Readability:* keeping the font size large, 20 pts. should normally be the minimum size. How many times have you heard a presenter say something like, 'I know you can't read this but …'? There is a better way but it takes time and effort because you have to figure out exactly what it is that you are trying to communicate (Kasser 2015a).
- *No typographical errors:* just because the spelling has bean [sic] checked does not mean that the spelling is correct. You may have accidently used a synonym which passed the check but is the wrong word in context. Did you

---

[13] Unless your audience consists on non-native English language speakers. In such a situation, read the slides word for word but also expand on the words.

The Seven Interdependent P's of a Project

note the spelling of 'bean' in this paragraph? Typographical errors in a presentation are distracting to the audience and easy to eliminate.
- *Lack of ambiguity:* ambiguity is a major contributor to miscommunications.
- *Contrast:* the contrast between text and background should be such that the slides can be read under all lighting conditions. Just because you can see it at the terminal when preparing the presentation does not mean that the audience will be able to see the projected version.
- *Appropriate to the presentation:* snazzy photographs can be distracters as the audience look at the photograph and wonder why it is being shown instead of listening to the presentation.
- *To the point:* reinforcing the words being spoken.
- *Consistency:* using the same word for the same concept each time it occurs. Synonyms can introduce ambiguity and misunderstandings.

2.1.6.2.3 Using Backup Slides

Time-limited presentations should flow without being bogged down in details. However, there are usually one or more persons attending the presentation, who will ask for details. You generally do not have enough time to go in to all the details, so go into none of them but anticipate the questions. Prepare backup slides providing detailed information that is likely to be requested. You might even prepare hyperlinks in the final slide to the backup slides and end the presentation with a reminder about the graphic and a question to the audience asking them which details they would like within the constraints of the remaining time. Information in the backup slides can include:

- *Anticipated technical questions:* the audience might want to know how something works.
- *Very important person hot buttons:* know your audience and their interests and anticipate the questions.
- *Supporting data:* in the form of charts and pictures together with source citations to assist in the credibility and indicate the timeliness of the information.

Before the era of electronic presentations, it was simple to create backup transparencies and select them in response to a question, and then display them. In PowerPoint and other presentation tools, it is more difficult when you first begin to use the software tool. However, you can create the backup information in various ways including:

- Using hidden slides and providing hyperlinks to and from the hidden slides.
- Locating the backup slides at the end of the presentation and:
  - Scanning though them to find the appropriate ones with which to respond to a question.
  - Using a menu slide in the backup section with hyperlinks to select the appropriate slide. In this situation, each backup slide should contain a hyperlink to return back to the menu slide.

2.1.6.2.4 Rehearsing the Presentation

Effective presentations are rehearsed especially when you are trying to communicate new ideas. Suggested activities for rehearsing the presentation are:

- *Timing the presentation:* keeping to time is the most important part of the presentation. If you go longer than the allotted time, the audience will start to think about where they should be and what they should be doing in that time and stop listening. If they are not too polite to leave, they may start to leave while you are still speaking.
- *Rehearsing in front of a video camera:* watch the video. Make sure you do not have any unconscious distracting behaviour such as waiving your hand about and fidgeting with the laser pointer.
- *Rehearsing in front of peers:* watch their body language to see where you start to lose their attention. As a rule of thumb if you speak for more than 2 min on a single slide you will start to lose the audience.
- *Using the feedback from your peers* to update the presentation both in terms of the contents of the slides (text and graphics) and in terms of what you say.
- *Keeping it current:* if you are going to be presenting at a symposium or conference, update the presentation during the event if you can to include something current from a pertinent presentation you attended. If you can't alter a slide, you can always make the reference verbally. In other types of presentations try to refer to some current pertinent event during the course of your presentation.
- *Rehearsing:* perform a final rehearsal the night before the presentation.
- *Finding the room beforehand:* this will
  - Help make sure that you will not be late to your own presentation, which can be very embarrassing.
  - Let you see if there is anything you should be aware of, and compensate for, such as bright lighting, size of room, etc.

2.1.6.2.5 Making the Presentation

Things you should know and consider when making your presentation include:

- *Keep to the allocated time:* the most important thing about the presentation is keeping to time.
- *Introduce the topic:* tell the audience what you are about to tell them during the rest of the presentation. In other words, give them an overview of the presentation as your introductions.[14] You might also want to thank the audience for attending.
- *Be enthusiastic!* How can you expect the members of the audience to be enthusiastic about your presentation if you aren't?
- *Expect people to leave in the middle:* they leave for various reasons such as the need to be somewhere else, lack of interest, waking up to the fact

---

[14] Notice the feed-forward concept.

they are in the wrong room and answering a call of nature. When someone leaves, don't comment, keep going but if everyone leaves you can stop talking.
- *Use of notes:* this is a personal issue. If you are comfortable speaking without notes, do so. If you need notes, use them. If you have some control over presentation location or computer you may be able to use two screens, the projected screen showing the presentation graphic and the local presenter's screen showing the notes part of the slide.
- *Eye contact with audience:* make eye contact with the audience. Do not focus on a single person, but slowly move around the audience making and holding eye contact for at least 5 s at a time. Watch the body language of the audience; if people nod as you speak, they probably agree with your points. If they are nodding with their eyes closed, they are probably falling asleep.[15]
- *Summarize the main concepts:* at the end of the presentation.

2.1.6.2.6 What Happens after the Presentation

After you have concluded your presentation, and invited questions or comments, the audience may or may not have a question.

2.1.6.2.6.1. No Questions In the event that nobody asks a question or makes a comment then perceptions from the *Continuum* HTP suggest that:

1. Nobody was interested.
2. You did a good job.
3. You overwhelmed the audience with...

Don't worry about it; you can't wind back the clock and redo the presentation. If you really want a question, then depending on the type of meeting, you can prevent the situation by passing a written question or two to the meeting manager to ask you in the event that nobody else poses a question before you make the presentation. The type of question could be to explain an issue you raised in the presentation (and have prepared backup slides) or to focus on something you wanted to mention but didn't fit into the main theme. Once the first question is asked and answered, the ice has been broken and other questions tend to follow.

2.1.6.2.6.2. Dealing with Questions Members of the audience generally make several types of questions or comments after a presentation, including:

- *Clarification of something you said:* in which case answer the question. If you don't know the answer, don't try to bluff. Admit that you don't know, but not by saying that you don't know the answer. After all you are supposed to be the expert. Instead, say that you do not have enough information to answer the question at this time and ask the questioner to send you an email

---

[15] This is an example of using a perception from the *Continuum* HTP to make poor jokes.

and you will provide an answer within a reasonable time. Don't accept a business card and promise to respond for two reasons:
1. You may not remember the question.
2. People often ask questions when they are not really interested in the answer, at least in the long term.

Asking for an email, means the person has to want the answer enough to send you an email. This approach used to cut out the dilettantes and reduce the speaker's workload. Unfortunately, these days with Wi-Fi, the questioner can send you an email from right there in the presentation room, so it is not much of a deterrent.

- *Request for additional information:* respond to these with a request for an email as discussed, or if you have the information on your computer with you, ask the questioner to see you later with a USB memory stick and you will download the information from your computer onto their memory stick.
- *Questioner wishes to make their point:* let the person speak and leave controlling the dialogue to the meeting manager.
- *Requests for copies of your presentation graphics:* respond to these in the same way as to a request for further information. However, be aware of intellectual property issues and don't pass out copies of copyrighted material when you don't own the copyright.

2.1.6.2.7 Learning from Other People's Presentations

When you watch other people presenting, watch what they are doing and presenting while thinking about the ideas they are presenting. You will see good and bad ways of presenting information. Adopt the good and make sure that your presentations are not as bad. If you like a presentation let the presenter know either in person or via an email. If you also ask a question in the email you may be on your way to developing a new colleague or even a new friend. If you disagree with the presenter or think the presenter was wrong, ask a polite question. If the presenter does not accept your opinion or answer your question, or even understand your question, you may find it better to remain silent. In any event, never make anyone lose face in a presentation. If you must ask embarrassing questions do so in private.

If the content does not interest you and you have no other place to be at, or cannot leave, then look at the way the information is presented (good and bad).[16]

2.1.7 MISCOMMUNICATIONS

One of the keys to project success (Section 4.4.1.12.1) is a common vision of what the project is going to achieve. Achieving a clear vision needs clear and concise communication. When trying to communicate with someone, you have to understand that

---

[16] I learnt this lesson many years ago when I went to see the London production of the musical *Cabaret* (Prince 1968). I was bored by the songs and dances and the slow plot development and felt like leaving. But then, I became interested in the use of the lighting (I trained as an electrical engineer) and the choreography and all of a sudden, the technical aspects of the play were interesting and I learnt a lot.

The Seven Interdependent P's of a Project

there are barriers that can block the transfer of meaning. Barriers to communications include the following (Kasser 2015a):

1. Cultural differences in perception discussed in Section 2.1.7.1.
2. Emotion discussed in Section 2.1.7.2.
3. Language discussed in Section 2.1.7.3.
4. Signal-to-noise ratio discussed in Section 2.1.7.4.

2.1.7.1   Cultural Differences in Perception

People in different cultures perceive the same situation differently. For example, in western Anglo-Saxon culture, it is polite to clear the plate when eating a meal as a guest. In Chinese culture, it is polite to leave some food on the plate to signal that you have had enough. Clearing the plate leaves an empty plate, which signals that you are still hungry and want more food. Any Anglo-Saxon visiting a Chinese home for dinner needs to be aware of this difference to avoid becoming fed up.[17] Cultures differ in organizations and nations as:

- *Organization:* Organizations have their own way of doing things, often called 'culture'. Communications may flow through channels in specific ways and your communication has to be adapted to enter the other organization and communicate the message in a clear and concise, non-ambiguous manner. Moreover, if the organization does things in a different way, or lacks a common frame of reference for the concept, the other organization may be unable to comprehend your concept. For example, try explaining the concept of colour to a blind person.
- *National:* Nations are an instance of a class or type of organization and inherit the properties of organizations.

2.1.7.2   Emotion

Emotion can be a barrier. If you are angry with someone, will you be willing to listen to what the person says? If you are in love with someone, will you listen and evaluate comments made about that person by someone else? If you determine that there is an emotional barrier, then be patient, and wait until the appropriate time for the communications.

2.1.7.3   Language

When people speak different languages, the barrier is obvious. However, the general solution of speaking louder is incorrect. Translators are employed to overcome this barrier. Even when you speak the same language, perceptions from the *Continuum* HTP indicate that you still may not communicate for the following reasons:

- *Different words for the same concept:* this situation gives rise to miscommunications and unnecessary complicated situations. This situation can arise when people in different disciplines address the same concepts and

---
[17] Pun intended.

the people in one discipline define new terminology rather than reuse the terminology for those concepts that exists in another discipline.[18]
- *Different concepts for the same word:* this situation can cause major communications problems when the concepts are close enough that the subtle differences are not noticed at the time, and each party subsequently proceeds in different directions. For example:
  - The word 'capability' has different meanings in different parts of the system or product lifecycle.[19]
  - There are subtle differences in the meaning of words used by systems engineers and software engineers and how to bridge that communications gap (Kasser and Shoshany 2000, 2001). These subtle differences may occur between all people communicating across other disciplines.
- *Opposite concepts represented by the same word:* an example of this situation is the notion that Great Britain and the United States are separated by a common language (Shaw 1925). The ramifications of this notion are subtle and it is not an easy concept to understand unless one has been sensitized to it. An example of the situation occurred in the mid-1970s during contract negotiations between personnel representing the U.S.-based Communications Satellite Corporation and British Aerospace (Kasser 1995). The language of the meeting was English. The meeting became stuck on one point. Someone then suggested tabling the issue. Both sides agreed and the meeting deteriorated even further. The situation was much improved when the interpreter who spoke both English and American pointed out that the verb 'to table' means:
  1. To place the subject *on* the table for immediate discussion – in English.
  2. To place the subject *under* the table for later discussion – in American.
- *Different perceptions of the meaning of the same word by different people:* people use words with a specific meaning in mind, which is different to the meaning perceived by another person. This is similar to the 'different concepts for the same word' example discussed above. This situation is akin to Humpty Dumpty telling Alice that, 'when he uses a word it means just what he chooses it to mean – neither more nor less' (Carroll 1872). The different

---

[18] On the other hand, in academic circles, the use of different terminology tends to result in research grants.

[19] As one more example, during a visit to Yamaguchi, Japan in 1980, I went into a department store to purchase a child's kimono for my 6-year old daughter as a present. It was a slack time in the store so as time went by, more sales assistants joined in the conversation. Nobody seemed to speak English but using a combination of pointing and phrases from the Berlitz phrasebook all went well. An appropriate kimono and obi were selected and then before concluding the purchase I wanted to know how to clean the garment. So, I looked up the Japanese word for 'cleaning' and stumbling a bit over the language asked in Japanese for instructions on how to clean the kimono. The sales assistants started whispering amongst themselves and looking at me strangely. After a minute or so, I opened the phrasebook and showed the chief sales assistant the word I had used. He looked, smiled and said something in Japanese to the remainder of the crowd. Everyone smiled and a number of people exclaimed, "dry clean"! Using a mixture of language and pantomime I was politely informed that the word I had used for 'clean' actually was normally only used in the context of 'cleaning the floor' (and they were wondering why I would want to wash the floor with this expensive kimono). In this situation (they agreed that foreigners were strange, but not that strange), since the context was obviously wrong, the unintentional misuse of the word 'clean' was recognized immediately and corrected.

meanings in the words may overlap the meanings of other words or contradict the meaning of the same word in another publication. For example, consider the word 'secure'. When told to 'secure' a building it is said that in the U.S. Department of Defense (DOD):
- *The Navy:* issues a purchase order for the building.
- *The Air Force:* locks the doors and turns on the alarm systems.
- *The Army:* first evacuates the personnel, then locks the doors and turns on the alarm systems.
- *The Marines:* assault the building using ground troops and air support, and then deploy squads in and around the building checking the credentials of all who aspire to enter and leave the building.

This example illustrates a subtle communications problem. When one hears unknown words, such as in a foreign language, the failure to communicate is obvious. However, when one hears words that sound correct in the context, the failure to communicate is not realized and sometimes produces serious consequences. This situation can happen when communications take place between different organizations, different national cultures and even different engineering specialties.
- *Use of the wrong word for a concept:* when the word is obviously wrong, this barrier is visibly present and can be corrected. However, this situation can cause major communications problems when the word is appropriate in context even though it is incorrect. The 'build a little test a little process' (Section 5.6.1) helps mitigate this communications barrier.

2.1.7.4 Signal-to-Noise Ratio

We are all subjected to information overload and it is sometimes difficult to separate pertinent information from the mass of information we received. Electronic engineers call this sorting the signals out of the noise.

2.1.8 MITIGATING THE EFFECT OF AND OVERCOMING MISCOMMUNICATIONS

The effect of the barriers can be:

- *Mitigated:* by planning a project using the 'build a little test a little process' (Section 5.6.1).
- *Overcome:* by dialogues using the question and answer approach.

This section discusses:

1. Questions in Section 2.1.8.1.
2. Answers in Section 2.1.8.2.
3. Active listening in Section 2.1.8.3.

2.1.8.1 Questions

In many cases, preparing or formulating and asking questions and using various techniques in a dialogue to verify that the concept has been communicated may overcome the barriers. Question and answer is most often thought of as taking the form of a dialogue. Questions and answers can take many forms including:

- Requirements elicitation, elucidation and validation.
- The dialogue between the lawyer and the witness in a courtroom.
- Examination questions and answers.
- Invitation to bid for a contract and the bid.
- Request for a tender or proposal and the tender or proposal.
- Feasibility studies which answer the questions concerning the feasibility of something.

This section discusses:

1. Types of questions in Section 2.1.8.1.1.
2. Attributes of a good question in Section 2.1.8.1.2.
3. The most useful types of questions in Section 2.1.8.1.3.
4. Preparing questions in Section 2.1.8.1.4.

2.1.8.1.1 Types of Questions

Questions can be closed or open where:

1. *Closed questions:* include all the solutions in the question. These are generally:
    1. Multiple-choice questions that limit the decision about the answer to the options in the question.
    2. Questions that require a 'yes' or 'no' response.
2. *Open questions:* do not include any solutions in the question. An example of an open question is, 'what are the three most important things you learnt from this chapter'?

2.1.8.1.2 Attributes of a Good Question

The attributes of a good question include:

- *Atomic:* contains only one main idea/concept, which prevents long multi-part questions where the early parts of the original question are forgotten. If you have to ask a multipart question, announce that fact and how many parts there will be in the question, then ask each part in turn and wait until it has been answered before asking the next part of the question.
- *Brevity:* you want to listen to the answer not talk. You learn more from listening than from talking.
- *Consistency:* applies within a single question and to a set of questions. Use the same word for the same meaning each time. Do not use synonyms or words with slightly different meanings. For example, do not use the words 'spade' and 'shovel' to refer to the same tool for digging holes in the ground.[20]

---

[20] They are different tools with different shaped blades, not synonyms for the same digging tool.

The Seven Interdependent P's of a Project

- *Clarity and conciseness:* maximizes the probability that you will be understood.
- *Grammatically correct:* use correct grammar, it not only helps to get your meaning across, it shows that you are educated.
- *Insensitiveness to errors in interpretation:* the question should be asked so that there is a low probability of error. For example, in an emergency situation after a disaster, when someone reports that they think someone else is dead, they should not be asked to 'make sure that the person is dead'.[21] They may interpret the question as an instruction to make sure the person is really dead by taking an appropriate action.
- *Unbiased:* don't bias the answer. Questions that bias the answer are known as leading questions in courtroom dialogue.

2.1.8.1.3 The Most Useful Types of Questions

The most useful questions to ask are the Kipling questions 'who, what, where, when, why and how' (Kipling 1912, Kasser 2018: Section 7.6). Each type of question initiates the exchange of concepts in different perspectives about different attributes of a concept. Consider a cup of instant coffee, for example, as the answers to the:

- *Who:* provide information about who is going to do operate, create or otherwise do something. In this instance, the answers provide information about who:
  1. Is going to prepare the cup of coffee.
  2. Serve the cup of coffee.
  3. Drink the cup of coffee.
  4. Dispose of the empty cup.
- *What:* information about the object associated with the concept. In this instance, the answers provide information about the ingredients and the equipment used to prepare the cup of coffee (cups, spoons, electric kettle, etc.).
- *Where:* provide information about where things happen. In this instance, the answers provide information about where the cup of coffee will be prepared and where it will be drunk. These answers may lead to further thoughts about transporting the cup of coffee from the place where it is prepared to the place where it is drunk.
- *When:* provide information about when things happen. In this instance, the answers provide information about when the cup of coffee will be prepared and when it will be drunk.
- *Why:*
  1. Provide information that helps you understand something, such as the cause of a problem, the need or how the situations arose. In this instance, the answers to the why questions provide the reasons why the cup of coffee is being drunk.

---

[21] See video joke on YouTube at www.youtube.com/watch?v=By0oe7BUDWQ last accessed on 26 July 2015.

2. Are useful for drilling down into the reasons for symptoms to determine the underlying cause, and are used in a concept called, 'The 5 why's' developed by Sakichi Toyoda for the Toyota Industries Corporation (Serrat 2009, Kasser 2018: Section 7.5).
- *How:*
  1. Often provide information about the process. In this instance, how the cup of coffee will be prepared and how it will be drunk. For example, if the coffee is drunk in sips over an hour or so, then you may need to serve the coffee in an insulated cup. If it is drunk within a few minutes, then a regular cup would be in order.
  2. Can also provide quantitative information such as 'how much will it cost'?

2.1.8.1.4 Preparing Questions

When preparing questions to ask, you have to consider several things including what you want the respondent to do. Do you want the person to:

1. Recall and repeat the information; demonstrate declarative knowledge (Section 2.1.3.1).
2. Apply information; demonstrate procedural knowledge (Section 2.1.3.1).
3. Analyse and use data or information; demonstrate procedural knowledge (Section 2.1.3.1).
4. Evaluate something; demonstrate conditional knowledge (Section 2.1.3.1).

So, you need to consider:

- *How easy it will be to recall and repeat the information:* you may have to allow for different response times and ways of delivering the answers depending on the nature of the topic and the type of question. Some types of questions may evoke an immediate written or verbal response, and some may need a study lasting for several months with written reports.
- *How easy it will be to answer the question:* this factor governs how much research needs to be done, how much time needs to be spent and possibly the cost of the resources used to determine the answer.
- *How difficult will it be to evaluate response:* one reason for multiple-choice questions in examinations is that they are very easy to evaluate.
- *How easy it will be to communicate the results* to those that need to make use of the information. This factor may govern how you want the responses delivered. Written responses are easier to distribute than verbal responses.
- *The validity of the question:* this factor produces a response containing the desired or pertinent information. Ambiguity reduces validity.
- *The reliability of the question:* this factor produces consistent similar responses from different respondents.

The Seven Interdependent P's of a Project

2.1.8.2 Answers

Questions and answers can be simple, and complex. The general rule in answering questions is to make it easy for the person who asked the question to realize that (1) the question has been answered and (2) to understand the answer.

Have you ever asked a question and instead of receiving an answer, you were faced with another question? Wasn't it annoying? Think about the process. The recipient received your question, thought about it and asked for further information or clarification as a prelude to answering the first question. Well you can defuse the annoyance of receiving a question instead of an answer is to use the words 'it depends' as the initial answer to most questions. Then go on and discuss what it depends on. This approach will open a dialogue, which will hopefully enable the concepts to be communicated. So, remember, never ever answer a question with another question, answer the question with the words 'it depends' instead.

2.1.8.3 Active Listening

Active listening (Rogers and Farson 1957, 2015) is a standard technique for applying the feedback principle to interpersonal communications to minimize errors in conveying meaning from one person to another. Active listening first recognizes that during a conversation, most people do not listen to what the other person is saying: they are too busy planning what they will say when the other person pauses. Standard active listening prevents that and comprises the following multi-step process:

1. When the other person speaks, give them your full attention and look them straight in the eyes. Then begin the following iterative loop.
    1. Listen to everything the other person says and try to understand it fully.
    2. Ask questions to clarify anything you don't understand and analyse the response.
    3. Watch the non-verbal communication cues. Sometimes the real message is in the emotion rather than the surface content. In these cases, you should respond to the emotional message.
    4. Rephrase what you have heard in your own words and ask the speaker if they meant what you are about to say. Use words such as, 'if I understand you, then ...', or 'Do you mean ...'. This is the principle of applying feedback.
    5. If, after you have rephrased what has been said and the person says, 'No that's not it!' or the equivalent, then go back to step 2. You may need to invoke the STALL technique (Kasser 2018: Section 3.2.9) at this time to regulate matters.[22] STALL is an original acronym and a mnemonic tool to help you remember how to deal with situations that range from being asked a question to avoiding panicking in a crisis. STALL stands for:
        - <u>S</u>tay calm: don't panic; wait until you understand what's going on. Then you can either panic or deal with the situation in a logical manner (*Continuum* HTP).

---

[22] Stalling for a while, but not too long, is a good way to initially deal with most undesirable situations.

- *Think:* think about what you're hearing, experiencing, being told or seeing, you're generally receiving symptoms; you need to understand the cause.
- *Analyse and ask questions:* gain an understanding what's going on. This can be considered as the first stage of the problem-solving process. And when you're asking questions and thinking, use idea-generating tools such as active brainstorming (Kasser 2018: Section 7.1) to examine the situation.
- *Listen:* you learn more from listening than from talking yourself.
- *Listen:* you learn more from listening than from talking yourself. Listening, analysing and asking questions are done iteratively until you feel you understand the situation. You have two ears and one mouth, that's the minimum ratio in which they should be used. Namely, that means do at least twice as much listening as talking.

2. When the speaker finally agrees with you, then you have (most probably) actually communicated and shared meaning.

In modifying active listening by the use of pattern matching, change step 3 to incorporate the pattern by adding words such as 'this reminds me of the [Type A Scenario]', and 'isn't this similar to [Type B scenario]' and explain why you find a similarity in the current situation. Use a metaphor appropriate to the other party such as sport.

Active listening can be used in other forms. For example, in the classroom instead of lecturing an overview of the readings or just requesting the students to read the readings, instruct the students to both read and present (Kasser 2013):

1. A summary of the reading.
2. The main points of the reading.
3. Some analysis of the reading.
4. A comparison of the reading with other literature.

This simple change to a class:

- Ensures that if the students misinterpret the readings the instructor can correct the misinterpretations by the next class session.
- Ensures that some of the students actually read the readings.
- Allows the students to see that different people/groups summarize readings slightly differently, in other words receive different messages from the same readings.

This form of active listening is used in the presentation exercises starting in Section 5.11.

2.2 POLITICS

Successful project management is directly linked to the ability of project managers and other key players to understand the importance of organizational politics and how to make them work for project success. While most of us view politics with distaste,

The Seven Interdependent P's of a Project

there is no denying that effective managers are often those who are willing and able to employ appropriate political tactics to further their project goals

Pinto (2000)

Perceptions of politics from the HTPs include:

- Big Picture: Politics is:
  - A major part of stakeholder management (Section 12.6).
  - The process of making decisions that applies to members of a group (Hague and Harrop 2013).
  - The use of intrigue or strategy in obtaining any position of power or control, as in business, university, etc. (Dictionary.com 2013).
  - A way of interacting with people to influence the achievement of goals by facilitating or impeding the achievement.
  - Distained by many people who have only been exposed to its negative use.

  Project management is generally performed in a political context because projects tend to exist outside the traditional organizational structure. Projects borrow people from functional departments so those people report to the managers of the functional departments who perform their evaluations and can reward good performance in various ways. The project manager who doesn't have this ability has to influence the project team to perform. Accordingly, one of the roles of the project manager is to use positive politics to facilitate the achievement of the project goals, via negotiation (Section 12.5) and conflict management (Section 12.3).

- *Operational:* note that projects produce changes, which tend to be resisted. For example, 500 years ago Machiavelli wrote, 'And it ought to be remembered that there is nothing more difficult to take in hand, more perilous to conduct, or more uncertain in its success, than to take the lead in the introduction of a new order of things. Because the innovator has for enemies all those who have done well under the old conditions, and lukewarm defenders in those who may do well under the new' (Machiavelli 1515). Sometimes change is resisted due to laziness, and sometimes due to fear of the unknown. You can overcome this hindering habit by showing people how the change will benefit them.[23] For example, it may:
  - Reduce the amount of work they will have to do once the change is implemented which will give them increased leisure or extra time to accomplish other tasks.
  - Increase their income.

  In other words, find out what is undesirable about the current situation or what problem people are having, then using positive politics show them how the change will help them (Crosby 1981: p. 92), namely, what's in it for them (Section 2.1.4.7). You can and should apply this idea to yourself.

---

[23] Especially if you want them to adopt your innovative idea.

- *Continuum:* note that politics is a tool that can be used:
  - *Negatively:* for self-serving purposes, revenge, spite, cover ups, etc.
  - *Positively:* to achieve the goals of a project, to improve the quality of life, etc.
- *Temporal:* note that even successful project managers can be hurt by negative politics and need to understand the nature of negative politics to prevent it from happening.

2.2.1 Positive Use of Politics

Some examples of using politics in a positive way are (Reardon 2005: p. 86) cited by Irwin (2007):

- *Creating a positive impression:* assuring that key people find you interesting and approachable.
- *Positioning:* being in the right place at the right time to advance your project.
- *Cultivating mentors:* locating experienced advisors.
- *Lining up the ducks:* making strategic visits to peers, senior people and support staff at which you mention your project and accomplishments and let them know what you can do for them.
- *Developing your favour bank:* favours usually require reciprocation; by agreeing to – or even offering – favours, you are making 'deposits' in anticipation that someday, when you need to call in a chit, you will have the currency to do so.

2.2.2 Negative Use of Politics

While the positive use of politics will advance your project or your organization, if it does this at the expense of others, it becomes negative politics. Some examples of using organizational politics in a genuinely negative manner include (Reardon 2005: p. 87) cited by Irwin (2007):

- *Poisoning the well:* fabricating negative information about others, dropping defaming information into conversation and meetings in the hope of ruining the target's career chances.
- *Faking left while going right:* leading others to believe you will take one action in order to increase the likelihood of succeeding via an entirely different manoeuvre, allowing or encouraging someone to think one condition exists when in reality another condition holds.
- *Deception:* lying, telling different people different things at the same time.
- *Entrapment:* steering or manipulating someone into a political position or action that results in embarrassment, failure, discipline or job loss.

People use negative politics for personal gain to the detriment of their organizations. They are successful because:

The Seven Interdependent P's of a Project 45

- Of the support given by their (middle) management who lack the perceptions from the *Big Picture* HTP to realize that the shenanigans they are supporting are harming the reputation of their organization.
- The organization does not seem to be able to detect and/or mitigate the use of negative politics.

2.2.3 Seven Steps That a Project Manager Can Take to Become Politically Astute

Seven steps that a project manager can take to become politically astute are (Pinto 2000):

1. *Understand and acknowledge the political nature of most organizations:* politics is a fact of life, exists in all organizations and has an impact on the success of the project. Accordingly, a project manager needs to understand how best to use positive politics to achieve a successful project.
2. *Learn to cultivate appropriate political tactics:* developing a reputation for keeping your word will assist negotiation and bargaining because you are often promising a future favour in return for something in the present. It is important to keep promises so do not make promises you cannot keep as part of the negotiation (Section 12.5) and bargaining. They will come back to haunt you. Other stakeholders outside the project may be concerned about the changes that will result from the project. They need to be reassured and their concerns need to be addressed.
3. *Understand and accept 'what's in it for me?':* the first thing to understand is 'what's in it for you – the project manager'. What's in it for you is a tool that if used properly will facilitate the success of the project and if misused can alienate influential stakeholders who can and probably will sabotage the project. Cultivate relationships with influential stakeholders *before* you need their aid. Develop positive personal relationships with stakeholders. Build this into your daily work plan (Section 12.1). When asking for help, you need to give your potential helper a reason as to why they should provide that help. Similarly, when dealing with project personnel, there needs to be a reason for the project personnel to perform. Do not be too blatantly obvious that you are attempting to influence the other person, keep it subtle. You should not wish to develop a reputation as a person playing politics.
4. *Try to level the playing before taking over the project:* many organizations do not allow project managers to influence performance evaluations of project personnel. Line managers often resist sharing the power of performance evaluation. However, since the project manager is most familiar with the performance of the personnel, it's to the benefit of the organization for the project manager to be involved in the performance evaluation. So, when taking over a project, ask for influence in the performance evaluation of the project personnel. It's probably not advisable to ask to take that role away from the line manager, it may not be possible to share it, but it

should be possible to provide input to the performance evaluation. When the performance evaluation is made, use positive politics; let the project team member have a copy of the evaluation before submitting it to the line manager and let the line manager know that the project team member has a copy.
5. *Learn the fine art of influencing:* use some of the leadership techniques discussed in Chapter 12. Understand the stakeholder you are dealing with, understand their concerns and the best way to influence them and then employ it. As a project manager you may not have financial reward power (Section 12.2.1), but you can deal with emotion (guilt, pride, shame, etc.). Recognize the type of critical thinker (Section 2.1.2.2) you are dealing with and tailor your arguments accordingly. Perceived from the *Generic* HTP, this is similar to understanding what motivates the person and then motivating the person accordingly.
6. *Develop your negotiating skills:* you need to understand the techniques of negotiation (Section 12.5) which:
   - Can be used against you.
   - You can use to achieve your goal.[24]

   The literature encourages the use of win–win techniques, but they can only be achieved with the collaborating style of negotiator. If you can recognize and match the style of the opposing negotiator, you stand a better chance of achieving your goal. Try and set up the negotiation as an example of mutual problem-solving looking for a win–win solution (Section 12.5.2).
7. *Recognize that conflict is a natural side effect of project management:* and can be dealt with constructively (Section 12.3).

2.3 PREVENTION

Prevention:

- Lowers costs and shortens schedules.
- May be thought of as the future value of work performed (or not performed) in the present. Perceptions from the *Generic* HTP show that this concept is similar to the concept of the future value of money which economists use to compare the value of money spent or saved in the future with the amount of money spent or saved in the present.
- Is difficult to prove after-the-fact. If something didn't happen, we cannot prove if it was prevented from happening or if the situation that would cause that something to happen did not arise.
- Is known as risk management when perceived from the *Generic* HTP.
- Is incorporated in the systemic and systematic approach to project planning and managing (Section 5.10).

---

[24] The same techniques used in reverse.

The Seven Interdependent P's of a Project

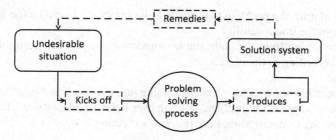

FIGURE 2.2 The problem-solving process as a causal loop.

2.4 PROBLEMS

Project management is about problem-solving. If a problem didn't need solving there wouldn't be any need for the project. The systemic and systematic way of identifying and resolving problems:

- Considers the problem-solving process as a causal loop as shown in Figure 2.2 (Kasser and Hitchins 2013).
- Is discussed in Chapter 3.

2.5 PROCESSES

Throughout management science - in the literature as well as in the work in progress - the emphasis is on techniques rather than principles, on mechanics rather than decisions, on tools rather than on results, and, above all, on efficiency of the part rather than on performance of the whole

Drucker (1973: p. 509)

Forty-five years later, according to the Project Management Institute, project management is accomplished through the appropriate application and integration of 47 logically grouped project management processes, which are categorized into five process groups (PMI 2013: p. 4), namely:

1. Initiating.
2. Planning.
3. Executing.
4. Monitoring and Controlling.
5. Closing.

Project management converts undesirable or problematic situations into desirable situations by means of people using resources for a period of time, namely:

1. Understanding the cause of undesirability in the situation.
2. Visualizing the situation without the undesirability (and improvements).

3. Formulating the problem that needs to be remedied to create the situation without the undesirability.
4. Transitioning from the undesirable situation to the planned desirable situation according to the plan.

This sequence is known as the problem-solving process (Section 3.6.1). Accordingly, the systematic approach to problem-solving uses iterative adaptations of the generic problem-solving or decision-making processes (Section 2.4).

2.5.1 THE PROCESS TIMELINE

A project takes place as a set of activities performed between milestones. Activities take time which cannot be recovered. Accordingly, it is important not to waste time doing the wrong thing or doing things ineffectively (Section 5.6.1).

2.5.2 THE THREE STREAMS OF ACTIVITIES

A project takes place as a set of activities performed between milestones. In development projects using one or more iterations of the waterfall approach, the milestones are predefined for management by objectives (MBO) (Mali 1972). The work between the milestones is split into three streams of activities (Kasser 1995, Kasser 2018: Section 8.14), which:

- Begins and ends at a major milestone.
- Is a template for planning a project workflow in conjunction with the Gantt chart (Kasser 2018: Section 2.12).
- Elaborates the Waterfall chart (Royce 1970, Kasser 2018: Section 14.9) view where the activities in each state of the waterfall are shown in Figure 2.3, where

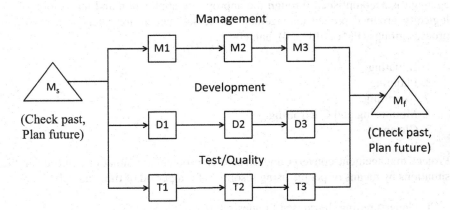

FIGURE 2.3 The three streams of activities in a project.

1. *Management:* the set of activities which include:
   - Monitoring and controlling the development and test stream activities to ensure performance in the state in accordance with the PP (Section 5.5).
   - Updating the PP to elaborate the work packages (WP) (Section 5.2.6) for the three streams of activities in the subsequent state in more detail.
   - Endeavouring to ensure that needed resources in the subsequent state will be available on schedule.
   - Providing periodic reports on the condition of the project to the customer and other stakeholders.
   - Being the contractual interface with the customer.
   - Performing the appropriate risk management activities on the process.
2. *Development:* the set of activities, which produce the products appropriate to the state by performing the design and construction tasks and appropriate risk management activities on the product.
3. *Test:* the set of activities known as quality control (QC) or quality assurance (QA), test and evaluation (T&E) and independent verification and validation (IV&V), which include:
   - Identifying defects in products.
   - Verifying the degree of conformance to specifications of the products produced by the development stream by performing appropriate tests or analyses.
   - Improving the process: 'Quality comes not from inspection, but from improvement of the production process' (Deming 1986: p. 29).
   - Prevention in the systems paradigm.

The development and management streams are tightly coupled; the quality/test stream and management streams are tightly coupled. The development and quality/test streams are independent between milestones, yet due to the interaction before, after and during, the milestone reviews they are interdependent over the project as a whole.

The interrelationships between the streams may be noted in Figure 2.3, which chunks the work between milestones (Section 5.6.1) into three parallel chunks where, in the first chunk:

- M1 is managing D1 and T1,[25] and planning how to manage D2 and T2.
- T1 is testing the products produced by D1 and planning to test the products to be produced in D2.

2.6 PROFIT

The profit made by a project is the return on investment (ROI) (Chapter 8). However, the ROI may not be financial, especially in government projects.

---

[25] Based on the updated PP at the starting milestone (M_s).

2.7 PRODUCTS

Products may be biological, chemical, procedural, technological and other, as well as a combination. Products may be produced during the performance of a project as well as by the project. For example, a requirements document is produced during the requirements state of the SDP and is used during the design and test states.

2.8 SUMMARY

This chapter helped you to understand projects by discussing seven interdependent P's of a project, which must be managed (or juggled) according to the PP (Section 5.5) to achieve the objectives of a project.

REFERENCES

Ackoff, Russel L. 1991. The future of operational research is past. In *Critical Systems Thinking Directed Readings*, edited by Robert L. Flood and Michael C. Jackson. Original edition. New York: John Wiley & Sons. Journal of the Operational Research Society, Volume 30, 1979.

Ackoff, Russel L., and Herber J. Addison. 2006. *A Little Book of f-Laws 13 Common Sins of Management*. Axminster: Triarchy Press Limited.

Ackoff, Russsel L., Herber J. Addison, and Carey Andrew. 2010. *Systems Thinking for Curious Manaers*. Axminster: Triarchy Press Limited.

Atkinson, Richard C., and Richard M. Shiffrin. 1968. Human memory: A proposed system and its control processes. In *The Psychology of Learning and Motivation: Advances in Research and Theory* (Vol. 2), edited by Kenneth W. Spence and Janet Taylor Spence. New York: Academic Press.

Augustine, Norman R. 1986. *Augustine's Laws*. New York: Viking Penguin Inc.

Carroll, Lewis. 1872. *Through the Looking Glass*. London: Macmillan.

Churchman, C. West. 1968. *The Systems Approach*. New York: Dell Publishing Co.

Covey, Steven R. 1989. *The Seven Habits of Highly Effective People*. New York: Simon & Schuster.

Crosby, Philip B. 1981. *The Art of Getting Your Own Sweet Way*. 2nd edition. New York: McGraw-Hill.

Deming, W. Edwards. 1986. *Out of the Crisis*. Cambridge, MA: MIT Center for Advanced Engineering Study.

Descartes, René. 1637, 1965. *A Discourse on Method*, trans. E. S. Haldane and G. R. T. Ross, Part V. New York: Washington Square Press.

Dictionary.com. 2013. *Dictionary.com* 2013. Available from http://dictionary.reference.com/.

Drucker, Peter F. 1973. *Management: Tasks, Responsibilities, Practices*. New York: Harper & Roe.

Eichhorn, Roy. 2002. *Developing thinking skills: Critical thinking at the army management staff college* [cited April 11, 2008]. Available from www.amsc.belvoir.army.mil/roy.html.

ETA. *General competency model framework*. The Employment and Training Administration 2010 [cited 9 September 2011]. Available from www.careeronestop.org/competencymodel/pyramid.aspx.

Facione, Peter. 1990. *Critical Thinking: A Statement of Expert Consensus for Purposes of Educational Assessment and Instruction*. Newark, DE: American Philosophical Association.

Facione, Peter. 2011. *THINK Critically*. Upper Saddle River, NJ: Pearson Education Inc.
Fayol, Henri. 1949. *General and Industrial Management*. London: Sir Isaac Pitman and Sons, Ltd.
Frank, Moti. 2010. Assessing the interest for systems engineering positions and the Capacity for Engineering Systems Thinking (CEST). *Systems Engineering* no. 13 (2):161–174.
Gallagher, Bill, Orvel Ray Wilson, and Jay Conrad Levinson. 1992. *Guerrilla Selling*. Boston, MA: Houghton Mifflin Company.
Gharajedaghi, Jamshid. 1999. *System Thinking: Managing Chaos and Complexity*. Boston, MA: Butterworth-Heinemann.
Goleman, Daniel. 1995. *Emotional Intelligence*. London: Bloomsbury.
Hague, Rod, and Martin Harrop. 2013. *Comparative Government and Politics: An Introduction*. London: Macmillan International Higher Education.
Herzberg, Frederick I. 1966. *Work and the Nature of Man*. Cleveland, OH: The World Publishing Company.
Irwin, Brian. 2007. Politics, leadership, and the art of relating to your project team. In *PMI® Global Congress 2007-North America*. Atlanta, GA. Newtown Square, PA: Project Management Institute.
Jones, Mari Riess, ed. 1955. *Nebraska Symposium on Motivation*. Lincoln, NE: University of Nebraska Press.
Kasser, Joseph Eli. 1995. *Applying Total Quality Management to Systems Engineering*. Boston, MA: Artech House.
Kasser, Joseph Eli. 2013. Introducing "knowledge readings": Systems engineering the pedagogy for effective learning. In *Asia-Pacific Council on Systems Engineering (APCOSE) Conference*, Yokohama, Japan.
Kasser, Joseph Eli. 2015a. *Holistic Thinking: Creating Innovative Solutions to Complex Problems* (Vol. 1). 2nd edition. Charleston, SC: Solution Engineering: Createspace Ltd.
Kasser, Joseph Eli. 2015b. *Perceptions of Systems Engineering* (Vol. 2). Charleston, SC: Solution Engineering: Createspace Ltd.
Kasser, Joseph Eli. 2018. *Systems Thinker's Toolbox: Tools for Managing Complexity*. Boca Raton, FL: CRC Press.
Kasser, Joseph Eli, and Derek K. Hitchins. 2013. Clarifying the Relationships between systems engineering, project management, engineering and problem solving. In *Asia-Pacific Council on Systems Engineering Conference (APCOSEC)*, Yokohama, Japan.
Kasser, Joseph Eli, Derek K. Hitchins, and Thomas V. Huynh. 2009. Reengineering systems engineering. In *3rd Annual Asia-Pacific Conference on Systems Engineering (APCOSE)*, Singapore.
Kasser, Joseph Eli, and Tim Mackley. 2008. Applying systems thinking and aligning it to systems engineering. In *18th INCOSE International Symposium*, Utrecht, Holland.
Kasser, Joseph Eli, and Sharon Shoshany. 2000. Systems engineers are from mars, software engineers are from venus. In *the 13th International Conference on Software & Systems Engineering & their Applications*, Paris.
Kasser, Joseph Eli, and Sharon Shoshany. 2001. Bridging the communications gap between systems and software engineering. In *INCOSE-UK Spring Symposium*.
Kast, Fremont E., and James E. Rosenzweig. 1979. *Organization and Management A Systems and Contingency Approach*. 3rd edition. New York: McGraw-Hill.
Kipling, Joseph Rudyard. 1912. The elephant's child. In *The Just So Stories*. Garden City: The Country Life Press.
Lawler III, Edward. 1973. *Motivation in Work Organizations*. Pacific Grove, CA: Brooks/Cole Publishing Company.
Lutz, S., and W. Huitt. *Information processing and memory: Theory and applications*. Valdosta State University 2003 [cited 24 February 2010]. Available from www.edpsycinteractive.org/papers/infoproc.pdf.

Luzatto, Moshe Chaim. circa 1735. *The Way of God*. Translated by Aryeh Kaplan. New York and Jerusalem, Israel: Feldheim Publishers, 1999.
Machiavelli, Nicollo. 1515. *The Prince* translated by W. K. Marriott, Electronically Enhanced Text. World Library, Inc. (1991).
Mali, Paul. 1972. *Managing by Objectives*. New York: John Wiley & Sons
Maslow, Abraham Harold. 1954. *A Theory of Human Motivation*. New York: Harper & Row.
Maslow, Abraham Harold. 1966. *The Psychology of Science*. New York: Harper and Row.
Maslow, Abraham Harold. 1968. *Toward a Psychology of Being*. 2nd edition. New York: Van Nostrand.
Maslow, Abraham Harold. 1970. *Motivation and Personality*. New York: Harper & Row.
McGregor, Douglas. 1960. *The Human Side of Enterprise*. New York: McGraw-Hill.
Merton, Robert King. 1948. The self-fulfilling prophecy. *The Antioch Review* no. 8 (2):193–210.
Miller, George. 1956. The magical number seven, plus or minus two: Some limits on our capacity for processing information. *The Psychological Review* no. 63:81–97.
Mills, Henry Robert. 1953. *Techniques of Technical Training*. London: Cleaver-Hume Press.
Morgan, Gareth. 1997. *Images of Organisation*. Thousand Oaks, CA: SAGE Publications.
Osborn, Alex F. 1963. *Applied Imagination Principles and Procedures of Creative Problem Solving*. 3rd Revised edition. New York: Charles Scribner's Sons.
Ouchi, William G. 1982. *Theory Z: How American Business Can Meet the Japanese Challenge*. London: Avon.
Paul, Richard, and Linda Elder. 2006. *Critical Thinking: Learn the Tools the Best Thinkers Use*. Concise edition. Upper Saddle River, NJ: Pearson Prentice Hall.
Peters, Tom, and Nancy Austin. 1985. *A Passion for Excellence*. New York: Warner Books.
Peters, Tom J., and Hayley R. Waterman. 1982. *Search of EXCELLENCE*. New York: Harper and Row.
Pinto, Jeffrey K. 2000. Understanding the role of politics in successful project management. *International Journal of Project Management* no. 18 (2):85–91. doi: 10.1016/S0263-7863(98)00073-8.
PMI. 2013. *A Guide to the Project Management Body of Knowledge*. 5th edition. Newtown Square, PA: Project Management Institute, Inc.
Prince, Harold. 1968. *Cabaret*. London: Palace Theatre.
Reardon, Kathleen. 2005. *It's All Politics: Winning in a World Where Hard Work and Talent aren't Enough*. New York: Doubleday.
Richmond, Barry. 1993. Systems thinking: critical thinking skills for the 1990s and beyond. *System Dynamics Review* no. 9 (2):113–133.
Rodgers, T.J., William Taylor, and Rick Foreman. 1993. *No-Excuses Management*. New York: Doubleday.
Rogers, Carl R., and Richard Evans Farson. 1957. Active listening. In *Communicating in Business Today*, edited by Ruth G. Newman, Marie A. Danzinger and Mark Cohen. Lexington: D.C. Heath & Company.
Rogers, Carl R., and Richard Evans Farson. 2015. *Active Listening, Communicating in Business Today*. Mansfeild Centre, CT: Martino Publishing.
Royce, Winston W. 1970. Managing the development of large software systems. In *IEEE WESCON*, Los Angeles.
Russo, J. Edward, and Paul H. Schoemaker. 1989. *Decision Traps*. New York: Simon and Schuster.
Schunk, Dale H. 1996. *Learning Theories*. Englewood Cliff, NJ: Pentice Hall.
Senge, Peter M. 1990. *The Fifth Discipline: The Art & Practice of the Learning Organization*. New York: Doubleday.
Serrat, Olivier. 2009. The five whys technique. In *Knowledge Solutions*. Singapore: Springer.
Shaw, George Bernard. 1925. England and America: Contrasts, "a conversation between Bernard Shaw and Archibald Henderson." *Reader's Digest*.

Sherwood, Dennis. 2002. *Seeing the Forest for the Trees. A Manager's Guide to Applying Systems Thinking*. London: Nicholas Brealey Publishing.
Skinner, Burrhus Frederic. 1954. *Science and Human Behavior*. Basingstoke: Macmillan.
Tittle, Peg. 2011. *Critical Thinking: An Appeal to Reason*. Abingdon: Routledge.
Vroom, Victor. 1964. *Work and Motivation*. Hoboken, NJ: John Wiley and Sons.
Wolcott, Susan K., and Charlene J. Gray. 2003. *Assessing and developing critical thinking skills* 2003 [cited 21 May 2013]. Available from www.wolcottlynch.com/Downloadable_Files/IUPUI%20Handout_031029.pdf.
Woolfolk, Anita E. 1998. Chapter 7 Cognitive views of learning. In *Educational Psychology*, 244–283. Boston, MA: Allyn and Bacon.
Wu, Wei, and Hidekazu Yoshikawa. 1998. Study on developing a computerized model of human cognitive behaviors in monitoring and diagnosing plant transients. In *1998 IEEE International Conference on Systems, Man, and Cybernetics*, San Diego, CA.
Yen, Duen Hsi. *The blind men and the elephant* 2008 [cited 26 October 2010]. Available from www.noogenesis.com/pineapple/blind_men_elephant.html.

3 Perceptions of Problem-Solving

This chapter deals with problems and problem-solving. If a problem didn't need solving there wouldn't be any need for management. The chapter:

- Helps you to understand the problem-solving aspects of management by discussing perceptions of the problem-solving process from a number of holistic thinking perspectives (HTP).
- Discusses the structure of problems.
- Discusses the levels of difficulty posed by problems.
- Discusses the need to evolve solutions using an iterative approach.
- Shows that problem-solving is really an iterative causal loop rather than a linear process.
- Discusses complexity and how to use the systems approach to manage complexity.
- Shows how to remedy well-structured problems.
- Shows how to deal with ill-structured, wicked and complex problems using iterations of a sequential two-stage problem-solving process.
- Discusses perceptions of problem-solving from the following HTPs:
  1. *Big Picture* in Section 3.1.
  2. *Quantitative* in Section 3.2.
  3. *Structural* in Section 3.3.
  4. *Continuum* in Section 3.4.
  5. *Functional* in Section 3.5.
  6. *Operational* in Section 3.6.
  7. *Scientific* in Section 3.7.

3.1 BIG PICTURE

Perceptions of problem-solving from the *Big Picture* HTP included:

1. Some assumptions underlying formal problem-solving in Section 3.1.1.
2. A number of myths about problem-solving in Section 3.1.2.

3.1.1 ASSUMPTIONS UNDERLYING FORMAL PROBLEM-SOLVING

Problem-solving like most other things is based on a set of assumptions. Waring provided the following four assumptions underlying formal problem-solving (Waring 1996):

1. The existence of the problem may be taken for granted.

2. The structure of the problem can be simplified or reduced so as to make its definition, description and solution manageable.
3. Reduction of the problem does not reduce effectiveness of the solution.
4. Selection of the optimal solution (decision-making) is a rational process of comparison.

However, while the existence of the problem may be taken for granted, it may take a while for the stakeholders to agree on the nature of the problem. Waring seems to be discussing well-structured problems (Section 3.4.5.1). The literature on decision-making, one of the key elements in problem-solving, has two schools of thought on Waring's fourth point.

1. *Agree:* decision-making is logical.
2. *Disagree:* decision-making is emotional.

Perceptions for the *Continuum* HTP indicate that some decisions are made emotionally and others are made logically (Kasser 2018a).[1] Accordingly, the two schools of thought perceive the decision-making from different single perspectives.

3.1.2 Selected Myths of Problem-Solving

There are a number of myths about problem-solving that hinder problem-solving and need to be exposed (Kasser and Zhao 2016a). These myths include:

1. All problems can be solved discussed in Section 3.1.2.1.
2. All problems have a single correct solution discussed in Section 3.1.2.2.
3. The problem-solving process is a linear time-ordered sequence discussed in Section 3.1.2.3.
4. One problem-solving approach can solve all problems discussed in Section 3.1.2.4.

3.1.2.1 All Problems Can Be Solved

One of the myths associated with the problem-solving process is that all problems can be solved. The reality is that:

1. *Problems are either solved, resolved, dissolved or absolved* (Ackoff 1978: p. 13), where only the first three actually remedy the problem. The word 'solve' is often misused in the literature to mean solved, resolved or dissolved, when a better word is 'remedy'. The four ways of dealing with a problem are:
    1. *Solving:* when the decision maker selects those values of the control variables which maximize the value of the outcome (satisfies the need, an optimal solution).

---

[1] And the same person can make the same decision emotionally at one time and logically at another time.

2. *Resolving:* when the decision maker selects values of the control variables which do not maximize the value of the outcome but produce an outcome that is good enough or acceptable (satisfices the need, Section 3.1.2.2).
3. *Dissolving:* when the decision maker reformulates the problem to produce an outcome in which the original problem no longer has any actual meaning. Dissolving the problem generally leads to innovative solutions.
4. *Absolving:* when the decision maker ignores the problem or imagines that it will eventually disappear on its own. Problems may be intentionally ignored for reasons that include:
   - They are too expensive to remedy.
   - The technical or social capability needed to provide a remedy is unavailable; it may not be known, affordable or available.

2. *Only well-structured problems* (Section 3.4.5.1) can be solved or resolved.

3.1.2.2 All Problems Have a Single Correct Solution

In school, generally, we are taught to solve problems using the simple problem-solving process by being given a problem and then asked to find the single correct solution. The assumption being that there is always a well-structured problem (Section 3.4.5.1) with a single well-defined correct solution. In some instances, such as in mathematics, there are single correct solutions to problems. However, a single correct solution to all problems is a myth that does not apply in the real world. For example, supposing that you are hungry; which is generally an undesirable situation. The well-structured problem is to figure out a way to remedy that undesirable situation by consuming some food to satisfy the hunger. There are a number of remedies to this problem including cooking something, going to a restaurant, collecting some takeaway food and telephoning for home delivery. Then there is the choice of what type of food; Italian, French, Chinese, pizza, lamb, chicken, beef, fish, vegetarian, etc. Now consider the vegetables, sauces and drinks. There are many solutions because there are many combinations of types of food, meat, vegetables and method of getting the food to the table. Which solution is the correct one? The answer is that the correct solution is the one that satisfies your hunger in a timely and affordable manner.[2] If several of the solution options can perform this function and you have no preference between them, then each of them are just as correct as any of the other ones that satisfy your hunger. The words 'right solution' or 'correct solution' should be thought of as meaning 'one or more acceptable solutions' on a continuum of solutions as shown in Figure 3.1. If you do have a preference for some of them, then they are 'optimal solutions'.

	Feasible solutions	
Unacceptable	Acceptable	Optimal
Range of potential solutions		

FIGURE 3.1 The continuum of solutions.

---

[2] And does not cause any gastric problems.

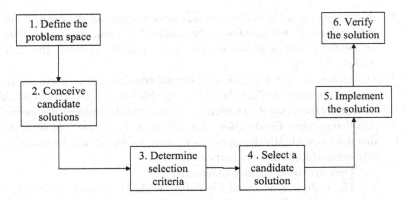

FIGURE 3.2 The traditional simple problem-solving process.

3.1.2.3 The Problem-Solving Process is a Linear Time-Ordered Sequence

The problem-solving process is taught as a linear process as shown in Figure 3.2. This is simple to teach but incorrect in the real world. For example, consider what happens if:

- None of the solutions meet the selection criteria in step 4 of the generic traditional simple problem-solving process shown in Figure 3.2.
- None of the solution options remedies the undesirable situation.
- All the solutions are too expensive, or will take too long to realize or are unacceptable for some other reason.

Then the choices to be made include:

- Absolve the problem (Section 3.1.2.1) for a while until something changes.
- Decide to remedy parts of the undesirable situation, sometimes known as reducing the requirements, until the remedy is feasible. Ways of doing this include:
  - Removing the lower-priority aspects of the undesirable situation and determining the new cost/schedule information until the solution option becomes affordable or can be realized in a timely manner. This is a holistic approach to the concept of designing to cost and used in conceptual design to cost (Section 8.3.1) and cost as an independent variable (CAIV).
  - Remedy the causes of undesirability with the highest priorities and absolve the problem posed by the remaining causes.
  - Continue to look for an acceptable feasible solution that will remedy the undesirable situation in a timely manner.

The reality is that the problem-solving process is iterative in several different ways including:

Perceptions of Problem-Solving

1. *Iteration of the problem-solving process across the system development process (SDP):* each state of the SDP contains the problem-solving process shown in Figure 3.2. The traditional waterfall chart view does not show the inside of the blocks in the waterfall; however, if the blocks inside the waterfall view of the SDP were visible, a partial waterfall would be drawn as in Figure 3.3.[3]
2. *Iteration of the problem-solving process within a state in the SDP:* each state in the SDP contains a problem-solving process with two exit conditions:
   1. The normal planned exit at the end of state.
   2. An anticipated abnormal exit anywhere in the state that can happen at any time in any state and necessitates either a return to an earlier state or a move to a later state and skipping intermediate states.

   In addition, the entire SDP may map into a single iteration of the problem-solving process.

See Chapter 19 for an example of how this process was used.

3.1.2.4 One Problem-Solving Approach Can Solve All Problems

The reality is that there are different types of problems that need different versions of the problem-solving process, e.g. research and intervention problems (Section 3.4.4).

3.2 QUANTITATIVE

Perceptions of problem-solving from the *Quantitative* HTP included the five components of a problem.

3.2.1 COMPONENTS OF PROBLEMS

The five components of a problem (Ackoff 1978: pp. 11–12) are:

1. *The decision maker:* the person faced with the problem.
2. *The control variables:* aspects of the problem situation the decision maker can control.
3. *The uncontrolled variables:* aspects of the problem situation the decision maker cannot control which constitute the problem environment.[4] The uncontrolled variables may give rise to unanticipated emergent properties of the solution often called undesirable outcomes.
4. *Constraints:* imposed from within or without on the possible values of the controlled and uncontrolled variables.

---

[3] Note that since the figure contains two levels of the hierarchy, it should only be used to show the repetition of the problem-solving process in each state of an ideal SDP, one in which no changes occur.

[4] There may be unknown uncontrolled variables, see Simpson's paradox (Savage 2009).

60   Systemic and Systematic Project Management

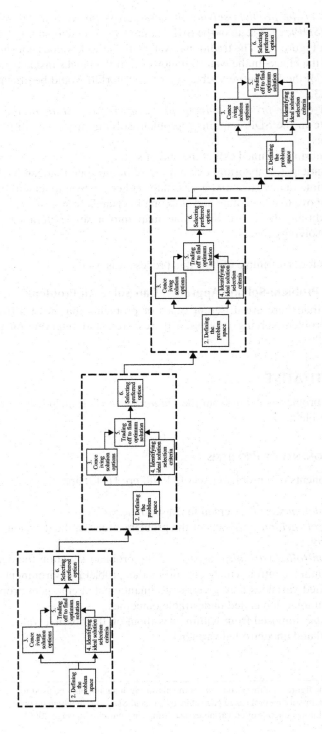

FIGURE 3.3 Iteration of the problem-solving process across the SDP in an ideal project.

5. *The possible outcomes:* desired and undesired produced jointly by the decision maker's choice and the uncontrolled variables.[5] The desired outcome may be represented in several ways including:
   - A specified relationship between the controlled variables and the uncontrolled variables.
   - A design or architecture.
   - A feasible conceptual future desirable solution (FCFDS) (Section 3.7.1).

3.3 STRUCTURAL

Perceptions of problem-solving from the *Structural* HTP included classifications of problems.

1. Classifications of problems discussed in Section 3.3.1.
2. The level of difficulty of problems discussed in Section 3.3.2.
3. The different structures of problems discussed in Section 3.4.5.

3.3.1 CLASSIFICATIONS OF PROBLEMS

Before trying to solve problems, it would be useful to have a classification of types of problems and ways to remedy them. The undesirable situation is the lack of such a classification, the FCFDS is a classification system, and the problem is to provide a classification of problems. So rather than inventing one, the literature was searched and several ways of classifying problems in various domains were identified including the level of difficulty of the problem.

3.3.2 THE LEVEL OF DIFFICULTY OF PROBLEMS

Ford introduced four categories of increasing order of difficulty for mathematics and science problems: easy, medium, ugly, and hard (Ford 2010). These categories may be generalized and defined as follows (Kasser 2015b):

1. *Easy:* problems that can be solved in a short time with very little thought.
2. *Medium:* problems that:
   - Can be solved after some thought.
   - May take a few more steps to solve than an easy problem.
   - Can probably be solved without too much difficulty, perhaps after some practice.
3. *Ugly:* problems are ones that will take a while to solve. Solving them:
   - Involves a lot of thought.
   - Involves many steps.
   - May require the use of several different concepts.

---

[5] Desired and undesired.

4. *Hard:* problems usually involve dealing with one or more unknowns. Solving them:
   - Involves a lot of thought.
   - Requires some research.
   - May also require iteration through the problem-solving process as learning takes place (knowledge that was previously unknown becomes known).

Classifying problems by level of difficulty is difficult in itself because difficulty is subjective since one person's easy problem may be another person's medium, ugly or hard problem.

3.4 CONTINUUM

Perceptions of problem-solving from the *Continuum* HTP included:

1. The difference between problems and symptoms discussed in Section 3.4.1.
2. The difference between the quality of the decision and the quality of the outcome discussed in Section 3.4.2.
3. The different decision outcomes discussed in Section 3.4.3.
4. The difference between research and intervention problems discussed in Section 3.4.4.
5. The different categories of problems discussed in Section 3.4.5.
6. The different domains of problems discussed in Section 3.4.6
7. The system implementation continuum discussed in Section 3.4.7.
8. The different levels of difficulty of a problem discussed in Section 3.3.2.

3.4.1 PROBLEMS AND SYMPTOMS

Perceiving the difference between problems, symptoms and causes from the *Continuum* HTP, the undesirable situation manifests itself as symptoms, which are used to diagnose the underlying problem. Having diagnosed the problem action is then taken to remedy the problem. The traditional problem-solving feedback approach is represented by the causal loop shown in Figure 3.4 (Kasser 2002a). An action can tackle the problem or a symptom. If the root cause of the problem is not found, a solution may not work or may only work for a short period of time. In addition, even if the implemented solution works it may introduce further problems that only show up after some period of time. Consider the implications of the time delay in the problem-solving feedback loop. Any action has an effect in the present and in the future. Group these effects in time as (Kasser 2002a):

- *First order:* noticeable effect within a second or less.
- *Second order:* noticeable effect within a minute or less.
- *Third order:* noticeable effect within an hour or less.
- *Fourth order:* noticeable effect within a day or less.
- *Fifth order:* noticeable effect within a week or less.
- *Sixth order:* noticeable effect within a month or less.
- *Seventh order:* noticeable effect within a year or less.

Perceptions of Problem-Solving

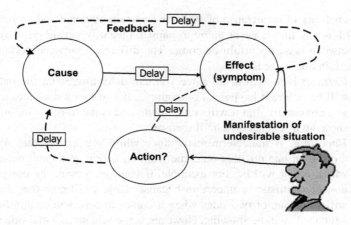

FIGURE 3.4 Problems, causes and effects (symptoms).

- *Eighth order:* noticeable effect within a decade or less.
- *Ninth order:* noticeable effect within a century or less.
- *Tenth order:* noticeable effect after a century or more.

The analysis of the requested change has to consider all of the above as applicable. While the higher-order effects may not be applicable in a computer-based system, they are applicable in long-lived systems such as those that affect the environment (dams, power plants, etc.). Sometimes a second action is taken before the effect of the first one is observed leading to the need for a further action to remedy the effect of the second one. Sometimes the action partially remedies the problem; sometimes the action only mitigates the symptoms and produces a new undesirable situation.

3.4.2 THE DIFFERENCE BETWEEN THE QUALITY OF THE DECISION AND THE QUALITY OF THE OUTCOME

Perceptions of decision-making from the *Continuum* HTP note, 'We need to differentiate between the quality of the decision and the quality of the outcome' (Howard 1973: p. 55). A good decision can lead to a bad outcome and conversely a bad decision can lead to a good outcome. The quality of the decision is based on doing the best you can to increase the chances of a good outcome hence the development and use of decision-making tools. Decisions can be made using quantitative and qualitative methods (Kasser 2018a: Chapter 4).

3.4.3 THE DIFFERENT DECISION OUTCOMES

All decision outcomes have consequences. Perceiving decision outcomes from the *Continuum* HTP:

- Outcomes and consequences lie on a probability of possibilities continuum as shown in Figure 3.1 ranging from 0% to 100% where an outcome with a probability of occurrence of 100% is a certain outcome and an outcome of 0% is one that is never going to happen (negative certainty). Anything in between is an uncertain outcome. The difference between certain and uncertain outcomes is:

- *Certain:* is deterministic since you can determine what the outcome will be before it happens. For example, if you toss a coin into the air, you are certain that it will come down[6] and come to rest with one side showing if it lands on a hard surface.
- *Uncertain:* is nondeterministic since while you know there may be more than one possible outcome from an action, you can't determine which one it will be. For example, if you toss a coin, the outcome is nondeterministic or uncertain because while you know that the coin will show one of two sides when it comes to rest, you cannot be sure which side will be showing. However, you could predict one side with a 50% probability of being correct.[7]
- Outcomes and consequences can be anticipated and unanticipated where:
  - Anticipated: can be:
    1. *Desired:* where the result is something that you want. For example, you want the coin to land showing 'heads' and it does.
    2. *Undesired:* where the result is something that you don't want. For example, you don't want the coin to land showing 'tails' and it lands showing 'tails'. This type of outcome and its consequences are known as risks before they occur and events once they have occurred.
    3. *Don't Care:* where you have no preference for the result. For example, if you have no preference as to which side is showing when the coin lands, you have a 'Don't Care' situation (Kasser 2018a: Section 3.2.2).
  - *Unanticipated:* can also be desired, undesired and don't care once discovered.

There can also be more than one outcome and consequence from an action; for example, each of the outcomes may be:

- Dependent on, or independent from, the other outcomes.
- Acceptable or not acceptable.
- Desired, undesired or 'Don't Care' (Kasser 2018a: Section 3.2.2).
- Unanticipated the first time that the action is taken.
- A combination of the above.

---

[6] Unless you toss it so fast that it escapes from the earth's gravity.
[7] So how can tossing a coin be certain and uncertain at the same time? It depends on the type of outcome you are looking for. Which side it will land on is uncertain; but that it will land on a side is certain. It is just a matter of framing the issue from the proper perspective.

Perceptions of Problem-Solving

TABLE 3.1
Decision Table for Known Outcomes of Actions

	Certain	Uncertain	Certain
Probability of occurrence	0% (will never happen)	0% < 100% (might happen)	100% (will always happen)
Desired	Need to conceptualize an alternative action	Opportunity that should be planned for, depending on probability of occurrence	Preferred outcome
Don't care	Ignore	Opportunity that might be considered depending on probability of occurrence	Opportunity that could be taken advantage of
Undesired	Can be ignored	Risk that should be mitigated depending on probability of occurrence and severity of consequences	Outcome that must be prevented or mitigated depending on severity of consequences

Table 3.1 shows the links between the known outcomes of decisions and uncertainty. Many of the decision-making tools in the literature deal with the decisions being made in the desired certain area. Don't Care outcomes should not be neglected but should be looked at as opportunities. For example, if you are considering purchasing a commercial-off-the-shelf (COTS) item and initially don't care about the colour then from the holistic perspective you might want to think about what additional benefits you might get from a specific colour. Typical active brainstorming questions (Kasser 2018a: Section 7.1):

- When considering risks include:
  - What if it is late?
  - What if it performs below specification?
  - What if it fails before the specified time?
- When considering opportunities, the questions would be the opposite of those asked from the perspective of risk, namely:
  - What if it is early?
  - What if it performs above specification?
  - What if it lasts longer than specified?

The answers and the resulting actions taken would depend on the situation.

3.4.3.1 Sources of Unanticipated Consequences or Outcomes of Decisions

Unanticipated consequences or outcomes of decisions need to be avoided or minimized.[8] In the systems approach, if we can identify the causes of unanticipated consequences, we should be able to prevent them from happening. A literature search

---
[8] This also applies to unanticipated emergent properties (*Generic* HTP).

found Merton's analysis, which discussed the following five sources of unanticipated consequences in social interventions (Merton 1936):

1. Ignorance.
2. Error.
3. Imperious immediacy of interest.
4. Basic values.
5. Self-defeating predictions.

These sources may be generalized as discussed below.

- *Ignorance:* deals with unanticipated consequences or outcomes due to the type of knowledge that is missing or ignored in making the decision. Ignorance in the:
  - *Problem domain:* may result in the identification of the wrong problem.
  - *Solution domain:* may produce a solution system that will not provide the desired remedy.
  - *Implementation domain:* may produce a conceptual solution that cannot be realized.
- *Errors:* there are two types of errors, errors of commission and errors of omission (Ackoff and Addison 2006) where:
  1. *Errors of commission:* do something that should not have been done. There are also two types of errors of commission: design errors and implementation errors.
     1. *A design error:* an error which produces an undesired outcome. For example, a logic error in a computer program.
     2. *An implementation error:* a mistake was made in creating the design. For example, a syntax error in a computer program, a failure to test something under realistic operating conditions, the wrong part was installed, or a part was installed backwards.
  2. *Errors of omission:*
     1. Fail to do something that should have been done such as in instances where only one or some of the pertinent aspects of the situation, which influences the solution are considered. This can range from the case of simple neglect (lack of systematic thoroughness in examining the situation) to 'pathological obsession where there is a determined refusal or inability to consider certain elements of the problem' (Merton 1936).
     2. Are more serious than errors of commission because, among other reasons, they are often impossible or very difficult to correct. 'They are lost opportunities that can never be retrieved' (Ackoff and Addison 2006: p. 20). Merton adds that a common fallacy is the too-ready assumption that actions, which have in the past led to a desired outcome, will continue to do so. This assumption often, even usually, meets with success. However, the habit tends to become automatic with continued repetition so that there is a

Perceptions of Problem-Solving

> failure to recognize that procedures, which have been successful in certain circumstances, need not be successful under any and all conditions.[9]
> - *Imperious immediacy of interest:* the paramount concern with the foreseen immediate consequences excludes the consideration of further or other consequences of the same act, which does in fact produce errors.[10]
> - *Basic values:* there is no consideration of further consequences because of the felt necessity of certain action enjoined by certain fundamental values. For example, the Protestant ethic of hard work and asceticism paradoxically leads to its own decline in subsequent years through the accumulation of wealth and possessions.
> - *Self-defeating predictions:* the public prediction of a social development proves false precisely because the prediction changes the course of history. Merton later conceptualized the, 'the self-fulfilling prophecy' (Merton 1948) as the opposite of this concept.

3.4.4 Research and Intervention Problems

Perceptions from the *Continuum* HTP indicate a difference between research and intervention problems (Kasser and Zhao 2016a). Consider both of them.

3.4.4.1 Research Problems

This type of problem manifests when the undesirable situation is the inability to explain observations of phenomena or the need for some particular knowledge. In this situation, using the problem formulation template (Section 3.7.1):

1. *The undesirable situation:* the inability to explain observations of phenomena or the need for some particular knowledge.
2. *The assumptions:* the research is funded and the researcher is an expert in the field.
3. *The FCFDS:* the outcome is the ability to explain observations of phenomena or the particular knowledge.
4. *The problem:* how to gain the needed knowledge.
5. *The solution:* use the scientific method, which works forwards from the current situation in a journey of discovery towards a future situation in which the knowledge has been acquired.

The scientific method:

- Is a variation of the generic problem-solving process.
- Is summarized in Figure 3.5.
- Is a systemic and systematic way of dealing with open-ended research problems.

---

[9] This assumption also applies to component reuse.
[10] Perceptions from the *Generic* HTP perceive the similarity to the decision traps in Section 4.2.

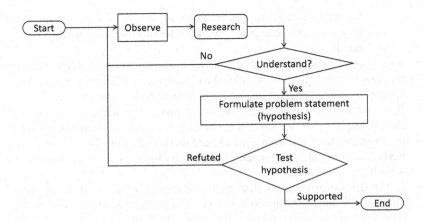

FIGURE 3.5 The scientific method.

- Has been stated as different variations of the following sequence of activities:
  1. Observe an undesirable situation.
  2. Perform research to gather preliminary data about the undesirable situation.
  3. Formulate the hypothesis to explain the undesirable situation using inductive reasoning (Kasser 2018a: Section 5.1.3.1.2).
  4. Plan to gather data to test the hypothesis. The data gathering may take the form of performing an experiment, using a survey, reviewing literature or some other approach depending on the nature of the undesirable situation and the domain.
  5. Perform the experiment or otherwise gather the data.
  6. Analyse the data (experimental or survey results) using deductive reasoning (Kasser 2018a: Section 5.1.3.1.1) to test the hypothesis.
  7. If the hypothesis is supported, then the researcher often publishes the research. If the hypothesis is not supported, then the process reverts to step 1.

In the real world, the hypothesis is often created from some insight or a 'hunch' in which the previous steps are performed subconsciously. The research then designs the data collection method, collects and examines the data to determine if the hypothesis is supported. In this situation:

- The publication is generally written as if the steps in the scientific method have been performed as described above.
- Half the data is used in defining the hypothesis and half the data is used in testing the hypothesis.
- There is also an unfortunate tendency to ignore or explain away data which does not support the hypothesis for reasons that include:

- The researcher may only be looking for data to support the hypothesis. See the decision traps (Russo and Schoemaker 1989); factors that lead to bad decisions in (Kasser 2018a: Section 8.3).
- The data sample may be defective. It is important to verify such data, because if the data are valid, they may indicate an instance of Simpson's paradox (Savage 2009) (Kasser 2018a: Section 4.3.2.6.12) and provide the opportunity for further research which could lead to the identification of one or more previously unknown and accordingly unconsidered variables in the situation which would then provide a better understanding and perhaps a Nobel Prize or equivalent in the specific research domain.

3.4.4.2 Intervention Problems

This type of problem manifests when a current real-world situation is deemed to be undesirable, and needs to be changed over a period of time into a FCFDS. In this situation, using the problem formulation template (Section 3.7.1):

1. *The undesirable situation:* may be a lack of some functionality that has to be created, or some undesirable functionality that has to be eliminated.
2. *The assumptions:* about the situation, constraints, resources, etc.
3. *The FCFDS:* one in which the undesirable situation no longer exists.[11]
4. *The problem:* how to realize a smooth and timely transition from the current situation to the FCFDS minimizing resistance to the change.
5. *The solution:* create and implement the transition process to move from the undesirable situation to the FCFDS together with the solution system operating in the situational context.

The decision maker or problem-solver is faced with an undesirable situation. Once given the authority to proceed:

1. The decision maker should use the relevant starter questions from the *Generic* HTP for active brainstorming (Kasser 2018a: Section 7.1) to determine if anyone else has faced the same or a similar problem, what they did about it, what results they achieved and the similarities and differences between their situation and the current situation. This is the concept behind TRIZ[12] (Barry, Domb, and Slocum 2007) and the Copy Cat systems thinking tool.
2. The decision maker uses the research problem-solving process to conceptualize a vision of the solution system operating in the FCFDS, which becomes the target or goal to achieve.
3. Then the problem the decision maker faces is to create the transition process and the solution system that will be operational in the FCFDS.

---

[11] And may have improvements to make it even more desirable.
[12] TRIZ is the Russian acronym for 'Teoriya Resheniya Izobretatelskikh Zadatch', which has been translated into English as 'Theory of Inventive Problem-Solving'.

4. The decision maker uses imagination, Gantt charts (Kasser 2018a: Section 2.14) and Program Evaluation Review Technique PERT charts (Kasser 2018a: Section 8.10) to work backwards from the FCFDS to the present undesirable situation creating the transition process.
5. The decision maker then documents the process in the project plan (PP) as a sequential process working forwards from the present undesirable situation to the FCFDS. This version of the problem-solving process is the SDP (Section 5.4).

3.4.5 The Different Categories of Problems

Perceived from the *Continuum* HTP, problems lie on a continuum of categories which range from 'well-structured' through 'ill-structured' to 'wicked'. Consider each of them.

3.4.5.1 Well-Structured Problems

Well-structured problems are problems where the existing undesirable situation and the FCFDS are clearly identified. These problems may have a single solution or sometimes more than one acceptable solution. Examples of well-structured problems with single correct solutions are:

- Mathematics and other problems posed by teachers to students in the classroom. For example, in mathematics, $1 + 1 = 2$ every time.
- Making a choice between two options. For example, choosing between drinking a cup of coffee and a cup of tea. However, the answer may be different each time.
- Finding the cheapest airfare between Singapore and Jacksonville, Florida if there is only one cheapest fare. However, the answer may be different depending on the time of the year.

Examples of well-structured problems with several acceptable but different solutions are:

- What food to eat to satisfy your hunger (Section 3.1.2.2).
- What brand of coffee to purchase? Although the solution may depend on price, taste and other selection criteria, there may be more than one brand (solution) that meets all the criteria.
- Which brand of automated coffee maker to purchase?
- What type of transportation capability to acquire?
- Finding the cheapest airfare between Singapore and Jacksonville, Florida if two airlines charge the same fare.

Well-structured problems:

- May be formulated using the five-part problem formulation template (Section 3.7.1).

Perceptions of Problem-Solving

- With single solutions tend to be posed as closed questions.
- With multiple acceptable solutions tend to be posed as open questions.

The traditional problem-solving approach for well-structured problems elaborates a complex problem into a number of simple problems so that when the simple problems have been solved, the complex problem has also (hopefully) been solved.

3.4.5.2 Ill-Structured Problems
Ill-structured problems:

- Sometimes called 'ill-defined' problems are problems where either or both the existing undesirable situation and the FCFDS are unclear (Jonassen 1997).
- Cannot be solved (Simon 1973); they have to be converted to one or more well-structured problems.

3.4.5.3 Wicked Problems
Wicked problems also known as 'messy' problems[13] are extremely ill-structured problems[14] first stated in the context of social policy planning (Rittel and Webber 1973). Wicked problems:

- Cannot be easily defined, therefore, all stakeholders cannot agree on the problem to solve.
- Require complex judgements about the level of abstraction at which to define the problem.
- Have no clear stopping rules (since there is no definitive 'problem', there is also no definitive 'solution' and the problem-solving process ends when the resources, such as time, money or energy, are consumed, not when some solution emerges).
- Have better or worse solutions, not right and wrong ones.
- Have no objective measure of success.
- Require iteration – every trial counts.
- Have no given alternative solutions – these must be discovered.
- Often have strong moral, political or professional dimensions.

3.4.6 The Different Domains of a Problem

Remedying a problem requires competency in three different domains, namely (Kasser 2015a):

1. *Problem domain:* the situation in which the need for the management activity has arisen.

---

[13] When complex.
[14] Technically, there is no problem since while the stakeholders may agree that the situation is undesirable, they cannot agree on the problem.

2. *Solution domain:* the situation and solution system that will be created as a result of the management activity.
3. *Implementation domains:* the environment in which the activity is being managed.

It is tempting to assume that the problem domain and the solution domain are the same, but they are not necessarily so. For example, the problem domain may be urban social congestion, while the implementation domain is tunnel boring and the solution domain may be a form of underground transportation system to relieve that congestion. Lack of problem domain competency may lead to the identification of the wrong problem, lack of implementation domain competency may lead to schedule delays due to preventable problems and lack of solution domain competency may lead to selection of a less than optimal, or even an unachievable, solution system.

3.4.7 THE TECHNOLOGICAL SYSTEM IMPLEMENTATION CONTINUUM

When considering candidate designs for a technological system, each candidate will lie on a different point on the implementation continuum with a different mixture of people, technology, a change in the way something is done, etc. at one end of the continuum is a completely manual solution, at the other is a completely automatic solution. The concept of designing a number of solutions and determining the optimal solution, which may either be one of the solutions or a combination of parts of several solutions, comes from the *Continuum* HTP. A benefit of producing several solutions is that one of the design teams conceptualizing the solutions may pick up on matters that other teams missed.

A benefit of recognizing the system implementation continuum is that the system can be delivered in Builds, wherein Build 1 is manual and the degree of automation increases with each subsequent Build (Section 10.15.8.3).

3.5 FUNCTIONAL

Perceptions of problem-solving from the *Functional* HTP included decision-making.

3.5.1 DECISION-MAKING

The most important use of systems thinking in problem-solving is in decision-making. When thinking about anything, we perceive it from different viewpoints, process the information and then infer a conclusion. That inference process is a decision-making process.

Since project managers spend a lot of their time making decisions, understanding the decision-making process will help the project manager to make better decisions.

The decision-making process is the front-end of the simple problem-solving process (steps 2–6) shown in Figure 3.2 based on Hitchins (2007: p. 173) perceived from the decision-making perspective. Figure 3.2 depicts the series of

Perceptions of Problem-Solving

activities which are performed in series and parallel to transform the undesirable situation into the strategies and plans to realize the solution system operating in its context.

1. The milestone to start the problem-solving process.
2. The authorization to make the decision.
3. The process to define the problem.
4. The process to conceive several solution options.
5. The process to identify ideal solution selection criteria.
6. The process to perform trade-offs to find the optimum solution.
7. The process to select the preferred option.
8. The process to formulate strategies and plans to realize the selected option.

For a more detailed discussion on decision-making and decision-making tools including decision traps (see Russo and Schoemaker 1989) and the *Systems Thinker's Toolbox* (Kasser 2018a: Chapter 5). Note:

- *Risks, opportunities and benefits are selection criteria for decisions:* the degree of risks and benefits are solution selection criteria allowing the degree of susceptibility to a specific risk or the possibility of taking advantage of a specific opportunity to be evaluated in conjunction with other criteria.
- *Make decisions using advice from appropriate personnel:* making decisions about the probability and severity of risks requires knowledge in the problem, solution and implementation domains. This means that even though the project manager makes the decision or approves and is responsible for the decision, the decision-making team needs to include personnel with the appropriate domain knowledge. It also helps to avoid the decision traps (Russo and Schoemaker 1989) and unanticipated negative consequences.

3.5.1.1 Decision-Making Tools

The literature tends to discuss each tool and decision-making approach as being used in an either-or case, namely use one tool or the other to make a decision. However, in the real world, we use perceptions from the *Continuum* HTP to develop a mixture of tools or parts of tools as appropriate. For example, determination of the selection criteria for the decision is often a subjective approach, even when those criteria are later used in a quantitative manner. For a more detailed discussion of decision-making, see the *System Thinker's Toolbox* (Kasser 2018a: Chapter 4).

3.6 OPERATIONAL

Perceptions of problem-solving from the *Operational* HTP included:

1. The simple problem-solving process discussed in Section 3.6.1.
2. The extended problem-solving process discussed in Section 3.6.2.

3.6.1 The Simple Problem-Solving Process

The traditional simple problem-solving process:

- Considers the problem-solving process as a linear sequence of activities, starting with a problem and ending with a solution as in Figure 3.2. The simple process contains six steps:
  1. Understand the problem and define the problem space or bound the problem.
  2. Conceive at least two candidate solutions.
  3. Determine selection criteria for choosing between the candidate solutions.
  4. Select a candidate solution (preferred option).
  5. Plan the implementation strategy.
  6. Implement the solution.
- Is recursive as each part contains an iteration of the problem-solving process.
- Is based on assumptions that include:
  - The problem is solved after one pass through the process, but often the outcome is 'oops' and the problem-solving process has to repeat. Accordingly, iteration is implied but not explicitly called out.
  - Someone has already defined the problem. This is often an incorrect assumption.
- Is accordingly incomplete. An additional step is required namely
  7. If the solution has not solved the problem, go back to step 1.

However, the modified simple problem-solving process is still incomplete.

3.6.2 The Extended Problem-Solving Process

> Problems do not present themselves as givens; they must be constructed by someone from problematic[15] situations which are puzzling, troubling and uncertain
>
> **Schön (1991)**

Accordingly, something is missing in the traditional simple problem-solving process shown in Figure 3.6 (Kasser and Zhao 2016c). Unlike the simple problem-solving process which begins with a problem and ends with a solution (Section 3.6.1), the systems approach takes a wider perspective and begins with an undesirable situation that needs to be explored (Fischer, Greiff, and Funke 2012). From this perspective, the observer becomes aware of an undesirable situation that is made up of one or more undesirable factors which has to be converted into a FCFDS by remedying a series of problems, for example:

---

[15] or undesirable.

Perceptions of Problem-Solving

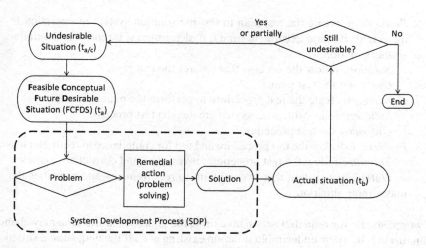

FIGURE 3.6 The extended problem-solving process. (*Source:* © 2016 IEEE. Reprinted, with permission, from Kasser, Joseph Eli, and Yang-Yang Zhao. 2016c. Wicked problems: Wicked solutions. In *11th International Conference on System of Systems Engineering*, Kongsberg, Norway.)

1. *Problem:* to do something about the undesirable situation.
   Solution: get the authorization of a project to do something about the undesirable situation together with adequate resources.
   Outcome: the authorization and the adequate resources.
2. *Problem:* to understand the undesirable situation, determine what makes the situation undesirable and define the problem space (bound the problem).
   Solution: follow the research problem-solving process (Section 3.4.4.1) to gain the understanding.
   Outcome: a statement of the correct problem.
3. *Problem:* to create a vision of a FCFDS that would remedy the undesirable situation.
   Solution: follow the problem-solving process to create the FCFDS.
   Outcome: the FCFDS.
4. *Problem:* how to transition from the undesirable situation to the FCFDS.
   Solution: create the specifications and project plan to realize the outcome.
   Outcome:
   1. A set of requirements (specifications) for a solution system that when operating in the situation would remove the undesirability.
   2. A PP for the remedial action that would realize the solution system specified by the requirements.
5. *Problem:* to perform the transition that creates the specified solution system according to the PP. This remedial action for complex problems often takes the form of the SDP.
   Solution: follow the SDP per the PP.
   Outcome: the undesirable situation has been transformed into what should be a desirable situation.

6. *Problem:* to create the test plan to test the solution system in operation in the actual situation existing at time t_1 to determine if it remedies the undesirable situation.
   Solution: follow the process that creates the test plan.
   Outcome: the test plan.
7. *Problem:* to create the test procedure to perform the planned tests.
   Solution: follow the process that creates the test process.
   Outcome: the test procedure.
8. *Problem:* to follow the test procedure and test for compliance to requirements.
   Solution: follow the test procedure, investigate and correct anomalies.
   Outcome: a system that is complaint to requirements and remedies the undesirable situation.

However, should the remedial action take time, the undesirable situation may change from that at t_0 to a new undesirable situation existing at t_2. If the undesirable situation is remedied, then the process ends; if not, the process iterates from the undesirable situation at t_2 to a new undesirable situation as shown in Figure 3.6.

In summary, in general:

- There is an undesirable or problematic situation.
- A FCFDS is created.
- The problem is how to transition from the undesirable situation to the FCFDS.
- The solution is made up of two parts:
  1. The transition process.
  2. The solution system operating in the context of the FCFDS.

It may require more than one iteration of Figure 3.6 to evolve a remedy.

3.7 SCIENTIFIC

Inferences from perceiving problem-solving from the *Scientific* HTP included:

- The activities in the problem-solving process relate to the five ways of approaching problems (Section 2.1.2.1) as shown in Table 3.2. Using the wrong type of project manager can contribute to the failure of a project.
- The five-part problem formulation template discussed in Section 3.7.1.

3.7.1 A Problem Formulation Template

A five-part problem formulation template (Kasser 2018a: Section 14.3) is:

- A tool to overcome the generic problem of 'poor problem formulation'.
- A tool to think through the whole problem-solving process.
- Based on the simple problem-solving process (Section 3.6.1).
- A way to assist the problem-solving process by encouraging the planner to think through the problem and ways to realize a solution when formulating the problem.

TABLE 3.2
How the Activity in the Problem-Solving Process Relates to the Five Ways of Approaching Problems

Activity in the Problem-Solving Process	Way of Approaching Problems
Defining the scope of the problem	IV
Gaining an understanding	IV
Conceiving several solution options	III
Performing trade-offs	III
Selecting the preferred option	III
Formulate strategies and plans	III
Implementing the solution system	II[a]
Validating the solution system	II[a]

[a] As long as everything goes according to plan. Should problems arise, types III and IV will be needed to determine the response to the problem (Section 2.4.3).

- Made up of following five parts:
  1. *The undesirable situation:* as perceived from the each of the pertinent descriptive HTPs. This is the 'as-is' situation in business process reengineering (BPR).
  2. *Assumptions:* about the situation, problem, solution, constraints, etc. that will have an impact on developing the solution. One general assumption is there is enough expertise in the group formulating the problem to understand the undesirable situation and specify the correct problem. If this assumption is not true, then that expertise needs to be obtained for the duration of the activity. Providers of the expertise include other people within the organization, consultants, and members of the group looking it up on the Internet.
  3. *The FCFDS (Scientific HTP)* or *desired outcome* as described by the appropriate descriptive HTPs. Something that remedies the undesirable situation and is to be interoperable with evolving adjacent systems over the operational life of the solution and adjacent systems (the outcome). This is generally:
     1. A conceptual 'as-it should be' situation. This is known as the 'to-be' situation in BPR.
     2. The undesirable situation without the undesirability and with improvements.
  4. *The problem:* how to convert the FCFDS into reality.
  5. *The solution:* two interdependent parts:
     1. *Process:* create and follow the SDP or transition process that converts the undesirable situation to a desirable situation[16] by first

---

[16] Or a less undesirable situation if the situation is complex and requires iterations of the problem-solving process.

visualizing the FCFDS and then realizing the conceptual system that will operate in the context of the FCFDS.
2. *Product or system:* follow the created process to conceptualize and create the solution system operating in the context of the FCFDS.

If the problem is objectively complex, the first version of the problem formulation template generally does not include a description of the solution. The problem-solving process creates the solution.

At the time the problem is being identified, the interdependent parts of the solution (at some time in the future) consist of two sets of functions;

1. *The product mission and support functions:* to be performed by the solution system once realized (F_s).
2. *The process functions:* which have to be performed to realize the solution system (F_w).

This concept can be represented by the following relationships:

$$\text{Total Solution } (S) = F_s + F_w \tag{3.1}$$

where

$$F_s = F_d - F_c \tag{3.2}$$

Which gives

$$S = (F_d - F_c) + F_w \tag{3.3}$$

In summary,

F_c = complete set of current functions; functionality provided in the existing situation which may range from zero (nothing exists) to some functionality in an existing system deemed as not providing a complete solution.
F_d = complete set of desired functions to be developed.
F_s = functions performed by solution system.
F_w = functions (needed to be) performed to realize the solution system (create the desired functions that do not exist at the time the project begins.

Moreover, the F_s and F_w can both consist of mission and support functions as discussed above.

3.8 COMPLEXITY

Consider perceptions of complexity from the following HTPs:

1. *Continuum* discussed in Section 3.8.1.
2. *Temporal* discussed in Section 3.8.2.
3. *Scientific* discussed in Section 3.8.3.

3.8.1 CONTINUUM

Perceptions of complexity from the *Continuum* HTP include:

1. The dichotomy discussed in Section 3.8.1.1.
2. Various definitions of complexity discussed in Section 3.8.1.2.
3. Partitioning complexity discussed in Section 3.8.1.3.

Consider each of them.

3.8.1.1 The Complexity Dichotomy

Perceptions of complexity from the *Big Picture* HTP (Kasser 2015a) indicate that there is a dichotomy on the subject of how to solve the complex problems associated with complex systems. There is literature that states:

1. *We have a problem:* there is a need to develop new tools and techniques to remedy complex problems.
2. *What's the problem?* There is no need for new tools and techniques. Complex problems are being remedied successfully.

3.8.1.2 Various Definitions of Complexity

The scientific community cannot agree on a single definition of a complex problem (Quesada, Kintsch, and Gomez 2005) cited by Fischer, Greiff, and Funke (2012). The literature contains many different definitions of complexity, for example:

> A complex system usually consists of a large number of members, elements or agents, which interact with one another and with the environment
>
> **ElMaraghy et al. (2012)**

According to this definition, the only difference between a system and a complex system is in the interpretation of the meaning of the undefined word 'large'.

> The classification of a system as complex or simple will depend upon the observer of the system and upon the purpose he has for constructing the system
>
> **Jackson and Keys (1984)**

> A simple system will be perceived to consist of a small number of elements, and the interaction between these elements will be few, or at least regular. A complex system will, on the other hand, be seen as being composed of a large number of elements, and these will be highly interrelated
>
> **Jackson and Keys (1984)**

> A complex system is an assembly of interacting members that is difficult to understand as a whole
>
> **Allison (2004: p. 2)**

The attributes associated with the different definitions of complexity include:

- Number of issues, functions or variables involved in the problem.
- Degree of connectivity among those variables.
- Type of relationships among those variables.
- Stability of the properties of the variables over time.

Since there are no specific numbers that can be used to distinguish complex systems from non-complex systems, it does seem that complexity is in the eye of the beholder (Jackson and Keys 1984).

3.8.1.3 Partitioning Complexity

Complexity can be partitioned in various ways including:

- A total of 32 different complexity types in 12 different disciplines and domains such as projects, structural, technical, computational, functional and operational complexity (Colwell 2005) cited by ElMaraghy et al. (2012).
- Subjective or objective complexity (Sillitto 2009) where:
  1. *Subjective complexity:* people don't understand it and can't get their heads round it, e.g. (Allison 2004: p. 2), discussed in Section 3.8.1.3.1.
  2. *Objective complexity:* the problem situation or the solution has an intrinsic and measurable degree of complexity, e.g. (ElMaraghy et al. 2012, Jackson and Keys 1984) discussed in Section 3.8.1.3.2.

3.8.1.3.1 Subjective Complexity

There do not appear to be unique words that uniquely define the concepts of 'subjective complexity' and 'objective complexity' in the English language. Hence, the literature accordingly uses the words 'complicated' and 'complex' both as synonyms to mean both subjective and objective complexity and to distinguish between subjective and objective complexity. To further muddy the situation, some authors use the word 'complex' to mean subjective complexity while other authors use the word complicated to mean subjective complexity and vice versa.

3.8.1.3.2 Objective Complexity

Perceptions from the *Continuum* HTP differentiate the various definitions of objective complexity in the literature into two types, as follows (Kasser and Palmer 2005):

1. *Real-world complexity:* elements of the real world are related in some fashion, and made up of components. This complexity is not reduced by appropriate abstraction; it is only hidden.
2. *Artificial complexity:* arising from either poor aggregation or failure to abstract out elements of the real world that, in most instances, should have been abstracted out when drawing the internal and external system boundaries, since they are not relevant to the purpose for which the system was

created. For example, in today's paradigm, complex drawings are generated that contain lots of information[17] and the observer is supposed to abstract information as necessary from the drawings. The natural complexity of the area of interest is included in the drawings; hence, the system is thought to be complex.

Using the analogy to complex numbers in mathematics (perceptions from the *Generic* HTP), objective complexity may be considered as the real part of complexity and subjective complexity may be considered as the imaginary and can be reduced by education and experience (Kasser and Zhao 2016a). This would allow problems to be plotted in a two-dimensional matrix with objective complexity along the vertical axis and subjective complexity along the horizontal axis (Section 3.8.3.2).

3.8.2 TEMPORAL

Perceptions of complexity from the *Temporal* HTP noted that the objective complexity of systems that humanity can manage has grown over the centuries as shown in Figure 3.7 (Kasser 2018b).

3.8.3 SCIENTIFIC

Perceptions of complexity from the *Scientific* HTP included resolving the complexity dichotomy.

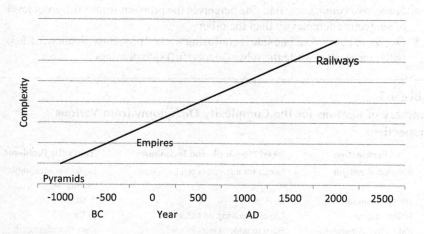

FIGURE 3.7 Perceptions of objective complexity from the *Temporal* HTP.

---

[17] The United States Department of Defense Architecture Framework (DODAF) Operational View (OV) diagrams can be wonderful examples of artificial complexity.

3.8.3.1 Resolving the Complexity Dichotomy

The complexity dichotomy (Section 3.8.1.1) may be resolved by observing that each side is focused on one or more different non-contradictory aspects of the situation as summarized in Table 3.3 where:

1. *The solution paradigm:* one side may be talking about the need to develop new tools and techniques to solve the problems associated with producing a single correct optimal solution that satisfies the problem, while the other (successful) side consists of those who are willing to settle for an acceptable solution that satisfices the problem.
2. *The Hitchins–Kasser–Massie Framework (HKMF) Column:* one side may be talking about developing new complex systems in the HKMF Columns A–F (Kasser 2018a: Section 5.2) and the other (successful) side may be talking about managing complex systems in operation in the HKMF Column G.
3. *The HKMF Layer:* one side may be positioned in the HKMF Layer 2 while the other side is positioned in the HKMF Layers 3–5. The theory of integrative levels (Needham 1937) cited by Wilson (2002) recognizes that system behaviour is different in the different levels of the hierarchy so that tools and techniques that work at one level may not work in others. Moreover, people in the Layer 2 side are used to dealing with their system in Layer 2, the meta-system in Layer 3 and the subsystem in Layer1. When they move up into Layer 3, they add Layer 4 to their area of concern but do not drop Layer 1, increasing the artificial complexity. Those in the other side of the dichotomy already in Layer 3 have dropped Layer 1 simplifying their area of concern.
4. *Subjective complexity:* one side perceives the problem from a different level of subjective complexity than the other.
5. *Degree of confusion:* one side is confusing wicked problems (Section 3.4.5.3) with complexity, while the other (successful) side does not.

TABLE 3.3
Summary of Reasons for the Complexity Dichotomy from Various Perspectives

	Perspective	Need New Tools and Techniques	What's the Problem?
1	Solution paradigm	Looks for a single correct solution	Looks for acceptable solutions
2	HKMF Column	B–F	A and G
3	HKMF Layer	Layer 2 moving up to Layer 3	In Layer 3
4	Subjective complexity	Hard to understand	Easy to understand
5	Degree of confusion	Confusing ill-structured problems with complexity	No confusion
6	Structure of the problem	Ill-structured, wicked	Well-structured
7	Boundary of knowledge	Outside	Inside

6. *Structure of the problem:* one side is successfully managing well-structured problems; the other is trying and failing to manage ill-structured and or wicked problems.
7. *Boundary of knowledge* (Kasser 2018b)*:* one (successful) side is:
   - Working *inside* the boundary of knowledge.
   - Working with well-structured problems (Section 3.4.5.1).

   The other (non-successful) side is:
   - Working *beyond* the boundary of knowledge.
   - Working with wicked and/or ill-structured problems, which cannot be solved (Simon 1973).

Recognition of items 4, 5, 6 and 7 provide a way to manage complexity. Donald Rumsfeld articulated the following three types of knowledge (Rumsfeld 2002):

1. Knowledge we know we know.
2. Knowledge we know we don't know.
3. Knowledge we don't know we don't know.

Mark Twain would have added a fourth type, as in, 'It ain't what you don't know that gets you into trouble. It's what you know for sure that just ain't so'.[18]

Let the line that represents the level of complexity in Figure 3.7 also represent the boundary of knowledge; Rumsfeld Type 1 knowledge lies below the line while Type 2 and Type 3 knowledge lie above the line.[19] This means that any complex system:

1. *Below that line:* can be managed posing well-structured problems using Rumsfeld Type 1 knowledge with deterministic results if all goes well. For example, problems faced by cruise ship companies, oil rigs, airlines and railway systems.
2. *Just above the line:* cannot be managed irrespective of the structure of the problem because it will take Rumsfeld Type 2 knowledge to manage it. This situation leads to applied research to convert Rumsfeld Type 2 knowledge to Rumsfeld Type 1 knowledge before the complex system may be managed. For example, the problem posed by sending a man to the moon and returning him to earth alive and well had elements just above and below the boundary of knowledge.
3. *Well above the line:* cannot be managed because it will take Rumsfeld Type 3 knowledge to manage it.

Questions arise as we convert Rumsfeld Type 2 knowledge to Rumsfeld Type 1 knowledge.[20] The act of posing these questions has converted some Rumsfeld Type 3 knowledge to Rumsfeld Type 2 knowledge. The systems that we can manage

---

[18] Also, Russo and Shumaker's *Decision Trap* (Russo and Schoemaker 1989, Kasser 2018a: Section 4.2).
[19] Referring to the types makes the argument clearer than repeating lots 'knowledge that you ...'.
[20] The learning process.

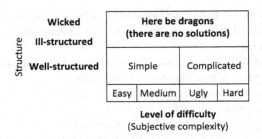

FIGURE 3.8 A problem classification framework.

successfully become more objectively complex as time passes and as the amount of Rumsfeld Type 1 knowledge increases.

3.8.3.2 The Problem Classification Framework

The structure of the problem and the level of difficulty of the problems are combined in the two-dimensional problem classification framework shown in Figure 3.8 (Kasser 2018a): Section 5.3). The two dimensions are:

1. *Structure of the problem:* ranges from non-complex through complex well-structured problems to complex ill-structured problems and wicked problems (Section 3.4.5.3).
2. *Level of difficulty:* four levels of subjective complexity ranging from easy to hard (Section 3.3.1), where simple problems can be easy and medium while complicated problems are those that are ugly and hard.

Different people may position the same problem in different places in the framework. This is because as knowledge is gained from research, education and experience, a person can reclassify the subjective difficulty of a problem down the continuum from 'hard' towards 'easy'.

3.9 REMEDYING WELL-STRUCTURED PROBLEMS

The traditional problem-solving approach manages large and objectively complex well-structured problems by breaking them out into smaller and simpler problems. Each of these problems is remedied in turn, which provides a remedy to the large and complex problem. When dealing with small problems, the process used to find a remedy is called the problem-solving process or the decision-making process. When faced with large and often complex problems the same generic process is known as the SDP (Section 5.4) as shown in Figure 3.9 (Kasser and Hitchins 2013). The words 'remedial action (problem-solving)' in Figure 3.6 have been replaced by the lower-level term, 'Series of (sequential and parallel) activities' in Figure 3.9. The figure shows that the lower-level term 'Series of (sequential and parallel) activities' is known as the SDP for large or complex systems which breaks up a complex problem into smaller less-complex problems (analysis) then solves each of the smaller problems and hopes that the combination of solutions

Perceptions of Problem-Solving

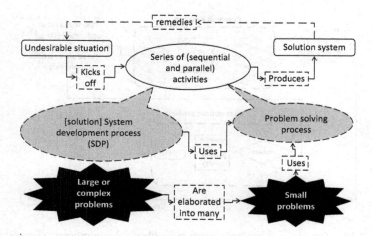

FIGURE 3.9 Activities in the context of the SDP and problem-solving.

to the smaller problems (synthesis) will provide a solution to the large complex problem. The SDP:

- Contains sub-processes for developing and testing products and parts of products as discussed in the three streams of activities (Section 2.5.2). Each of these processes needs to be under configuration control so that the version of the process used at any time to develop or test any product is known and the processes are repeatable.
- May be considered as
  - Two sequential problem-solving processes (Section 3.9.1).
  - One iteration of the multiple-iteration problem-solving process for remedying complex problems (Section 3.9.2).

3.9.1 THE TWO-PART SDP

The top-level SDP is a two-part sequential problem-solving process as shown in Figure 3.10 (Kasser and Zhao 2016a): planning and doing or implementing.

The first problem-solving process:

- Takes place in the needs identification state of the SDP.
- Shown in Figure 3.2 is the front end of the simple problem-solving process. The process contains the following major milestones (identified in triangles) and activities or processes (shown in rectangles):
  1. The milestone to provide authorization to proceed.
  2. The process for defining the scope of the problem and gaining an understanding.
  3. The process for conceiving several solution options.
  4. The process for identifying ideal solution selection criteria.
  5. The process for performing trade-offs to find the optimum solution.

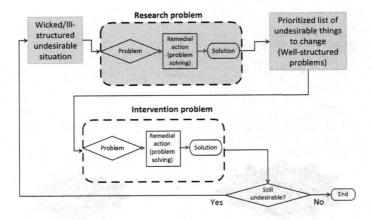

FIGURE 3.10 The two-part multiple-iteration problem-solving process. (*Source:* © 2016 IEEE. Reprinted, with permission, from Kasser, Joseph Eli, and Yang-Yang Zhao. 2016b. Simplifying Solving Complex Problems. In *the 11th International Conference on System of Systems Engineering*, Kongsberg, Norway.)

6. The process for selecting the preferred option.
7. The process for formulating strategies and plans to implement the preferred option.
8. The milestone to confirm consensus to proceed with implementation.

The second problem-solving process:

- Begins once the stakeholder consensus is confirmed at Milestone 8 at the end of Figure 3.11 (Kasser and Zhao 2016a).

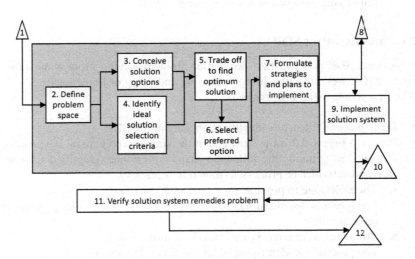

FIGURE 3.11 Modified Hitchins' view of the problem-solving decision-making process.

Perceptions of Problem-Solving

- Covers the remaining states in the SDP; the project can move on to the implementation states shown from the *Functional* HTP in Block 9 of Figure 3.11 where the additional following major milestones and activities are:
  9. The process for implementing the solution system often using the SDP.
  10. The milestone review to document consensus that the solution system has been realized and is ready for validation.
  11. The process for validating the solution system remedies the evolved need in its operational context, often known as operational test and evaluation (OT&E) for complex systems.
  12. The milestone to document consensus that the solution system remedies the evolved need in its operational context.

3.9.2 THE MULTIPLE-ITERATION PROBLEM-SOLVING PROCESS

The various problem-solving processes in the literature are parts of a meta-problem-solving process that starts with ill-defined problems, converts them to well-defined problems and evolves a remedy to the set of well-defined problems recognizing that the problem may change while the remedy is being developed

Kasser and Zhao (2016a)

From the perspective of the problem-solving paradigm of project management, the standard approach to evolving a solution can be reworded to become 'an evolutionary approach to remedying the undesirability in a situation by turning an ill-structured problem into a number of well-structured problems, remedying the well-structured problems and then integrating the partial remedies into a whole remedy' (Kasser and Hitchins 2012).

By observing the management of complex systems in industry from the HTPs, it was possible to infer a meta-process for solving the problems associated with complex systems based on a modified version of the extended problem-solving process (Section 3.6.2). Let this meta-process shown in Figure 3.10 be called the multiple-iteration problem-solving process. The multiple-iteration problem-solving process consists of two sequential problem-solving processes (Section 3.9.1) embedded in an iterative loop.

1. *The first problem-solving process:* converts the ill-structured problem posed by the situation into one or more well-structured problems (Section 3.9.1).
2. *The second problem-solving process:* is tailored to remedy specific type of problems since one problem-solving approach does not fit all problems (Section 3.1.2.4)

Choice of which of the problems identified by the first problem-solving process to tackle in the second problem-solving process will depend on a number of factors including urgency, impact on undesirable situation, the need to show early results and available resources.

This sequential evolutionary process is sometimes known as the 'build a little test a little' (Section 5.6.1) and evolves the solution from a baseline or known state to the subsequent milestone which then becomes the new baseline.

The causes of the undesirability may be remedied at different levels and locations in the situation hierarchy, simultaneously or sequentially.

3.9.2.1 The First Problem-Solving Processes

The first problem-solving process in the multiple-iteration problem-solving process remedies a research problem (Section 3.4.4.1) using an adaptation of the scientific method. This process:

- Takes place in the needs identification state of the SDP (Section 5.4.1).
- Figures out the nature of the problematic situation and what needs to be done about it.
- Creates a prioritized list of things to change. The problem-solvers create the prioritized list of things to change by:
  - Gaining a thorough understanding of situation.
  - Identifying the undesirable aspects of the situation.
  - Performing research to gather preliminary data about the undesirable situation.
  - Formulating the hypothesis to explain causes of the undesirable situation.
  - Estimating the approximate contribution of each cause to the undesirable situation.
  - Developing priorities for remedying the causes.
  - Deciding which causes to remedy.
  - Conceptualizing the FCFDS.
  - Performing feasibility studies.

Even when consensus on the causes of the undesirability cannot be achieved, it is often possible to achieve consensuses on what is undesirable and on the FCFDS.

Feasibility studies (Kasser 2019b: Section 9.5) are performed on the FCFDS because there is no point in creating a FCFDS if it is not feasible. Examples include:

- *Operational feasibility:* at least one solution or combination of partial solutions is achievable.
- *Quantitative feasibility:* Cost is affordable, risk and uncertainty are acceptable.
- *Structural (technical) feasibility:* suitable technologies exist at the appropriate technology readiness levels (Section 10.4).
- *Temporal feasibility:* Schedule (the solution system will be ready to operate in the FCFDS when needed).

As an example, the complex problem may be associated with undesirable traffic congestion in an urban area. The mayor, feeling under pressure to do something about the growing traffic congestion in her city, provides the trigger to initiate the problem-solving process, which begins with the ill-structured problem of how to remedy the undesirable effects of traffic congestion. An understanding of the situation might produce a number of causes, e.g. commuting to work and school, deliveries and tourists, etc. The analysis would also provide quantitative information such as an estimate of the degree of the contribution by the cause to the undesirable situation, e.g. commuting to work (40%) and school (30%), deliveries (20%) and tourists (10%). The need to remedy each cause would then be prioritized according to selection criteria that might include cost, schedule, political constraints, performance and robustness.

The often-forgotten domain knowledge needed to gain consensus on, and prioritize the cause is 'human nature'. Each of the stakeholders needs to know 'what's in it for me?' in implementing the change (Section 2.1.4.7). So, it is the responsibility of the project manager to identify and communicate that information.

Tools developed for gaining an understanding of the system (situation) and the nature of its undesirability include:

- Checkland's soft systems methodology (SSM) (Checkland 1993: pp. 224–225).
- Avison and Fitzgerald's interventionist methodology (Avison and Fitzgerald 2003).
- The Nine System Model (Kasser 2019b: Chapter 10).
- Active brainstorming (Kasser 2018a: Section 7.1).
- The HTPs (Section 2.1.1.4).

3.9.2.2 The Second Problem-Solving Process

The second problem-solving process in the multiple-iteration problem-solving process is:

- The traditional SDP (Section 5.4).
- An intervention problem (Section 3.4.4.2) and remedies (some of) the aspect(s) of the undesirable situation identified by the first problem-solving process. Since one problem-solving approach does not fit all problems (Kasser and Zhao 2016a), the second problem-solving process is tailored to remedy the specific type of problems. Once the second problem-solving process is completed the process may iterate back to the beginning for a new cycle as shown in Figure 3.12 (Kasser and Zhao 2016c) because the second problem-solving process may:
  - Only partially remedy the original undesirable or problematic situation.
  - Contain unanticipated undesirable emergent properties from the *solution system* and its interactions with its adjacent systems.
  - Only partially remedy new undesirable aspects that have shown up in the situation during the time taken to develop the solution system.
  - Produce new unanticipated undesired emergent properties of the solution system and its interactions with its adjacent systems, which in turn produce new undesirable outcomes.

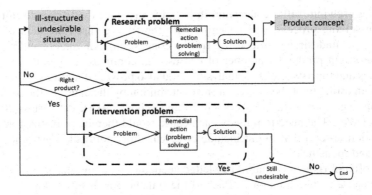

FIGURE 3.12 A new product development variation of the multiple-iteration problem-solving process. (*Source:* © 2016 IEEE. Reprinted, with permission, from Kasser, Joseph Eli, and Yang-Yang Zhao. 2016c. Wicked Problems: Wicked Solutions. In *the 11th International Conference on System of Systems Engineering*, Kongsberg, Norway.)

3.9.3 A New Product Development Process

New product development has been mapped into the learning process as a sequence of problems known as design thinking (Beckman and Barry 2007, Brown 2008, Leifer and Steinert 2011). The undesirable situation is the need to develop a new desired product. The first well-structured problem is what product (or service) to provide to users, which is not necessarily the first product that comes to mind. After working on the first product, the learning process often produces a finding that the first product is not what the users need and identifies an alternative product, so the process iterates and the new product development team learns about the need of the user and how the product will be used in its context (Beckman and Barry 2007, Brown 2008). The second well-structured problem is how to create the product that was the outcome from remedying the first well-structured problem.

This two-problem sequential process can be mapped into a modified version of the multiple-iteration problem-solving process shown in Figure 3.12 where the output from the first research process is not a list of problems to solve, but is instead a product concept or prototype. The first process ends at a stage gate which determines if the product is indeed, what the user needs and if the product should proceed to the second process which in this situation is the production process.

Having mapped new product development into the multiple-iteration problem-solving process, the benefits of applying thinking in improving design thinking (Zhao 2015) in dealing with complexity can be seen. Design thinking has a number of definitions including:

- A systems approach to design visualizing the product operating in its context (Cross 2011).
- An iterative learning process by multidisciplinary experts from sociology, psychology, engineering, science, design, etc., working together and evolving the new product (Leifer and Steinert 2011).

- A human-centred approach enabling the engineers to think outside of the box (Collins 2013, Brown 2008).

Design thinking takes place in needs identification state of the SDP. This new product development process reduces the number of iterations in the traditional new product development process. When the systems comprising the undesirable situation and the FCFDS (the proposed new product operating in its context) are examined systemically from the HTPs, a better understanding of the user's as well as other stakeholders' needs is achieved as a result of observing the situation from the different perspectives. This better initial understanding reduces the number of iterations of the first process.

3.10 REMEDYING ILL-STRUCTURED PROBLEMS

Ill-structured problems cannot be remedied; they must first be converted to well-structured problems (Simon 1973). Perceptions from the *Generic* HTP show that is what we do in the problem-solving process depicted in Figure 3.12. The undesirable or problematic situation poses an ill-structured problem. As we gain an understanding of the situation, we convert the ill-structured problem into one or more well-structured problems.

3.11 REMEDYING COMPLEX PROBLEMS

A complex problem may be defined as, 'one of a set of problems posed to remedy the causes of undesirability in a situation in which the solution to one problem affects another aspect of the undesirable situation' (Kasser and Zhao 2016b). Consequently, remedying a complex problem will depend on the structure of the problem.

3.11.1 Remedying Well-Structured Complex Problems

Well-structured complex problems:

- Consist of a set of interconnected well-structured problems. Choice of which of the symptoms identified by the first problem-solving process (Section 3.9.2) to tackle in the second problem-solving process will depend on a number of factors (selection criteria) including urgency, impact on undesirable situation, the need to show early results and available resources.
- Are remedied using the multiple-iteration problem-solving process (Section 3.9.2). This is a time-ordered-multi-phased evolutionary approach that provides remedies to one or more of the well-structured problems, integrates the remedies, re-evaluates the situation and then repeats the process for the subsequent set of problems (Kasser 2002b).

3.11.2 Remedying Ill-Structured Complex Problems

The undesirable situation causing the ill-structured complex problem cannot be remedied until the ill-structured complex problem has been transformed into one or more well-structured problems. Consequently, finding a solution requires converting the ill-structured complex problem into a well-structured complex problem or series of well-structured complex problems. Determining the real cause(s) of the undesirable situation and finding solutions sometimes means doing both functions in an iterative and interactive manner. In this situation, initially:

1. *The undesirable situation:* an ill-structured problem.
2. *The assumptions:* there may be than one cause of undesirability. Ill-structured problems may be a result of stakeholders perceiving the situation from single different HTPs in the manner of the fable of the blind men perceiving the elephant (Yen 2008)[21]; they identified different animals, in this situation, the stakeholders are identifying different causes. Even if the stakeholders cannot agree on the causes of the undesirability, they can usually agree on the nature of the undesirability.
3. *The FCFDS:* (desired outcome) one or more well-structured problems.
4. *The problem:* how to convert the ill-structured problem into one or more well-structured problems.
5. *The solution:* follow the multiple-iteration problem-solving process (Section 3.9.2).

The research problem-solving process (Section 3.4.4.1) is used to convert the ill-structured problem into one or more well-structured problems which can then be remedied either singly or as a group (Section 3.11.1). However, take care when converting ill-structured problems into a series of well-structured problems because you can end up with different and sometimes contradictory well-structured problems, which would generate different and sometimes contradictory solutions.

As an example, consider the ill-structured complex problem of how to win a war. This problem is broken out into two lower level ill-structured problems (1) Defence: how to defend the nation and (2) Attack: how to destroy the other side and end the war. Each of these problems is then further broken out into a set of well-structured problems, which if remedied successfully in a timely manner[22] will end the war successfully.

The first approach to create the set of well-structured problems goes beyond systems thinking (Section 2.1.1.4) and uses perceptions from the *Continuum, Temporal* and *Generic* HTPs often using lessons learned in Command School. Questions to be researched include[23]:

- Who has faced this situation before?
- What did they do about it?
- Why would that solution work/not work in this case?
- How can we improve on the previous instance?

---

[21] They identified different animals.
[22] Before the enemy does the same.
[23] These are standard questions to be posed in similar problematic situations.

3.11.3 REMEDYING WICKED PROBLEMS

Wicked problems (Section 3.4.5.3) are considered impossible to solve using the current problem-solving paradigm. When faced with insolvable problems, the best way to approach them is to dissolve the problems (Section 3.1.2.1) or bypass them by finding an alternative paradigm (Kuhn 1970). Using inferences from the *Scientific* HTP, instead of trying to solve or resolve wicked problems, dissolve the problem by changing the paradigm from 'problem' to 'situation' (Kasser and Zhao 2016c). Instead of dealing with wicked problems, deal with wicked situations. Then, nondeterministic behaviour:

1. Is not a characteristic of complex problems.
2. Is a characteristic of:
   1. Ill-structured and wicked situations.
   2. The initial state in the scientific method (Section 3.4.4.1) as perceived from the *Generic* HTP due to:
      1. A lack of understanding of the situation which precludes determining the behaviour.
      2. Being beyond the boundary of the current body of knowledge, the line in Figure 3.8.

One of the characteristics of wicked problems (Section 3.4.5.3) is 'Cannot be easily defined so that all stakeholders cannot agree on the problem to solve'. Accordingly, assume multiple unknown causes of undesirability in the undesirable situation.[24] The assumption of multiple causes leads to perceiving that there may be multiple solutions (perhaps even at different levels in the hierarchy of systems) one or more for each cause.

1. Elaborate the undesirable situation into one or more undesirable situations.
2. Use the multiple-iteration problem-solving process (Section 3.9.2) to create wicked solutions (Kasser and Zhao 2016c) which have similar characteristics to wicked situations.

When creating wicked solutions, the initial solution may not be the needed solution, since wicked solutions:

- Evolve via the multiple-iteration problem-solving process (Section 3.9.2).
- Each instance of the wicked solution:
  1. May only remedy part of the undesirability in the whole wicked situation.
  2. May satisfice and not necessarily satisfy the problem in a single pass through the multi-pass problem-solving process.
  3. May apply simultaneously in the wicked situation hierarchy at more than one level and more than one location at a particular level.

---

[24] The hypothesis in the scientific method.

3.12 SUMMARY

This chapter helped you to understand the problem-solving aspects of management by discussing perceptions of the problem-solving process from a number of HTPs, then discussing the structure of problems and the levels of difficulty posed by problems and the need to evolve solutions using an iterative approach. After showing that problem-solving is really an iterative causal loop rather than a linear process, the chapter then discussed complexity and how to use the systems approach to manage complexity. The chapter then showed how to remedy well-structured problems and how to deal with ill-structured, wicked and complex problems using iterations of a sequential two-stage problem-solving process.

Reflecting on this chapter, it seems that iteration is a common element in remedying any kind of problem other than easy well-structured ones irrespective of their structure.

REFERENCES

Ackoff, Russel L. 1978. *The Art of Problem Solving*. New York: John Wiley & Sons.
Ackoff, Russel L., and Herber J. Addison. 2006. *A Little Book of f-Laws 13 common Sins of Management*. Axminster: Triarchy Press Limited.
Allison, James T. 2004. *Complex System Optimization: A Review of Analytical Target Cascading, Collaborative Optimization, and Other Formulations*. Ann Arbor, MI: The University of Michigan.
Avison, David, and Guy Fitzgerald. 2003. *Information Systems Development: Methodologies, Techniques and Tools*. Maidenhead: McGraw-Hill Education (UK).
Barry, Katie, Ellen Domb, and Michael S. Slocum. 2007. *TRIZ—What is TRIZ?* 2007 [cited 31 October 2007]. Available from www.triz-journal.com/archives/what_is_triz/.
Beckman, Sara L., and Michael Barry. 2007. Innovation as a learning process: Embedding design thinking. *California Management Review* no. 50 (1):25–56.
Brown, Tim. 2008. Design thinking. *Harvard Business Review* no. 6 (June):84.
Checkland, Peter. 1993. *Systems Thinking, Systems Practice*. Chichester: John Wiley & Sons.
Collins, Harper. 2013. Can design thinking still add value? *Design Management Review* no. 24 (2):35–39.
Colwell, Bob. 2005. Complexity in design. *IEEE Computer* no. 38 (10):10–12.
Cross, Nigel. 2011. *Design Thinking: Understanding How Designers Think and Work*. Oxford: Berg.
ElMaraghy, Waguih, Hoda ElMaraghy, Tetsuo Tomiyama, and Laszlo Monostori. 2012. Complexity in engineering design and manufacturing. *CIRP Annals—Manufacturing Technology* no. 61 (2):793–814.
Fischer, Andreas, Samuel Greiff, and Joachim Funke. 2012. The process of solving complex problems. *The Journal of Problem Solving* no. 4 (1):19–42.
Ford, Whit. *Learning and teaching math* 2010 [cited 8 April 2015]. Available from http://mathmaine.wordpress.com/2010/01/09/problems-fall-into-four-categories/.
Hitchins, Derek K. 2007. *Systems Engineering. A 21st Century Systems Methodology*. Chichester, England: John Wiley & Sons.
Howard, Ronald A. 1973. Decision analysis in systems engineering. In *Systems Concepts*, edited by Ralph F. Miles Jr, pp. 51–85. Hoboken, NJ: John Wiley & Son
Jackson, Michael C., and Paul Keys. 1984. Towards a system of systems methodologies. *Journal of the Operations Research Society* no. 35 (6):473–486.

Jonassen, David H. 1997. Instructional design model for well-structured and ill-structured problem-solving learning outcomes. *Educational Technology: Research and Development* no. 45 (1):65–95.

Kasser, Joseph Eli. 2002a. Configuration management: The silver bullet for cost and schedule control. In *IEEE International Engineering Management Conference (IEMC 2002)*, Cambridge, UK.

Kasser, Joseph Eli. 2002b. Isn't the acquisition of a system of systems just a simple multi-phased time-ordered parallel-processing process? In *11th International Symposium of the INCOSE*, Las Vegas, NV.

Kasser, Joseph Eli. 2015a. *Holistic Thinking: Creating Innovative Solutions to Complex Problems* (Vol. 1). 2nd edition. Charleston, SC: Solution Engineering: Createspace Ltd.

Kasser, Joseph Eli. 2015b. *Perceptions of Systems Engineering* (Vol. 2). Charleston, SC: Solution Engineering: Createspace Ltd.

Kasser, Joseph Eli. 2018a. *Systems Thinker's Toolbox: Tools for Managing Complexity*. Boca Raton, FL: CRC Press.

Kasser, Joseph Eli. 2018b. Using the systems thinker's toolbox to tackle complexity (complex problems). In *SSSE Presentation at Roche*. Zurich: Swiss Society of Systems Engineering.

Kasser, Joseph Eli. 2019b. *Systemic and Systematic Systems Engineering*. Boca Raton, FL: CRC Press.

Kasser, Joseph Eli, and Derek K. Hitchins. 2012. Yes systems engineering, you are a discipline. In *22nd Annual International Symposium of the INCOSE*, Rome, Italy.

Kasser, Joseph Eli, and Derek K. Hitchins. 2013. Clarifying the relationships between systems engineering, project management, engineering and problem solving. In *Asia-Pacific Council on Systems Engineering Conference (APCOSEC)*, Yokohama, Japan.

Kasser, Joseph Eli, and Kent Palmer. 2005. Reducing and managing complexity by changing the boundaries of the system. In *Conference on Systems Engineering Research*, Hoboken, NJ.

Kasser, Joseph Eli, and Yang-Yang Zhao. 2016a. The myths and the reality of problem-solving. In *Systemic and Systematic Integrated Logistics Support: Course notes*, edited by Joseph Eli Kasser. Singapore: TDSI.

Kasser, Joseph Eli, and Yang-Yang Zhao. 2016b. Simplifying solving complex problems. In *The 11th International Conference on System of Systems Engineering*, Kongsberg, Norway.

Kasser, Joseph Eli, and Yang-Yang Zhao. 2016c. Wicked problems: Wicked solutions. In *11th International Conference on System of Systems Engineering*, Kongsberg, Norway.

Kuhn, Thomas S. 1970. *The Structure of Scientific Revolutions*. 2nd edition, Enlarged ed. Chicago, IL: The University of Chicago Press.

Leifer, Larry J., and Martin Steinert. 2011. Dancing with ambiguity: Causality behavior, design thinking, and triple-loop-learning. *Information Knowledge Systems Management* no. 10 (1):151–173.

Merton, Robert King. 1936. The unanticipated consequences of social action. *American Sociological Review* no. 1 (6):894–904.

Merton, Robert King. 1948. The self-fulfilling prophecy. *The Antioch Review* no. 8 (2):193–210.

Needham, Joseph. 1937. *Integrative Levels: A Revaluation of the Idea of Progress*. Oxford: Clarendon Press.

Quesada, Jose, Walter Kintsch, and Emilio Gomez. 2005. Complex problem solving: A field in search of a definition? *Theoretical Issues in Ergonomic Science* no. 6 (1):5–33.

Rittel, Horst W., and Melvin M. Webber. 1973. Dilemmas in a general theory of planning. *Policy Sciences* no. 4:155–169.

Rumsfeld, Donald. 2019. *DoD news briefing—Secretary Rumsfeld and gen. Myers*, February 12, 2002. DoD 2002 [cited 18 January 2019]. Available from http://archive.defense.gov/Transcripts/Transcript.aspx?TranscriptID=2636.

Russo, J. Edward, and Paul H. Schoemaker. 1989. *Decision Traps*. New York: Simon and Schuster.

Savage, Sam L. 2009. *The Flaw of Averages*. Hoboken, NJ: John Wiley and Sons.

Schön, Donald A. 1991. *The Reflective Practitioner*. Farnham: Ashgate.

Sillitto, Hillary. 2009. On systems architects and systems architecting: Some thoughts on explaining and improving the art and science of systems architecting. In *19th International Symposium of the INCOSE*, Singapore.

Simon, Herbert Alexander. 1973. The structure of ill structured problems. *Artificial Intelligence* no. 4 (3–4):181–201. doi:10.1016/0004-3702(73)90011-8.

Waring, Alan. 1996. *Practical Systems Thinking*. London, England: International Thompson Business Press.

Wilson, Tom D. 2002. Philosophical foundations and research relevance: Issues for information research (Keynote address). In *4th International Conference on Conceptions of Library and Information Science: Emerging Frameworks and Method*, July 21 to 25, 2002, University of Washington, Seattle, WA.

Yen, Duen Hsi. *The blind men and the elephant* 2008 [cited 26 October 2010]. Available from www.noogenesis.com/pineapple/blind_men_elephant.html.

Zhao, Yang-Yang. 2015. Towards innovative systems development: A joint method of design thinking and systems thinking. In *25th International Symposium of the (INCOSE)* Seattle, WA.

4 Management
General and Project Management

The purpose of management is to accomplish a goal by getting other people to do the work. This chapter helps you to understand how to achieve that purpose by discussing management, general management and project management, and how to accomplish that goal. After a brief discussion on general management, the chapter focuses on the attributes of projects and project management as a problem-solving activity. This chapter explains:

1. The purpose of management in Section 4.1.
2. Perceptions of management from the holistic thinking perspectives (HTP) in Section 4.2.
3. General management in Section 4.3.
4. Project management in Section 4.4.
5. Taking over a project in Section 4.5.
6. Research projects in Section 4.6.

4.1 THE PURPOSE OF MANAGEMENT

The purpose of management is to accomplish a goal by getting other people to do the work. One way of achieving this purpose is to delegate all the tasks to other people and let them get on with doing the work. They should be monitored but in general should not be micromanaged. The only personnel that should be micromanaged are the Type I apprentices (Section 2.1.2.1). While the work is being done, management should:

- Assign credit and recognition to those who did the work (Section 12.2.1).
- Communicate accomplishments, what is happening, why it's happening as well as problems that have been identified and are being mitigated or prevented.

4.2 PERCEPTIONS OF MANAGEMENT

Perceptions of management from the HTPs include:

4.2.1 BIG PICTURE

Perceived from the *Big Picture* HTP, management is using positive politics (Section 2.2.1) to motivate people to use resources to achieve objectives by taking advantages of opportunities and overcoming adversity in the context of (constrained by) an organizational environment and implementation domain.

Three systems which form the context of management in the implementation domain (Section 3.4.6) are:

1. *The organization in which the project/process is performed:* provides the resources as well as the constraints on management. The context is the environment in which management is taking place. The constraints come from the rules, regulations and standard operating procedures in the organization. This system is not within the manager's sphere of control, but may be influenced by the actions of the manager and relationships with other members of the organization.
2. *The process achieving the objective:* is directly within the manager's sphere of control. The role of the manager is to enable the process to meet its objectives in the most effective manner at that time and place.
3. *The objective being achieved:* is only indirectly within the manager's sphere of control. The objective is generally being met by the project team who have the requisite skills and competencies to achieve the objective.

4.2.2 OPERATIONAL

Perceptions from the *Operational* HTP note:

- Managers performing the traditional activities of applied management (Section 4.2.4.3) which consists of planning, organizing, directing and controlling (Fayol 1949: p. 8) as well as staffing.
- There seems to be one very common practical traditional management paradigm – the Calamity Jane or John Wayne Paradigm expressed as (Kasser 1995):
  1. Shoot first; ask questions later.
  2. Ready, fire! aim.[1]

This paradigm is reactive; the project manager (PM) jumps into a situation, shoots in all directions and completes the task. However, very soon the same problems arise again, so the PM reacts by shooting in all directions until the task seems completed. Sometime later, the same problem comes back, so the task is never ended. This paradigm is characterized by the following:

- *Closed communications:* Jane/John act as information sponges gathering information about the product and state of the process but do not share it. Consequently, people are never sure about what to do. This results in:
- *Continually fighting crises:* problems arise all the time. Many times, the effort spent in dealing with the crisis was not scheduled, so time passes and the main development effort does not progress. As a consequence, they all experience:

---

[1] Reactive; proactive management is 'ready, aim, fire'.

- *Frenzied activity:* Everybody is very busy working hard, but somehow the work does not lead to deliverable results. Nobody has time to plan, so they are under:
- *High pressure:* When a deadline approaches, John/Jane exhorts the troops to deliver the specific item. Much overtime is worked, and the product is delivered in spite of John/Jane due to the diligence of the workers. This leads to:
- *Schedule and budget overruns:* All the extra work took time away from doing the job, so the job takes longer than estimated. Consequently, John/Jane has to explain the situation to top management, leading to:
  - *High visibility:* Management monitor the situation closely and 'help' get the product delivered. This situation, in extreme cases can lead to 'fraud' on contracts. Senior management puts pressure on John/Jane to close out tasks. Jane/John complies and closes out a cost account when the work is not completed. If the work is reported (claimed) as being completed, when in fact it wasn't, then that work will have to be performed later and charged to a different cost account, a situation which could be construed as a false claim (Section 13.5.2.1).
  - *High employee turnover:* Effective people tend to leave these kinds of tasks because they get frustrated with the waste inherent in the paradigm. They'd rather work more productively. Sometimes they transfer to another project, more often they move to another organization. As the good people leave, their work is shared among the existing project team or they are replaced by less effective people. The replacements need time to get up to speed. This, in engineering terms is a positive feedback situation, since as the level of expertise decreases due to people transferring out, the amount of ineffective work increases, leading to further departures.

4.2.3 FUNCTIONAL

Perceptions from the *Functional* HTP show many managers using the cognitive thinking skills of pure management to perform applied management (Section 4.2.4.3).

4.2.4 CONTINUUM

Perceptions from the *Continuum* HTP include:

1. The differences between a manager and a leader discussed in Section 4.2.4.1.
2. The difference in competence discussed in Section 4.2.4.2
3. The three types of management discussed in Section 4.2.4.3.
4. The difference between the requirements for the project and the requirements for the product produced by the project discussed in Section 4.2.4.4.
5. The different styles of management in the role of leaders discussed in Section 12.4.3.

4.2.4.1 The Differences between a Manager and a Leader

The differences have been expressed in various ways and include those shown in Table 4.1. Most managers combine some degree of management and leadership.

4.2.4.2 The Difference in Competence

There is a wide range of competence demonstrated by managers. Some managers are extremely competent, some merely competent while others can be characterized by Scott Adams' Dilbert cartoon strip (Adams 1996).

4.2.4.3 The Three Types of Management

Perceptions from the *Generic*[2] and *Continuum* HTPs note that management can be divided into the following three types:

1. *Pure management:* the cognitive thinking skills (Section 2.1.1) used to perform applied management.
2. *Applied management:* the traditional activities of managers: planning, organizing, directing and controlling (Fayol 1949: p. 8) as well as staffing.
3. *Domain management:* the three domains in which management is performed (Section 3.4.6).

4.2.4.4 The Difference between the Requirements for the Project and the Requirements for the Product Produced by the Project

Perceptions from the *Continuum* HTP note the following difference. The requirements for the:

1. *Project:* are developed during the project planning state (Section 5.3.1.2)
2. *Product the project produces:* are developed during the project performance state (Section 5.3.1.3) in the system requirements state of the system development process (SDP) (Section 5.4.1).

TABLE 4.1

The Differences between Leaders and Managers

	Leaders	Managers
Goals and objectives	Set goals and objectives	Ensure the goals and objectives are met
The process	Are administrators making sure that resources are available on schedule, and that the process is being followed according to the plan	Motivate people to perform the process
People	Care about the people they lead	Managers may not care about the people they manage

---

[2] Similar to the three types of systems engineering (Kasser and Arnold 2016).

Management: General and Project Management

4.2.5 QUANTITATIVE

Perceptions from the *Quantitative HTP* include:

1. The three types of management discussed in Section 4.2.4.3.
2. The seven interdependent P's of a project (Chapter 2).
3. About 80%–85% of all organizational problems are caused by management (Juran 1992).
4. About 94% of the problems faced by workers belong to the system (Deming 1986) (i.e. are the responsibility of management).

4.2.6 TEMPORAL

The delegation of work goes back to ancient times. For example, when Jethro visited Moses and saw how Moses managed (micromanaged?) the people, he recommended appointing leaders of tens, hundreds and thousands to lighten his workload (Exodus 18:13–23). This advice:

- Made Jethro the world's first documented management consultant.
- Has been interpreted as recommending a hierarchical organization structure.

4.3 GENERAL MANAGEMENT

Perceive general management from five HTPs:

- *Big Picture:* general management deals with the management of organizations or parts of organizations that exist on a permanent basis, such as departments in a large organization.
- *Operational:* the general manager is responsible for the work performed by a part of an organization, for example, a department or a shop.
- *Functional:* the general manager uses the functions of pure management to perform the applied management of running the part of the organization.
- *Structural:* a general manager is the person at the top of the organization chart for that part of the organization.
- *Continuum:* there is a difference between a general manager and general management. A general manager is a role or a job description that applies to the person running an organization. General management is the function performed by the general manager.

4.4 PROJECT MANAGEMENT

Project management is defined as 'the planning, organizing, directing, and controlling of company resources (i.e. money, materials, time and people) for a relatively short-term objective. It is established to accomplish a set of specific goals and objectives by utilizing a fluid, systems approach to management by having functional personnel (the traditional line-staff hierarchy) assigned to a specific project (the horizontal hierarchy)' (Kezsbom, Schilling, and Edward 1989).

4.4.1 Projects

A project is a temporary endeavor undertaken to create a unique product, service or result

PMI (2013)

All projects should contain the following attributes:

1. Purpose
2. Activities
3. Funding
4. Milestones
5. Need
6. Priority
7. Sponsor
8. Stakeholders
9. Customers
10. Timeline
11. Return on investment (ROI)
12. Outcomes
13. Risks

Consider each of them.

4.4.1.1 Purpose

The purpose of the project is to create the product, service or result for which it was established, on or before the scheduled delivery date within the resource budget.

4.4.1.2 Activities

Activities are the basic building blocks of projects. All work in the project is done by activities that take time to produce a product using resources. Activities may take place in series or in parallel. Most activities contain risks; undesirable events that, should they occur, will increase the cost or schedule to the detriment of the project. The time the activity takes is shown in a schedule in the form of a Gantt chart. Each activity has:

- *Customer(s):* to whom the product produced by the activity is delivered.
- *Prerequisite(s):* something that has to happen, or be created before the activity can commence.
- *Resources:* people, materiel, time, money, etc.
- *Supplier (s):* who supply something that is transformed by the activity into products.
- *Value:* by transforming the input from the supplier into a product that is used by a customer.

Management: General and Project Management 103

Perceived from the *Continuum* HTP, each activity uses the resources to produce:

1. *Product(s):* the desired output.
2. *Waste:* the undesired output, e.g. defective products and parts, leftover parts and transformed materials, etc. Waste incurs an unnecessary cost and must be:
   1. Minimized.
   2. Disposed of carefully.

4.4.1.3 Funding

Adequate funding is critical to project success. Underfunded projects tend to fail. However, many projects are knowingly underfunded by the project sponsors (Section 4.4.1.7) sometimes because if the true costs were known the project would never be approved; politics (Section 2.2) in action. The project sponsors hope that there will be changes during the performance state of the project that will inflate the cost of the project to the realistic expected amount.

4.4.1.4 Milestones

Projects incorporate major and minor milestones. The number of milestones depends on the duration of the project. Major milestones may be the milestones in the SDP (Section 5.4); minor milestones may be reporting meetings that take place periodically, e.g. weekly or monthly depending on the scope and direction of the project.

4.4.1.5 Need

All projects should have a need. If there is no need to deliver the output of the project, the effort put into the project is wasted and should be usefully applied elsewhere.[3]

4.4.1.6 Priority

All projects should have a priority within the organization in which the project is performed. This priority allows upper management to determine which projects should be delayed or even cancelled, should shared resources not be available when needed.

4.4.1.7 Sponsor

There are various opinions on what constitutes a sponsor. In some instances, the sponsor represents the customer, in other instances the activities assigned to the sponsor overlap the activities assigned to the PM. Perceptions from the *Continuum* HTP show that there is a difference between the roles of the sponsors in projects that are internal to organizations and projects that are undertaken under contract for an external customer. The different roles are:

1. *Internal projects:* 'project sponsors champion the project and use their influence to gain approval for the project. Their reputation is tied to the success of the project, and they need to be kept informed of any major

---

[3] On the other hand, there might be a need to keep people busy even if the product is not going to be used.

developments. They defend the project when it comes under attack and are a key project ally' (Larson and Gray 2011: p. 343). The project sponsor is often the project customer.
2. *External projects:* the customer is the project sponsor.

4.4.1.8 Stakeholders

A stakeholder is an individual, group, or organization who may affect, be affected by, or perceive itself to be affected by a decision, activity, or outcome of a project. Stakeholders may be actively involved in the project or have interests that may be positively or negatively affected by the performance or completion of the project. Different stakeholders may have competing expectations that might create conflicts within the project. Stakeholders may also exert influence over the project, its deliverables, and the project team in order to achieve a set of outcomes that satisfy strategic business objectives or other needs

(PMI 2013: p. 30)

Stakeholders need to be managed over the life of the project (Section 12.6).

4.4.1.9 Customers

The most important stakeholder is the one who funds the project. In this book, that stakeholder is known as the customer.

4.4.1.10 Timeline

Projects have fixed timelines or at least they are fixed when the project begins. The timeline extends from the start of the project to the end of the project passing through a number of pre-planned milestones (Section 5.4.2). This facilitates management by objectives (MBO) (Mali 1972) (Section 11.7), if the prerequisite for achieving a milestone is the accomplishment of one or more objectives. The timeline may be shown as a single line or in Gantt and Program Evaluation Review Technique (PERT) charts, depending on which aspect of the timeline is being presented.

If the project is developing a new product, then in general the shorter the timeline, the faster the product will get to market and the greater will be the profit or ROI.

4.4.1.11 The ROI

The traditional approach to estimate the ROI is to divide the financial benefit of doing the project by the cost of doing project. However, perceptions from the *Continuum* HTP indicate that there may also be:

- *A cost of starting late:* starting late means finishing late, which results in missing the opportunity. The missed opportunity cost is there even if the project takes the scheduled amount of time. For example, the late start delays the date of a product introduction into the market allowing a competitor to gain market leadership.
- *A cost of finishing late:* the cost of the resources used after the planned finishing time such as salaries. For example, project salary budget is $10,000 a month, then for every month the project is late, the cost goes up by $10,000. And this is in addition to any penalty cost clauses in the contract. Accordingly,

the systems approach also considers the effect of that month's delay in completing the project (Malotaux 2010). For example, if the project was to develop a new product, and the monthly revenues that the product would bring in are estimated as $500,000, the cost of the delay would be $510,000 not $10,000.

4.4.1.12 Outcomes

Projects have two kinds of outcomes: successful and unsuccessful.

4.4.1.12.1 Successful Projects

The keys to project success generally include:

- Clear and concise communications (Section 2.1.6).
- A common vision of what the project is going to achieve.
- Adequate funding.
- Good time management (Section 12.1).
- The absence of the top ten risk-indicators portending a project failure (Section 11.2.1).

A traditional measure of success is on-time delivery. On-time delivery is important because there are often penalty costs for late delivery; however, that must be coupled with the product being delivered meeting its requirements. For example:

- There is an apocryphal story about an overseas company that was building a product for a U.S. company. When it came time to deliver a shipment, the company knew they would be unable to deliver on the scheduled date. The PM telephoned his customer and asked for a delay in the scheduled delivery explaining why they would be unable to make the delivery. The purchasing officer replied, 'no way, and if you are late, the penalty clause goes into effect'. The PM was desperate, his project would not be able to deliver and he needed a way to avoid the penalty clause. So, he read the contract very carefully and found that while the contract specified a delivery had to be made on that date, it did not specify what had to be delivered. So, he shipped an empty box containing a note to that effect to the purchasing officer. He had met the letter of the contract rather than the intent.[4]
- Consider the Sydney Opera House. The development project was originally scheduled to take 4 years to complete and cost $7 million. However, it took 14 years to muddle through and cost $102 million. A classic example of a failed project! Yet today Sydney Opera House is considered as a (SOH 2007):
  - Masterpiece of late modern architecture admired internationally and proudly cherished by the people of Australia.
  - Masterpiece of human creative genius and a daring and visionary experiment that has had an enduring influence on the emergent architecture of the late 20th century.

---

[4] The lesson learned is to specify exactly what is to be delivered each time there is a delivery. Alternatively, there is a need for the customer and supplier to cooperate for a project to be successful.

4.4.1.12.2 Unsuccessful or Failed Projects

The characteristics of unsuccessful or failed projects in the:

- *People dimension:* are summed up in the top risk-indicators for predicating project failure (Section 11.2.1). They are well known, yet continue to manifest themselves: an instance of Cobb's Paradox (VOYAGES 1996). Cobb's Paradox states 'We know why projects fail, we know how to prevent their failure; so why do they still fail?' Now a paradox is a symptom of a flaw in understanding the underlying paradigm. Perhaps Juran and Deming provided the remedy. Juran as quoted by Harrington (1995) stated that management causes 80%–85% of all organizational problems. Deming (1993) stated that 94% of the problems belong to the system (i.e. were the responsibility of management).
- *Process dimension* include:
  - The wrong management style for the type of technical project (Section 10.4).
  - Assuming a success-oriented schedule.
  - Planning insufficient iterations of the SDP for the project.

4.4.1.13 Risks

Risks are discussed in Chapter 10.

4.4.2 THE TRIPLE AND QUADRUPLE CONSTRAINTS OF PROJECT MANAGEMENT

In traditional project management, the effect of a change on a project is said to impact the triple constraints of project management shown in Figure 4.1, namely, scope, cost and schedule. This is because a change in any one causes changes in the other two. Accordingly, it is possible to:

- Shorten the schedule of a project but that will increase the cost.
- Lengthen the schedule of a project to reduce the cost.
- Lower the cost of a project but that will increase the schedule or reduce the scope.
- Change the scope of a project but that will affect cost or schedule, or both.

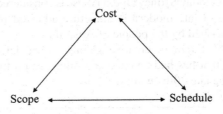

FIGURE 4.1 The traditional triple constraints.

Management: General and Project Management   107

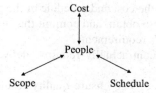

FIGURE 4.2   The systemic quadruple constraints.

The systems approach considers the quadruple constraints[5] shown in Figure 4.2 by adding the people element to the triple constraints because people are one of the seven interdependent P's of a project and closely coupled to the others (Chapter 2). Experience has shown that the right people can make a project and the wrong people can break a project (Augustine 1986), and this factor needs to be considered during the life of the project since people can:

- Lower costs and shorten schedules by preventing problems and taking advantage of opportunities.
- Increase costs and schedules by causing problems or not mitigating problems (risks) in a proactive manner.

4.4.3 PROJECT ORGANIZATION

Project management and leadership are about making optimal decisions in a timely manner. Consequently, the leadership and/or management of a task can only be accomplished effectively if the leader/manager understands the technical implications of decisions, as well as the people, schedule and budget implications (Section 2.1.1.4). For this reason, the systems approach to project organization uses situational leadership and allows leaders with the relevant competencies to emerge for specific activities. The functions performed by the leadership and management roles can be classified as administrative and technical. The administration functions for the team are performed by an administrative assistant[6]; the technical functions by the technical personnel and the leader/manager takes care of the entire project. This arrangement reduces the cost of the project by not having the administrative activities performed at PM salary rates.

4.4.3.1 Project Hierarchy

The hierarchy of the three leadership/management roles in a project can be as follows:

1. *The PM:*
    - Is the focal point of the project.
    - Is the principal contractual interface with the customer.
    - Is the project interface to the organization's marketing and procurement departments.

---

[5] A change in one will produce a change in the others.
[6] The administrative activities are not spilt between the PM and the administrative assistance, they are portioned to avoid duplication.

- Is responsible for the cost and schedule of the project.
- Has the authority to obtain and commit the necessary resources to satisfy the customer's requirements.
- Ensures that all contractually required deliverables are provided on schedule.
- Has the responsibility to ensure quality results and fulfil contractual requirements.
- Has direct access to corporate management.
- Has approval authority over all personnel actions, procurement and subcontracting activities.

2. *The Task Leader (TL)*:
   - Reports to the PM.
   - Is responsible for the cost, quality, schedule and technical performance[7] of all development stream of activities (Section 2.5.2) in a work package (WP).
   - Is a working supervisor.
   - Is usually the most experienced team member in the technical or programmatic discipline needed for the WP.
   - Has the authority to provide technical direction to all team members assigned to the WP.

3. *The Quality Leader (Q):*
   - Is responsible for the cost, quality and schedule of all the quality/test stream of activities in a WP.

Depending on the size of the project, one person may have more than one role. For example:

- The PM and a task leader (TL) could be the same person.
- One person could serve as the task lead for several WPs.

The literature generally recommends that Q not report to PM, but report to a higher level in the organization. In an organization in which on scheduled delivery counts more than the quality of the product, there is little purpose in this arrangement because the schedule set by the PM will always be followed. One way to overcome this anomaly is to let Q have an information interface to the customer describing exactly what is being delivered at any delivery milestone.

4.4.3.2 Customer–Project Interfaces

From the contracting perspective, there are two interfaces to the customer, namely:

1. *The contractual interface:* between the customer and the PM. The PM can access higher levels of team company management and has been given a high level of independent authority. Thus, problems and issues that may

---

[7] If the TL is also the technical lead person.

Management: General and Project Management

surface should be resolved quickly. As a result, customer satisfaction should remain at a high level throughout the duration of the project (or off with his head!). This interface is for:
- Managing change requests.
- Reporting.
- Billing.
- Compliance with legal requirements.
- Other contract management functions.

2. *The information interface:* between the customer and the TL. This interface is primarily for WP management in the following areas:
   - Joint or customer resolution of problems in the WP.
   - Ensuring WP performance as an integral part of the project/customer team effort.

The information interface is not for accepting informal change requests from the customer bypassing the formal change control process.

4.5 TAKING OVER A PROJECT

Taking over a project poses a problem irrespective of whether you are taking over a project because it is in trouble or because it is doing so well that the PM has been promoted or assigned to take over a different project that is in trouble and try to get it out of trouble. In all situations, a variation of the generic problem-solving process provides a systemic and systematic way to take over the project, decide what changes need to be made and implemented:

1. Gain an understanding of the situation discussed in Section 4.5.1.
2. Decide if changes need to be made or do not need to be made discussed in Section 4.5.2.
3. If changes do not need to be made, allow the project to continue as planned.
4. If changes do need to be made, then:
   1. Prepare a change management plan discussed in Section 4.5.3.
   2. Present the change management plan to the important stakeholders discussed in Section 4.5.4.
   3. Make the changes discussed in Section 4.5.5.

4.5.1 Gain an Understanding of the Situation

Gain an understanding of the situation by examining the project from a number of perspectives. For example, perceptions from the following HTPs:

- *Big Picture:* will show you the context of the project, its goals and aims, what it is supposed to be doing and the assumptions behind it.
- *Temporal:* will let you know the state of the project and how it got into that state.

- *Structural:* will show you the organization structure.
- *Operational:* will show you what the project is doing.
- *Functional:* will show you the functions being performed by the people in the project.
- *Scientific:* will infer reasons for the state of the project.

A good way of doing this is to talk to the people in the project and the external stakeholders. Visit them in their work places rather than inviting them to attend to you in your workplace. Not only does it show respect, but it also provides you with contextual information. When meeting with project personnel who you will be managing, a starting set of open-ended questions to ask each person is (Kasser 2011, 2015: pp. 376–388):

1. What does the individual like?
2. What does the individual dislike?
3. What does the team do?
4. What does the person feel the team should be doing but isn't?

4.5.2 Decide If Changes Need to Be Made

Decide if changes need to be made. If changes need to be made, decide what changes need to be made and prioritize them. The priority will depend on the situation. For example, do you have to show results quickly? If so, changes that will show results quickly should have the highest priority. If you have to reduce the cost and/or the schedule, then the changes that will have the greater impact on cost and schedule should have the highest priority.

4.5.3 Prepare a Change Management Plan

Plan to make the changes one at a time allowing time for the stakeholders to see the beneficial effect of the change.[8] The change management plan should include:

1. A summary of the current situation.
2. The reason for the change.
3. The benefits of the change.
4. The nature of the change.
5. The schedule for the change.

The degree of formality of the change management plan will depend on the project. For example, it may be:

- A PowerPoint presentation.
- A document and a PowerPoint presentation.
- Some notes on the back of an envelope.

---

[8] The build a little test a little process (Section 5.6.1).

4.5.4 Present the Change Management Plan to the Important Stakeholders

The important stakeholders in this instance are those who will benefit from the change and those who will probably be in a position to resist it. Explain it to them as well as what's in it for them (Section 2.1.4.7) and use other techniques for overcoming resistance to change (Kasser 1995).

4.5.5 Make the Changes

Make the changes one at a time using the build a little test a little process (Section 5.6.1) allowing time for the stakeholders to see the beneficial effect of the change. Examine the effect of the change to be sure that the correct issue was changed and then move on to the next change. Publish the effect of the change in the appropriate management review meetings.

4.6 RESEARCH PROJECTS

This book focuses on managing development projects, which produce a specific product according to an estimated cost and schedule. Research projects on the other hand are generally performed according to an activity-based schedule for whatever amount of time a certain amount of money will buy. Accordingly, management of research projects has some differences when compared to the management of development projects.[9] This section discusses the following three differences:

1. Types of research projects.
2. Sponsor management.
3. The need for a pollinator.

4.6.1 Types of Research Projects

There are three types of research projects (Jain and Triandis 1990):

1. *Basic research:* produces significant advances across the broad front of understanding of natural and social phenomena. Basic research projects produce information rather than products. The output tends to be publications and patents. The research ends when the funds are exhausted.
2. *Applied research:* produces prototypes, which are generally not suitable for turning into products, fosters inventive activity to produce technological advances.
3. *Innovation:* combines understanding and invention in the form of socially useful and affordable products and processes.

---

[9] The *Continuum* HTP.

In addition, some development projects may rely on the completion of applied research and or innovation to create the technology that would allow the development project to be successful. For example, if a development project is going to produce some kind of system that has to weigh less than 1 kg in 2 years and the current state-of-the-art can only produce that system with a minimum weight of 2 kg, that project will only be successful if the state-of-the-art of technology advances to the point where that 1 kg requirement is met. In such a situation, the development PM needs to use the Technology Availability Window of Opportunity (TAWOO) to lower the risk of not being able to meet that requirement (Section 10.4.1).

4.6.2 Sponsor Management

Sponsors of these types of projects generally want speedy outcomes so they can justify spending the funding. In managing research project sponsors the PM needs to:

- Understand the needs of the sponsors.
- Respond to sponsor's questions and comments in a positive manner demonstrating the understanding of the sponsor's need and indicating progress towards meeting the sponsor's needs.
- Educate the sponsor about the nature of the research and why it's in the sponsor's interest to follow a systemic and systematic time-consuming process to produce a prototype that will meet the sponsor's needs without any short- or long-term negative side effects.

4.6.3 The Need for a Pollinator

Every research project needs a pollinator. The pollinator or catalyst role is that of a key communicator who (Chakrabarti and O'Keefe 1977):

- Is invaluable to an organization.
- Is in a supervisory position only about half the time.
- Reads the literature in the research field, particularly the 'hard' papers.
- Talks frequently with outsiders and insiders to the team.
- Provides desired information.
- Locates written sources.
- Participates in the generation of ideas.
- Puts people in contact with each other.
- Evaluates ideas.
- Offers support.
- Briefs key decision makers about recent developments in fields.
- Makes contacts inside and outside the organization to promote ideas.

If the PM does not have the technical or scientific expertise, then the role of the pollinator should be split between the PM and a team member who does have the appropriate scientific or technical expertise.

4.7 SUMMARY

The purpose of management is to accomplish a goal by getting other people to do the work. This chapter helped you to understand how to achieve that purpose by discussing management, general management and project management, and how to accomplish that goal. After a brief discussion on general management, the chapter focused on the attributes of projects and project management as a problem-solving activity.

REFERENCES

Adams, Scott. 1996. *The Dilbert Principle*. New York: HarperBusiness.
Augustine, Norman R. 1986. *Augustine's Laws*. New York: Viking Penguin Inc.
Chakrabarti, Alok K., and Robert D. O'Keefe. 1977. A study of key communicators in research and development laboratories. *Group and Organizational Studies* no. 2 (3):336–345.
Deming, W. Edwards. 1986. *Out of the Crisis*. Cambridge, MA: MIT Center for Advanced Engineering Study.
Deming, W. Edwards. 1993. *The New Economics for Industry, Government, Education*. Cambridge, MA: MIT Center for Advanced Engineering Study.
Exodus 18:13–23. The Five Book of Moses. Mt Sinai.
Fayol, Henri. 1949. *General and Industrial Management*. London: Sir Isaac Pitman and Sons, Ltd.
Harrington, H. James. 1995. *Total Improvement Management the Next Generation in Performance Improvement*. New York: McGraw-Hill.
Jain, Ravinder Kumar, and Harry C. Triandis. 1990. *Management of Research and Development Organisations*, edited by Dundar F. Kocaoglu, ETM Wiley Series in Engineering & Technology Management. New York: John Wiley & Sons.
Juran, Joseph M. 1992. *Juran on Quality by Design*. New York: The Free Press.
Kasser, Joseph Eli. 1995. *Applying Total Quality Management to Systems Engineering*. Boston, MA: Artech House.
Kasser, Joseph Eli. 2011. An application of Checkland's soft systems methodology in the context of systems thinking. In *5th Asia-Pacific Conference on Systems Engineering (APCOSE 2011)*, Seoul, Korea.
Kasser, Joseph Eli. 2015. *Holistic Thinking: Creating Innovative Solutions to Complex Problems* (Vol. 1). 2nd edition. Charleston, SC: Solution Engineering: Createspace Ltd.
Kasser, Joseph Eli, and Eileen Arnold. 2016. Benchmarking the content of Master's Degrees in systems engineering in 2013. In *the 26th International Symposium of the INCOSE Edinburgh*, Scotland.
Kezsbom, Deborah S., Donald L. Schilling, and Katherine A. Edward. 1989. *Dynamic Project Management. A practical guide for managers and engineers*. New York: John Wiley & Sons.
Larson, Erik W., and Clifford F. Gray. 2011. *Project Management the Managerial Process*. 5th edition. The McGraw-Hill/Erwin series operations and decision sciences. New York: McGraw-Hill.
Mali, Paul. 1972. *Managing by Objectives*. New York: John Wiley & Sons
Malotaux, Neils. 2010. Predictable Projects delivering the right result at the right time. EuSEC 2010 Stockholm: N R Malotaux Consultancy.
PMI. 2013. *A Guide to the Project Management Body of Knowledge*. 5th edition. Newtown Square, PA: Project Management Institute, Inc.
SOH. 2007. *Sydney Opera House*. Sydney: Sydney Opera House.
VOYAGES. *Unfinished voyages, A follow up to the CHAOS report* 1996 [cited January 21, 2002]. Available from www.pm2go.com/sample_research/unfinished_voyages_1.asp.

5 Project Planning

Successful projects are planned. This chapter helps you to understand planning by focusing on product-based planning, the project and system lifecycles (SLC) and planning methodologies. The chapter discusses plans, the difference between generic and project-specific planning, how to incorporate prevention into the planning process to lower the completion risk and shows how to apply work packages (WP) instead of work breakdown structures (WBS). The chapter also explains why planning should iterate from:

1. Project start to finish at the conceptual level.
2. Project finish back to start at the detailed level.

The chapter explains:

1. The project planning paradox in Section 5.1.
2. Project planning and managing tools in Section 5.2.
3. The project lifecycle in Section 5.3.
4 The SLC in Section 5.4.
5. The project plan (PP) in Section 5.5.
6. Generic planning in Section 5.6.
7. Specific planning in Section 5.7.
8. The planning process in Section 5.8.
9. The systems approach to project planning in Section 5.9.
10. Using 'prevention' to lower project completion risk in Section 5.10.

5.1 THE PROJECT PLANNING PARADOX

There are a number of prerequisites for a project to be approved. These include a sponsor, a budget and schedule. Yet you can't have a budget and a schedule unless someone has planned what the project is going to do and what resources it needs. The paradox can be resolved by realizing that there are two versions of PPs:

1. A draft PP that contains rough estimates of budget and schedule to show cost and timeliness feasibility produced in the project initialization state (Section 5.3.1.1). The draft is based on:
   • A template associated with the project development methodology. For example, the waterfall methodology contains a set of milestones and products that are delivered at those milestones. This means that the generic activities and products for any project using the waterfall are the same before being customized for the scope and type of the specific project.
   • Experience on similar projects and provides a working baseline for estimating cost, schedule and resources.
2. A finished PP produced in the project planning state (Section 5.3.1.2) and updated prior to each major milestone during the project performance state.

5.2 PROJECT PLANNING AND MANAGING TOOLS

The project manager's toolbox contains office tools and specialized databases. Specific project management tools include:

1. The *problem formulation template:* a tool to think through the issues associated with remedying the undesirable situation discussed in Section 3.7.1
2. The *product-activity-milestone (PAM) chart:* a thinking tool to create the project activities, identify milestones, activities and products situation discussed in Section 5.2.1.
3. *Gantt charts:* tools to show activity schedules using the information in the WPs discussed in Section 5.2.2.
4. *Programme evaluation review technique (PERT) charts:* tools to show activity dependencies and the critical path using the information in the WPs situation discussed in Section 5.2.3.
5. *Timelines:* tools to show the duration of project activities discussed in Section 5.2.4.
6. *The Gantt-PERT cross-check:* a tool to find missing elements when creating the PP, discussed in Section 5.2.5.
7. *WP:* a tool that contains the interdependent planning and controlling data associated with the activity situation discussed in Section 5.2.6. The WP is used in this book for planning and controlling a project.
8. *Earned value analysis (EVA):* a tool for cost and schedule control situation discussed in Section 11.4.
9. *Categorized requirements in process (CRIP) chart:* a tool that provides a window into the project to allow problems to be mitigated as soon as they arise discussed in Section 11.5.
10. *Enhanced traffic light (ETL) chart:* a tool that also provides a window into the project to allow problems to be mitigated as soon as they arise discussed in Section 11.6.2.
11. *Management by objectives (MBO):* discussed in Section 11.7.
12. *Management by exception (MBE):* discussed in Section 11.8.
13. *WBS:* a tool that displays information in the WPs in a hierarchical format (Kasser 2018: Section 8.18). The systems approach *does not use* the WBS as an input tool.

5.2.1 THE PAM CHART

The PAM chart (Kasser 2018: Section 2.14):

- Is a tool to facilitate project planning by:
    - Defining a point in time (milestone).
    - Defining the product(s) or goals to be achieved by the milestone.
    - Determining the activities to produce the product(s).
    - Defining the resources needed to produce the product(s).
- Is designed to facilitate thinking backwards from the answer/solution to the problem (Kasser 2018: Section 11.8).

Project Planning

- Is shown in Figure 5.1.
- Has been found to be a very useful project planning tool for thinking about the relationships between the product, the activities that realize the product and the milestone by which the product is to be completed.
- Is a concept map (Kasser 2018: Section 6.1.2) linking products, activities and milestones.
- Can be used to think about the inputs to PERT charts (Section 5.2.3). Note that the PAM chart milestones are in triangles, while the PERT chart milestones are in circles.
- Consists of four parts:
  1. *The milestone:* shown as a triangle.
  2. *The product(s) produced or delivered at the milestone:* drawn as a sloping line(s) leading towards the milestone. Two products (A.1 and A.2) are shown in the Figure 5.1.
  3. *The activities that produce the products:* drawn as horizontal lines leading to the product line. They are listed above the line. Labelling reflects the activities associated with the product, so activities A.1.1 and A.1.2 are associated with producing Product A.1, and activity A.2 is associated with producing product A.2. All activities shall start and end at milestones.
  4. *The resources associated with each activity:* shown as labels below the activity lines. They are listed below the line. Labelling reflects the resource associated with the activities, so resources for A.1.1 are listed below A.1.1, resources for A.2 are listed below A.2, etc.

5.2.1.1 Creating a PAM Chart

Use the following process to create a PAM chart.

1. Start with a blank page.
2. Position a milestone at the right side of the paper.
3. Draw diagonal arrows for each product to be delivered at the milestone.
4. Draw horizontal activity lines that end at the product lines for each activity that creates the product. The starting point of each activity will be a

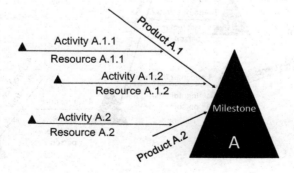

FIGURE 5.1 The PAM chart.

previous (in time) milestone when the PAM network chart (Kasser 2018: Section 2.14.2) is completed.
5. Number the milestones, products and activities where each milestone, product and activity has an identical numeral component as shown in Figure 5.1. The PAM triptych numbers at each milestone must match which facilitates identifying missing products and activities. The letter 'A' identifies activities, 'P' identifies products and 'R' identifies resources. Thus, product P1 is produced by activity A1 using resource R1, which may consist of R1-1, R1-2, R1-3, etc.

The PAM chart is a node in a project network because there is more than one milestone within a project. For example, consider the partial PAM network chart linking the products activities, resources and milestones in making a cup of instant coffee shown in Figure 5.2. Working back from the last milestone (4) the product is a stirred cup of instant coffee ready to drink. The activity between milestones 3 and 4 is 'stir the mixture' and the necessary resources are a spoon, the mixture and a person to do the stirring. At milestone 3, the product is the 'mixed ingredients' and there are two activities, adding hot water (between milestones 1 and 3) and adding the ingredients (between milestones 2 and 3). The product produced at milestone 1 is the hot water and the resources needed consist of water, the kettle, electricity and a person to do the job. The product produced at milestone 2 is the set of ingredients (instant coffee, creamer and sugar) purchased separately or as a 3-in-1 packet. Look at Figure 5.2; can you see what is missing? No, then what are you thinking of putting the water and ingredients in before stirring the mixture? The answer is, of course, the cup.

5.2.2 GANTT CHARTS

Gantt charts:

- Are a specialized type of horizontal bar chart which were invented by Henry Gantt to compare the time planned for activities with the time taken by the activities (promises with performance) in the same chart (Clark 1922).

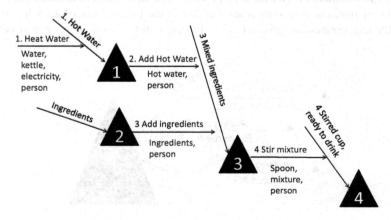

FIGURE 5.2 A partial PAM network chart for making a cup of instant coffee.

Project Planning

- Are thinking and communications tools widely used in project management.
- Are tools with which to view a project from the *Temporal* HTP.
- Are not the schedules; are the representations of the schedules.
- Represent the calendar time of a project in an easy to understand graphic. Activities start at the beginning of a bar and end at the end of the bar. While most software allows any number of bars to be shown in a Gantt chart, the systems approach limits the number of bars on any one Gantt chart to nine in accordance with Miller's rule (Miller 1956) and uses a hierarchical set of Gantt charts to represent a project where the top-level chart represents the top-level WPs in the timeline.
- Contains activities and milestones as shown in Figure 5.3 where:
  - *ID number and labels:* identify the activities.
  - *Activities:* horizontal bars which indicate the duration of activities from a start date to a finish date. The length of the bar represents the length of time for an activity and the colour or thickness of the bar can be used to represent things about the tasks (completed, level of difficulty, etc.). The colours of the horizontal bars are often assigned as follows:
    - *Green:* estimated schedule and on schedule once the project has started.
    - *Red:* behind schedule once the project has started.
    - *Blue:* ahead of schedule once the project has started.
  - *Milestones:* triangles or diamonds which represent points in time where activities start and end. Milestones can be major or minor (Section 5.4.2) and may take time. For example, a major milestone review for a complex project may take several days and needs to be shown as such in the Gantt chart.
- Show when activities start and finish and how long they take.
- Should not show dependencies between one activity and another. Dependencies should be shown in PERT charts.[1]

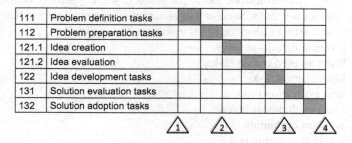

FIGURE 5.3 The initial project schedule shown in a Gantt chart.

---

[1] Some project management software does provide the capability to draw arrows between the end of one task and the start of another in the Gantt chart view. I recommend that you do not use that capability if you have more than a few tasks on the chart since the additional lines will cause clutter and may be misinterpreted. PERT charts (Section 5.2.3) are designed for the purpose of showing dependencies and can usually be drawn by the same software.

5.2.2.1 Creating a Gantt Chart

Traditionally, Gantt charts are created when creating a timeline (Section 5.2.4.1) during the project planning process (Section 5.8). The adapted traditional process for manually creating a Gantt chart is as follows:

1. Identify the starting and ending milestones for the activities to be shown in the Gantt chart.
2. Working back from the ending milestone; creating a list (Kasser 2018: Section 9.4) of all the major milestones (Section 5.4.2) between the starting and ending milestones.
3. For each pair of milestones:
   1. Create a list (Kasser 2018: Section 9.4) of all the activities to be shown in the Gantt chart. Use a Gantt chart (Section 5.2.1) to minimize missing activities.
   2. Combine low-level activities into higher-level activities, until there is only one high-level activity between that pair of milestones.

For example, consider creating a Gantt chart for the problem-solving process according to Osborn (1963). In this instance:

1. The starting and ending milestones are:
   1. The starting milestone is where the problem-solving activity begins.
   2. The ending milestone is the one in which the solution is working and remedying the problem.
2. The major milestones are:
   1. The ending milestone in which the solution is working and remedying the problem.
   2. Ideas have been developed.
   3. The problem has been defined.
   4. The starting milestone where the problem-solving activity begins.
3. The list (Kasser 2018: Section 9.4) of activities at some intermediate level, from the starting milestone to the ending milestone is:
   1. Problem definition tasks.
   2. Problem preparation tasks.
   3. Idea creation tasks.
   4. Idea evaluation tasks.
   5. Idea development tasks.
   6. Solution evaluation tasks.
   7. Solution adoption tasks.

A Gantt chart of the activities drawn at this time would look like Figure 5.3. Each activity has a number associated with the milestone. In the real world if there are a large number of tasks, it would be very complex (objective complexity) and difficult to understand (subjective complexity) (Section 3.8.1.3). Accordingly, the tasks need to be aggregated and grouped into higher-level tasks. Recognizing that the Gantt chart is an example of

Project Planning

111	Problem definition tasks		o					
112	Problem preparation tasks			o				
121.1	Idea creation				o			
121.2	Idea evaluation					o		
122	Idea development tasks						o	
131	Solution evaluation tasks							o
132	Solution adoption tasks							

△1　△2　△3　△4

Notes: o' is shown for educational purposes, not normally used in Gantt charts

FIGURE 5.4　Aggregating activities in a Gantt chart.

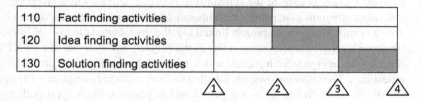

FIGURE 5.5　The top-level schedule.

the N^2 chart (Lano 1977, Kasser 2018: Section 2.10), the activities can be aggregated as shown in Figure 5.4 where the zeros have been inserted for educational purposes and are not normally used in project Gantt charts. The zeros in the N^2 chart show that when an activity in one row inputs to an activity in a following row and no other, the two activities can potentially be combined. When the activities are combined, and the higher-level activities renamed, the resulting simplified Gantt chart showing the top-level schedule is as shown in Figure 5.5. The top-level chart shall contain no more than nine bars in accordance with Miller's Rule (Miller 1956, Kasser 2018: Section 3.2.5).

In the systems approach to project management, Gantt charts are a timeline view of the set of project WPs (Section 5.2.6) and should not be created independently.

5.2.3 PERT Charts

A PERT chart:

- Is a project management tool used to help thinking about scheduling and coordinating tasks within a project.
- Is a specialized version of a flow chart (Kasser 2018: Section 2.7).
- Was developed by the U.S. Navy in the 1950s to manage the timing of research activities for the Polaris submarine missile programme (Stauber et al. 1959) at about the same time that the critical path method (CPM) was developed for project management in the private sector to perform the same function.

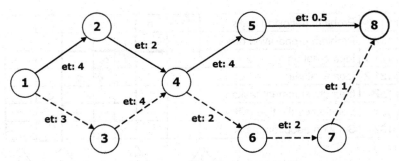

Shows expected completion times (et) for tasks between Milestones 1 and 8

FIGURE 5.6 A part of a typical PERT chart.

- Presents a concept map of the relationship between activities in a project as a network diagram consisting of numbered nodes representing the dates of events or milestones in the project linked together by labelled directional lines (lines with arrows) representing tasks in the project. For example, the PERT chart shown in Figure 5.6 has eight nodes, of which Nodes 1, 4 and 8 are major milestones. The tasks (arrows) are labelled to show expected completion times (et) of the tasks. When tasks are performed in parallel, the longest path is known in project management as the critical path and identified in the figure by dotted lines. For example, in the time between Milestones 1 and 4, the critical path is the route formed by Tasks 1-3 and 3-4 because it takes 9 time units as opposed to the 8 time units for the path formed by Tasks 1-2 and 2-4.
- May be considered as a PAM network chart (Kasser 2018: Section 2.14.2) with the product information abstracted out, hidden or removed (whichever way you care to think about it).

While the PERT chart is a very effective management tool, it can also be used in its flow chart incarnation (Kasser 2018: Section 2.7) as a communications tool to show people:

- How their work relates to the project as a whole.
- Whom their input comes from (suppliers).
- Whom their output goes to (customers).
- When their output is needed.

Once upon a time, a systems engineer used a PERT chart as a communications tool in the following manner. The context was in 1989 in the Systems Engineering and Services (SEAS) contract supporting the National Aeronautics and Space Administration (NASA) Goddard Spaceflight Center (GSFC) in a multitasking environment supporting the upgrading of a number of interdependent data processing and command and control facilities in which the subcontractor was responsible for systems engineering while the contractor was responsible for management and software engineering.

One day, the systems engineer who worked for the subcontractor tried to visualize how all the tasks were related. So, he gathered information from the different groups

Project Planning

on what they were doing, who supplied them with inputs and who used their outputs and printed[2] a PERT chart and coloured the systems engineering, hardware engineering, software engineering and test engineering activities in different distinctive colours using highlighters. He also coloured the arrows between the activities in the corresponding colours and highlighted the transfer points (i.e. a point in time where development transfers the system to the test team). He then placed the coloured PERT chart on the wall in the corridor outside his office.[3]

It didn't take very long for people to come out of their offices and study the chart. For the first time, everyone had a picture of the project as a whole, and how their activities were related to those of the other groups working on the project. A few days later similar PERT charts appeared on the walls of other departments showing other project activities. That was the upside of the situation. A week or so later he received an informal reprimand because he was a systems engineer and PERT charts were management tools, which, as a systems engineer, he should not be using.

5.2.3.1 Creating a PERT Chart

Remember:

- Milestones are dates and are shown as circles; activities take place over time and are represented as the arrows between the circles as shown in Figure 5.6.
- The PERT chart focuses on the activities and milestones. The products and resources associated with the activities are not included in the PERT chart. The schedule, dates and other time information is also not included, other than the nominal time to perform the activity. This time information is used to depict the critical path.
- PERT charts are created during the planning process (Section 5.8)

The non-systems activity-based manual approach for creating a PERT chart assuming there is no timeline (Section 5.2.4) is as follows:

1. Start at the initial milestone.
2. Create the timeline (Section 5.2.4).
3. Think of the activities that will commence at that milestone and draw them out to milestones.
4. Repeat step 3 at those milestones.
5. Repeat steps 3 and 4 until the activity lines come together at the final milestone.

The systems approach for manually creating a PERT chart assuming there is no timeline (Section 5.2.4) is as follows:

1. Start at the final milestone, the last milestone in the project.
2. Create the timeline (Section 5.2.4) by working backwards to the initial milestone (Kasser 2018: Section 11.8).

---

[2] Using a dot matrix printer and sticking the strips of paper together with clear tape.
[3] It was too long to go on the wall in the office.

3. Use a Gantt chart (Section 5.2.1) to figure out what products are to be delivered at that milestone, and what activities produce the products.
4. Draw the activities leading to the milestone as arrows. Each activity starts at another milestone.
5. For each of these new milestones, use a Gantt chart (Section 5.2.1) to figure out what products are to be delivered at that milestone.
6. Draw the activities leading to these milestones as arrows in the PERT chart. Each activity starts at a previous milestone.
7. Repeat steps 5 and 6 until you reach the milestone that kicks off the project.

By applying systems thinking to this process, it is obvious that repeating steps 5 and 6 will produce a complex and complicated chart that violates the KISS principle (Kasser 2018: Section 3.2.3). Accordingly, the systems approach creates a set of PERT charts (mostly abstracting out the activities that create the WPs (Section 5.2.6)) as follows:

1. Determine the major milestones for the project (Section 5.4.2). This is done as part of the planning process (Section 5.8).
2. Create the timeline (Section 5.2.4) by working backwards to the initial milestone (Kasser 2018: Section 11.8).
3. Determine the objectives that have to be met at each milestone from the WP (Section 5.2.6). These will generally be the products that have to be produced by the date of the milestone. The objective may also include an activity that takes place at the milestone. For example, if the activities that take place at the milestone include presentations, then creating and delivering the presentation is also an objective. By determining the objectives for each milestone, the systems approach provides for MBO (Section 11.7) into the management process in a transparent manner.
4. Draw the single line of arrows and milestones starting the kick-off major milestone to the final major milestone. The notional reference major milestones (Section 5.4.2) provide a template which can be adjusted to suit the project.
5. Start at the final milestone, the last milestone in the project. If you don't have a WP (Section 5.2.6), use a Gantt chart (Section 5.2.2) or timeline (Section 5.2.4) to figure out what products are to be delivered at that milestone, and what activities produce the products.
6. Draw the activities leading to the major milestone as arrows. Each activity starts at another milestone. Consider these milestones as minor milestones, or lower-level milestones between the major milestone and its previous major milestone. For example, if the major milestone is critical design review (CDR), the previous major milestone will be the preliminary design review (PDR). The minor milestones:
    1. Will be milestones where intermediate products are transferred from one part of the organization to another between the two major milestones.
    2. May be in any of the three streams of activities (Section 2.5.2).
7. Each activity has a component in the three streams of activities (Section 2.5.2) with the appropriate products and resource requirements.

Project Planning

8. For each of these new milestones, use a Gantt chart (Section 5.2.2) to figure out what products are to be delivered at that milestone.
9. Draw the activities leading to these milestones as arrows in the PERT chart. Each activity starts at another, previous in time, milestone.
10. Repeat steps 8 and 9 until the arrows come together at the previous major milestone.
11. Start at this major milestone. Use a Gantt chart (Section 5.2.2) to figure out what products are to be delivered at that milestone, and what activities produce the products.
12. Repeat steps 6–10 until the arrows come together at the first or kick-off major milestone.

The systems approach to drawing the PERT chart has the following advantages:

- It keeps each drawing understandable in accordance with Miller's Rule (Miller 1956, Kasser 2018: Section 3.2.5).
- It ensures that the three streams of activities (Section 2.5.2) come together at each major milestone.
- It provides for MBO (Section 11.7) in a transparent manner.

In the systems approach to planning, risk and opportunities are considered when estimating times for activities in the planning process (Section 5.8). These contribute to the early and late times of the time estimates and allow 'what-if?' questions to be asked at planning time to mitigate risks or take advantage of opportunities.

5.2.4 TIMELINES

A timeline is a tool:

1. To quickly visualize a sequence of events, for example, in a project.
2. To clearly convey the sequence to interested parties.

A timeline:

- Consists of a sequence of activities and milestones: in series, parallel or a combination of series and parallel activities.
- Is often shown from two perspectives:
  1. *Gantt charts:* (Section 5.2.2) which show the timing and are often called schedules.
  2. *PERT charts:* (Section 5.2.3) which show the dependency between the activities and identify the critical path.
- Is sometimes shown as a horizontal line with milestone markers.

5.2.4.1 Creating a Timeline

Timelines are created as part of the project planning process (Section 5.8) working backwards from the final milestone to the start of the project (Kasser 2018: Section 11.8).

There are software packages that combine the calendar relationships and the dependencies by showing an arrow from the end of one task to the start of the next task. These are useful when there are a small number of tasks. However, when trying to understand the relationships between a large number of tasks in a project unless the combined chart shows nine or less lines in compliance with Miller's Rule (Miller 1956, Kasser 2018: Section 3.2.5), they should not be used; the two separate charts should be used. This is in accordance with the dictum of viewing something from more than one perspective (Kasser 2018: Chapter 10). In this case, the two perspectives are calendar time and task dependencies.

5.2.5 THE GANTT–PERT CROSS-CHECK

While Gantt charting software provides dependency information, the systems approach *does not use* that functionality for the following reasons:

1. It increases the subjective complexity (Section 3.8.1.3.1) of the chart making it difficult to understand.
2. It does not allow the two different views to be used to find missing activities and dependencies.

Consider the Gantt and PERT charts shown in Figure 5.7 (Kasser 2017). When the dependencies are easy to see as in the PERT chart, the lack of an activity being dependent on activity 2–5 is obvious. However, the PERT chart does not provide a reason, which needs to be investigated. Reasons include:

- A missing activity, 5-6.
- Milestone 5 is an error and should be either milestone 4 or 6.
- There is an error in hand drawing the chart; it is not a view of the database in a software tool.

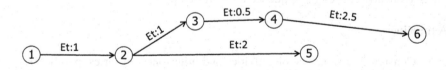

FIGURE 5.7 The Gantt and PERT cross-check.

5.2.6 THE WP

The systems approach to project management uses WPs instead of a WBS to create and control a project by considering the WBS, product breakdown structure (PBS), and timeline, Gantt and PERT charts as outputs from a project database. A WP is:

- A tool used in the systems approach to project management to plan the work, identify and prevent risks from happening.
- A set of information associated with an activity.

The WP is based on the quality system elements (Kasser 2000) and contains the following information:

1. *Unique WP identification number:* the key to tracking.
2. *Name of activity:* a succinct summary statement of the activity or activities.
3. *Priority:* the priority associated with the task linked back to the source requirement.
4. *Narrative description of the activity:* A brief description of the activity or activities.
5. *Estimated schedule for the activity:* estimated time taken to perform the activities. This item contributes to the baseline schedule for the project.
6. *Accuracy of schedule estimate:* median, shortest and longest times to complete. This item is used to create the initial estimate of the critical path and shows up in the PERT charts (Kasser 2018: Section 8.10).
7. *Actual schedule:* filled in as and after the activity is performed. This item is compared with the estimated schedule during the performance of the project.
8. *Products:* (outputs) produced by the activities in the WP.
9. *Acceptance criteria for products:* the response to the question, 'how will we know that the product is working as specified', as agreed between the supplier and the customer.
10. *Estimated cost:* the estimated fixed and variable costs of the activities and materiel. This item project contributes to the baseline budget at completion (BAC) for the project.
11. *The level of confidence in the cost estimate:* or the accuracy.
12. *Actual cost (AC):* filled in as and when the activities are performed. This item is compared with the estimated schedule during the performance of the project as part of EVA (Section 11.4).
13. *Reason activity is being done:* important because personnel leave and join and the reasons may be forgotten. Some activities are preventative and the reason may not be immediately obvious. This information is also useful when considering change requests during the operations and maintenance states of the SLC (Section 5.4).
14. *Traceability:* (source of work) to requirements, the concept of operations (CONOPS), laws, regulations, etc.
15. *Prerequisites:* the products that must be ready before the activity can commence.

16. *Resources:* the people, equipment, materiel, etc.
17. *Name of person responsible:* for the activity or task.
18. *Risks:* description and estimate of probability and seriousness of consequences.
19. *Risk mitigation information:* mitigation WP ID(s) for each risk listed in Item 18.
20. *Lower-level WP ID's:* if any.
21. *Decision points:* if any.
22. *Internal key milestones:* if the activity is broken out into lower-level WPs.
23. *Assumptions not stated elsewhere:* so as to be available for checking by cognizant personnel.

A typical WP at planning time is shown in Table 5.1. The actual values of cost and schedule will be inserted into the WP during the performance of the activity and summarized as part of the EVA (Section 11.4). Comments on some of the entries are:

- *Item 1:* The ID is 124D showing it is an activity in the development stream of work (citation) in WP 124. With this numbering system, the other two WPs would be 124M for the corresponding management activities and 124T for the corresponding test activities in the three streams of activities (Section 2.5.2).
- *Item 2:* The name has a verb in it to show that it is an activity.
- *Item 4:* CONOPS (P123P) is listed to define the product that will be used in the activity.
- *Item 7:* the product produced by activity 124D is 124P to correlate the product to the activity in accordance with Gantt chart numbering (Section 5.2.1).
- *Item 9:* the fixed and variable costs associated with the activity.
- *Item 10:* the accuracy of cost estimate is same as accuracy of schedule. For example, if the schedule accuracy is ±3 days on 10 days, the cost accuracy is ±30%.
- *Item 13:* the prerequisites that have to be completed before the task can begin are the availability of product 123P.
- *Item 15:* the personnel should be named to avoid overloading them with too many tasks. The amount of time spent on the task should be listed. This information is used to determine the labour costs.
- *Item 20:* must match descriptions in narrative Item 4. Each lower-level WP has corresponding activities in the other two streams of activities (Section 2.5.2).

5.2.6.1 The Benefits of Using WPs

The benefits of using WPs include:

- Collecting information that is interdependent between product and process provides the ability to create and use attribute profiles (Kasser 2018: Section 9.1). For example, the risk mitigation plan would be based on an abstracted view of the risk elements in the WP database.
- New perspectives on the system based on the attribute profiles.

TABLE 5.1
Typical (Planned) WP (Spreadsheet)

1	Unique WP identification number	124
2	Name of activity	Feasibility study
3	Priority	2
4	Narrative of the activity	A systems engineer will perform the feasibility study on the selected CONOPS to verify realization of CONOPS is feasible within project constraints. Cognizant personnel will be identified and interviewed
5	Estimated schedule	2 weeks
6	Accuracy of schedule estimate	± 3 days
7	Products	Feasibility study report (124P)
8	Acceptance criteria for products	Consensus that study findings are correct
9	Estimated cost	$3,000
10	The level of confidence in the cost estimate	±10%
11	Reason activity is being done	To ensure feasibility of the CONOPS
12	Traceability	An inherent part in the problem-solving process
13	Prerequisites	Completion of CONOPS (product ID P123P)
14	Resources	Task leader (4 h)
		systems engineer (1 full-time equivalent)
		Engineering specialist (1 full-time equivalent)
15	Person responsible for task	Mark Time
16	Risks	Cognizant personnel not available when needed
17	Risk mitigation	Telephone ahead of time to make appointments
		Identify alternate candidates for interviewing as part of 124-01
18	Lower-level WP ID's (if any)	124-01 Identify cognizant personnel
		124-02 Interview cognizant personnel
		124-03 Perform rest of feasibility study
		124-04 Write up report
19	Decision points (if any)	None
20	Internal key milestones	End of 124-01
		End of 124-02
21	Assumptions not stated elsewhere	Personnel will be available to perform the activity

- The project information resides in a single integrated information database that contains information from the three streams of activities (Section 2.5.2), which dissolves the problem of updating separate databases to keep them current in the non-systems approach to project management.

5.3 THE PROJECT LIFECYCLE

Since the PP covers the activities in the project lifecycle which may take place at any point in the SLC, this section discusses the project lifecycle, the SLC and the differences between them. The SLC may be considered as the states a system exists in, starting from its conception through development and operation, and ending once it has been disposed of. The system development process (SDP) contains the system development states of the SLC. Projects, being temporary endeavours, may appear in any one state of the SLC or in multiple states crossing state boundaries. For example, if a project is to:

- Develop a new product it will take place over the whole SDP of that product.
- Improve the way requirements are elicited and elucidated it will take place in the system requirements state of the SDP.

5.3.1 THE FOUR-STATE PROJECT LIFECYCLE

The project lifecycle contains the following four states:

1. The project initialization state discussed in Section 5.3.1.1.
2. The project planning state discussed in Section 5.3.1.2.
3. The project performance state discussed in Section 5.3.1.3.
4. The project closeout state discussed in Section 5.3.1.4.

The project sequences through the lifecycle states as shown in Figure 5.8. Each state:

- Starts and ends at a milestone.
- May take an amount of time that ranges from minutes to months depending on the complexity of the project. For example, if the project is to:
  - *Develop a new government policy:* the project initialization state may take months to determine what the policy should be (and may be a project in itself).
  - *Build an airport:* the project planning state may take months if it includes figuring out where to locate the airport.
  - *Organize a party:* the project planning state may take 20 min or less.

5.3.1.1 The Project Initiation State

The project begins in the project initiation state as a proposed project. The product produced in the project initiation state is the project initiation document (PID). This document contains the guiding information for the project including:

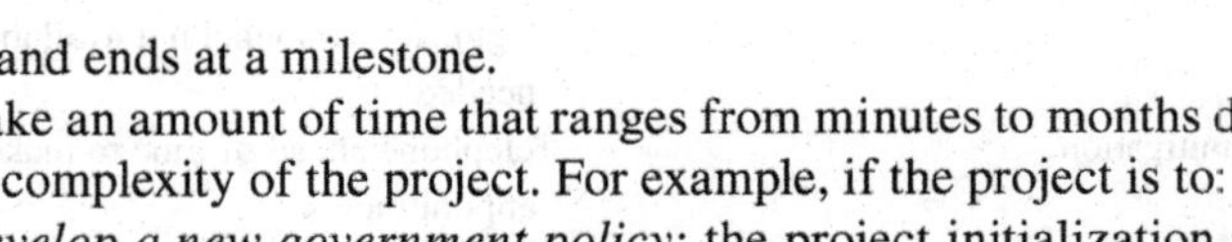

Project Initialization State			
Project Planning State			
Project Performance State (SDP)			
Project Closeout State			

FIGURE 5.8 The project lifecycle.

Project Planning

- The sponsor.
- The business case for the project.
- The undesirable situation that needs to be remedied by the project.
- The priority of the project with respect to other projects taking place in the organization.
- The stakeholders in the project and their order of importance and influence, if known.
- The proposed project organization.
- The customer who funds the project and who must be satisfied with the deliverables.
- An estimate of the budget for the proposed project.
- An estimate of the completion date: the date the product to be produced by the proposed project is needed.
- An estimate of the other types of resources needed by the proposed project.

The PID may be created using annotated outlines (Kasser 2018: Section 14.1) and the documentation process (Kasser 2018: Section 11.4).

The project initiation state terminates at a milestone which reviews the information in the PID and provides consensus to proceed with the project. The review is not the place to adjust the scope of the project to lower the cost, change the schedule or reduce the scope. This should be done as part of creating the PID. If the project is complex, the project initiation state may be a project in itself.

5.3.1.2 The Project Planning State

The product produced in the project planning state is the initial draft of the PP (Section 5.5). Many projects combine all the information in the PID into the PP. On the other hand, many projects don't have a PID or even a PP and so don't have clear goals, and accordingly fail (Section 4.4.1.12.2). The project planning state terminates at a milestone which reviews the information in the draft PP and provides consensus to proceed with the project.

5.3.1.3 The Project Performance State

The project performance state is the longest state in the project. It is the state in which the project is managed and it produces the products or product that will remedy the undesirable situation using the SDP. If the project takes time, it may not completely remedy the undesirable situation because:

1. The product operating in its context does not remedy the entire original undesirable situation.
2. New undesirable aspects have shown up in the situation during the time taken to develop the product.
3. Unanticipated undesired emergent properties of the product and its interactions with its adjacent systems may produce new undesirable aspects of the situation.

Errors of omission and commission (Section 3.4.3.1) made in each part of SDP in the project performance state can produce undesirable outcomes. For example, if the wrong cause of the undesirable situation is identified, the wrong problem will be stated, and not only will the solution not remedy the undesirable situation, the solution may make the situation even more undesirable. Similarly, if the correct problem is identified but the wrong solution conceptualized and realized, then again, the undesirable situation will not be remedied and may become even more undesirable. Even if the correct solution is conceptualized, errors in the realization process may produce an incorrect solution. Accordingly, the project performance state has to use a methodology that will increase the probability of delivering the product or products that will remedy the undesirable situation as it exists at the time of delivery, rather than as it existed at the time the project began.

5.3.1.4 The Project Closeout State

The project closeout state is the reverse of the project performance state. The project performance state started with nothing and created a project. The project closeout state starts with a project and reduces it to nothing. The project closeout state forms a WP of its own and contains the following process:

1. Delivering the products produced by the project to the customer.
2. Getting delivery acceptance from customer.
3. Shutting down resources and releasing them to new users.
4. Rewarding the project team; a certificate of recognition of contributions is a minimum.
5. Reassigning project team members: they should be reassigned according to a published schedule so there is no uncertainty as to where people will go when the project ends. This minimizes reduction in morale. Ideally, the people have a place to go within the organization, perhaps returning to their home department or moving on to the next project.
6. Closing accounts and seeing all bills are paid.
7. Arranging the project records in a logical manner and storing them in an accessible but secure location. The records need to be available for technical, procedural, liability and other legal reasons.
8. Updating the project lessons learned into an accessible database.
9. Producing a formal project closure notification to higher management.
10. Creating and delivering a final report to the appropriate stakeholders. A useful list of contents for a project final report is (Larson and Gray 2011: pp. 510–511):
    1. *Executive summary:* highlighting the key findings on facts relating to the implementation of the project including:
       - The degree of meeting project goals for the customer.
       - Stakeholder satisfaction.
       - User reaction to the quality of the deliverables.
       - How the deliverables are being used; as intended and providing the expected benefits or not.
       - Summaries of final cost, schedule and scope.
       - Major problems encountered and addressed.
       - Key lessons learned.

Project Planning

2. *Review and analysis:* distinct and factual review statements of the project including:
   - Project mission and objectives.
   - Procedures.
   - Systems used.
   - Organizational resources used.
   - schedules, cost, scope and risk information
3. *Recommendations:* recommendations for improvements on future projects both technical and non-technical. They may include changing or reasons to keep vendors and subcontractors.
4. *Lessons learned:* a summary of the major lessons learned on the project. This is why each management review at a major milestone should include a lessons-learned component. Capturing the lessons learned at the time will provide a more extensive and usable set that can be applied before the project ends.
5. *Appendices:* critical information supporting the previous sections.

5.4 THE SLC

The SLC may be considered as the set of states a system exists in, starting from its conception through development and operation, and ending once it has been disposed of.

5.4.1 THE STATES IN THE SLC

While there is no generally accepted set of system states in the SLC, the states are normally thought of as a linear time-ordered sequential process and shown as a Gantt chart or a N^2 or waterfall chart (Royce 1987) as shown in Figure 5.9. The waterfall chart is a N^2 chart with the row and column lines blanked out. States A–F can be considered as

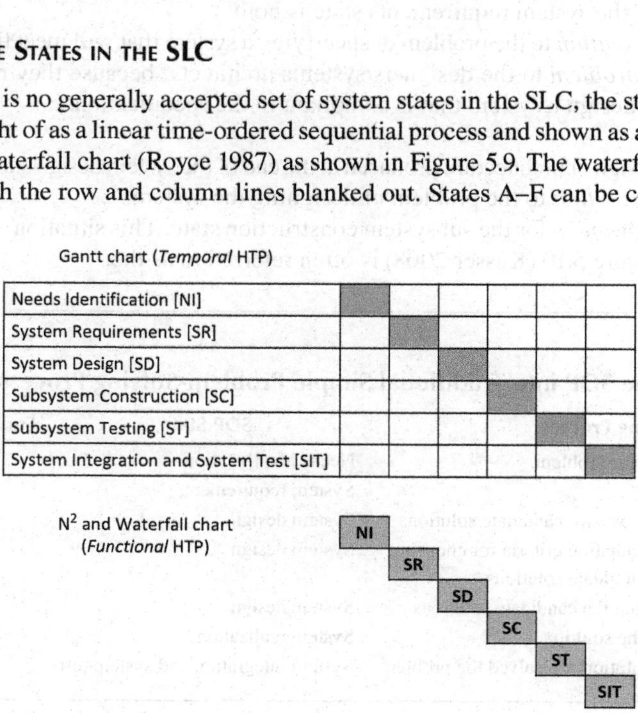

Gantt chart (*Temporal* HTP)

Needs Identification [NI]						
System Requirements [SR]						
System Design [SD]						
Subsystem Construction [SC]						
Subsystem Testing [ST]						
System Integration and System Test [SIT]						

N^2 and Waterfall chart (*Functional* HTP)

NI
  SR
    SD
      SC
        ST
          SIT

FIGURE 5.9 The SDP: A planning view.

TABLE 5.2
The SLC, SDP and Hitchins–Kasser–Massie Framework (HKMF) States

State	HKMF State	SDP	SLC
Needs identification	A	Yes	Yes
System requirements	B	Yes	Yes
System design	C	Yes	Yes
Subsystem construction	D	Yes	Yes
Subsystem testing	E	Yes	X
System integration and system test	F	X	Yes
Operations and maintenance	G		Yes
Disposal	H		Yes

the SDP as shown in Table 5.2. The SDP maps into the simple problem-solving process shown in Figure 3.6 as shown in Table 5.3. Each state in the SLC may be considered as starting with a problem and ending with a solution. Accordingly, the solution output of any state becomes the problem input to the subsequent state. For example:

- The matched set of specifications for the system and subsystems produced during the system requirements state is both:
  - *A solution* to the problem of specifying a system that will meet the needs.
  - *A problem* to the designers/systems architects[4] because they now have to design a system that is compliant to the specifications.

- The design or architecture produced at the end of the system design state is both:
  - *A solution* to the problem of designing the system.
  - *A problem* for the subsystem construction state. This situation, shown in Figure 5.10 (Kasser 2008) is often referred to as the:

TABLE 5.3
Mapping the SDP into Traditional Simple Problem-Solving Process

Problem Solving Process	SDP State	Ending Milestone
Understanding the problem.	Needs identification	OCR
	System requirements	SRR
Conceiving at least two candidate solutions	System design	PDR
Determining evaluation criteria for choosing between the candidate solutions	System design	PDR
Deciding between the candidate solutions	System design	CDR
Implementing the solution	System realization	IRR
Verifying the solution has solved the problem	System integration and system test	DRR

---

[4] In the system design state.

Project Planning

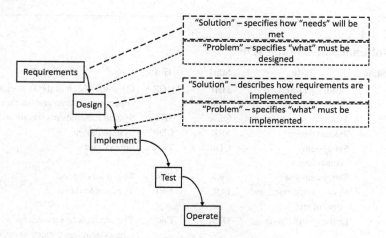

FIGURE 5.10 The waterfall view—problem-solving perspective (partial).

- *'What's':* which refer to what needs to be done, or the problem.
- *'How's':* which refer to how it is done, or the solution.

The activities performed in each state are discussed in *Systemic and Systematic Systems Engineering* (Kasser 2019).

5.4.2 Milestones

Each state starts and ends at a major milestone. While there is also no generally accepted set of major project milestones, the major project milestones used in most system and software development project processes can be mapped into the following template:

1. Start-up meeting (which formally starts the SDP).
2. Operations concept review (OCR).
3. Systems requirements review (SRR).
4. Preliminary Design Review (PDR).
5. Critical Design Review (CDR).
6. Subsystem test readiness review (TRR).
7. Integration readiness review (IRR).
8. Delivery readiness review (DRR).
9. Acceptance test review which formally terminates the project.

Each milestone review covers two sets of work:

1. Work accomplished prior to the milestone.
2. Work to be accomplished before the next milestone review.

The relationship between the states, milestones and major products produced during each state for delivery at the milestones in the 'A' paradigm (Kasser 2012) is shown in Table 5.4.

TABLE 5.4
The Notional States in the SDP

HKMF State	State	Start	End	Produces
A	Needs identification	Start	OCR	Common vision of the system in operation, system architecture
B	System requirements	OCR	SRR	System requirements specifications, PP
C	System design	SRR	CDR	System design
D	Subsystem construction	CDR	TRR	Subsystems
E	Subsystem test	TRR	IRR	Tested subsystems
F	System integration and system test	IRR	DRR	The tested system
	Delivery and handover	DRR	End	The products or services for which the system was created in operation

The waterfall model was developed as a planning tool in a time when the requirements didn't change very quickly. This may have been because microcomputers had not been invented and systems were created in hardware which was difficult to change. The waterfall model is easy to teach and understand and is still applicable and controllable and can cope with changes in requirements. The waterfall model copes with changes in requirements:

1. As long as the time taken to progress through the states in the waterfall is shorter than the time between changes.
2. The project has a working change management system and can adjust the workload to meet the changes in the need. This may require reverting to previous stages in the waterfall depending on the nature the change request (Section 11.3.2).

Just because the SDP has been discussed as a single waterfall as a teaching example, in the real world there may be multiple iterations of the waterfall which allow the solution system to evolve as in the multiple-iteration problem-solving process (Section 3.9.2) or the cataract methodology (Section 5.4.3). Ideally, the length of each waterfall SDP should be shorter than the time it takes for the undesirability being remedied by the project to change for the worse. Perceptions from the *Generic* HTP note a similarity to the Observe–Orient–Decide–Act (OODA) loop (Boyd 1995). However, perceptions from the *Continuum* HTP note that while the focus of the OODA loop is on the speed of completing the loop to get inside the opponent's decision-making cycle; here the focus is getting inside the rate of change of undesirability.

As inferred from the *Scientific* HTP, the relationship between the type of project manager (the project manager's way of approaching problems) (Section 2.1.2.1) and the state in the SDP is shown in Table 5.5. During the needs identification state the project is trying to understand the problem and so the project manager needs to be

TABLE 5.5
Type of Project Manager Needed in Each State of the SDP

State in Project Lifecycle	Type of Project Manager
Needs identification	Type IV (V)
System requirements	Type III (V)
System design	Type II
Subsystem construction	Type II
Subsystem test	Type II
System integration and system test	Type II

a Type IV. During the system requirements state, the solution is being conceptualized and the PP created or finalized, so the project manager needs to be a Type III. Once the PP is in existence, a Type II manager can manage the requirements design construction test and integration states as long as there is no deviation from the PP, namely as long as there are no problems (Section 2.1.1.4).

5.4.2.1 Informal Reviews
Informal reviews tend to be to be in-process tests and working meetings (Section 2.1.5) such as:

- Concept reviews
- Implementation (management) plan reviews
- Requirements reviews
- Design reviews
- Code walkthroughs
- Test plan reviews
- Test procedure reviews

5.4.2.2 Formal Reviews
The purpose of a formal review is:

1. To provide visibility into the state of a project.
2. To formalize approval of decisions previously made.
3. To transition from one state of the SDP to the subsequent in an orderly manner, namely to form a stage gate.
4. Not to present surprises or controversial changes.

5.4.2.2.1 Template for a Management Review Presentation
This generic template for a management review presentation contains the following four parts:

1. Section 1 Overview
    1. Cover slides stating title, date and agenda (list of topics to be covered).

    2. The overall health and status of the project or projects. An effective way of presenting this is to use the ETL chart (Section 11.6.2). If the presentation is for a single project, the chart will only contain a single row and may be combined with the following charts.
    3. Situation summary:
        1. The financial state of the project, presented in EVA graphical format. The information in graphs is easier to comprehend than the same information in tables of numbers.
        2. The current schedule of the project, presented in the form of a Gantt chart. The chart should be at a high level and contain no more than nine bars in accordance with Miller's Rule (Miller 1956).
        3. CRIP Charts showing the summary of the workflow (Section 11.5).
2. Section 2 Details:
    1. Work that was planned for this reporting period.
    2. The activities actually performed since the last review presentation starting with the cost (EVA charts) and schedule (Gantt charts) information showing the planned for the reporting period and the actual values.
    3. Explanation of variations (excuses).
    4. Risk report showing the status of the outstanding risks and the mitigation activities.
    5. Supporting information presenting details about any cost and schedule variations.
    6. Any other pertinent information relating to activities performed.
3. Section 3 Looking ahead:
    1. Next major milestone(s).
    2. Work planned for next reporting period.
    3. Problems and issues pending (summary, details as backups).
    4. Plans for the activities to be performed during the next reporting period.
4. Section 4 Closing material:
    1. Brief summary.
    2. Lessons learned (Kasser 2018: Section 9.3) during the reporting period.
    3. Any additional information that is relevant to the presentation (i.e. date and time of next meeting, due dates for inputs).

5.4.3 THE CATARACT METHODOLOGY

The Cataract[5] Methodology (Kasser 2002b) relies on two factors:

1. The waterfall methodology works very well over a short period of time.
2. Implementation and delivery of systems and software are often performed in partial deliveries, commonly called 'Builds' in which each successive Build provides additional capabilities (Section 5.6.1). Build planning is not

---
[5] Mini waterfalls.

a new concept. It has been used in software maintenance for many years. For example, it was also incorporated in the U.K. Defence Evaluation and Research Agency (DERA) Reference Model (DERA 1997).

In the software world, a Build means a defined software configuration. Successive Builds enhance the capability of the software. In the hardware world, Builds can comprise subsystems, or the integration of two or more subsystems. The work associated with each Build takes place in the three parallel streams of activities (Section 2.5.2).

The cataract approach to Build planning may be likened to a rapid prototyping scenario within the spiral in which the requirements for each Build are frozen at the start of the Build. This approach, however, is more than just grouping requirements in some logical sequence and charging ahead. Build plans must be optimized on the product, process and organization axis to:

- Implement the highest-priority requirements in the earlier Builds. Then, if budget cuts occur during the implementation phase, the lower-priority portions are the ones that can readily be eliminated because they were planned to be implemented last.
- Make use of the insight that, typically, 20% of the application will deliver 80% of the capability (Arthur 1992) by providing that 20% in the early Builds.
- Allow the waterfall approach to be used for each Build. This tried-and-true approach works on a small project over a short timeframe.
- Produce a Build with some degree of functionality that if appropriate can also be used by the customer in a productive manner. For example, the first Build should generally, at a minimum, provide the user interface and shell to the remainder of the functions. This follows the rule of designing the system in a structured manner and performing a piecemeal implementation.
- Allow a factor for the element of change.
- Optimize the amount of functionality in a Build (features versus development time).
- Minimize the cost of producing the Build.
- Level the number of personnel available to implement the Build (development, test and systems engineers) over the SDP to minimize staffing problems during the SDP.

5.4.3.1 Build Zero

Once the initial set of requirements has been signed off, the system architecture designed and the implementation allocated into a series of Builds, the implementation phase embodying the cataracts begins. The cataract methodology incorporates an initial Build, Build Zero, which contains the same initial two phases, requirements and design, of the waterfall methodology with the exception that there is recognition that

- All the requirements are not finalized at SRR.
- Additional requirements will become known as the project progresses.
- Design and implementation decisions will be deferred so as to maximize the 'Don't Care' situations and made in a just in time (JIT) manner.

The work in Build Zero is to:

1. Identify the highest-priority requirements.
2. Baseline an initial set of user needs and corresponding system requirements.
3. Develop the requirements database.
4. Complete the first draft of the PP and CONOPS document (Kasser and Schermerhorn 1994).
5. Design the architecture framework for the system.
6. Perform risk assessment to determine if the proposed architecture framework can meet all the highest-priority requirements.
7. Document the assumptions driving the architecture framework and a representation of operational scenarios (use cases) that the architecture framework prohibits. This activity also helps identify missing and non-articulated requirements early in the SDP. The design of the architecture framework for the entire system in Build Zero introduces a risk that it may not be suitable for changes years later in its operations and maintenance phase (or even earlier). This is why part of the Build Zero effort is to determine scenarios for which the system is not suitable. The customer is then aware of the situation. The goal of the cataract methodology is to achieve convergence between the customer's needs and the operational system. In the course of time, one can expect that the need will change to something for which the system cannot provide capability. At that time, a revolutionary Build will be needed to replace the system. However, it will be done with full knowledge in a planned manner, rather than the ad hoc manner of today's environment.
8. Develop the WPs to level the workload across the future Builds and implement the highest-priority requirements in the earlier Builds as described above.

From Build One inclusive, each subsequent Build is a waterfall in itself. The requirements for the Build are first frozen at the Build SRR. Then the design effort begins. Once the design is over, the Build is implemented and when completed turned over for integration. While the design team does assist with the integration, their main effort is to start to work on the design of the next Build. Once the first Build has been built and is working, the requirements for the second Build are frozen and the design-integrate-test-transition and operate states of the SLC commences for the second Build. This cycle will continue through subsequent Builds until the system is decommissioned although the contract may change from the development organization to the maintenance organization. Each Build is an identical process but time delayed with respect to the previous one. Each successive Build provides additional capabilities. When the Builds are placed under configuration control, the waterfall may initially be drawn as shown in Figure 5.11 (Kasser 2002b); however, this figure is misleading. Externally driven changes are requested and problems tend to show up during the integration and test phases. When a problem is noticed, a Discrepancy Report (DR) is issued against the symptom. This DR is analysed and the cause identified. A change request is then issued by the Change Control Board (CCB) to resolve the defect either in the current Build before delivery, or by assigning it to be fixed in a subsequent Build.

Project Planning

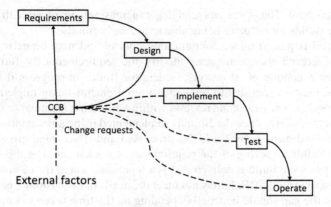

FIGURE 5.11 The configuration control view of waterfall.

Thus, Figure 5.11 should be replaced with Figure 5.12 (Kasser 2002b) showing that the cataract methodology explicitly modifies the DERA evolutionary lifecycle approach (DERA 1997) to incorporate:

- The effect of changing external requirements.
- The management of changing requirements.

The change request, if accepted, is assigned to be implemented in the appropriate future Build. Think of each Build as being completed a little behind the arrowhead of the advancing requirements. From this perspective, the gap between the user's need and the completed section of the system converges over time. Project personnel move from one Build to the next; the development team moves from one Build to the next, as does the testing team. Ideally, the Builds are sequential with no wasted

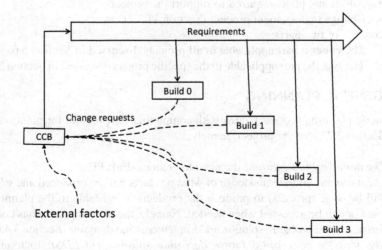

FIGURE 5.12 The cataract methodology.

time between them. The customers tend to get increasingly involved with the system during later Builds by virtue of being able to use early Builds.

Each Build is placed under configuration control and may be delivered to the customer. Accepted change requests modify the requirements for future Builds, with the sole exception of 'stop work' orders for Builds-in-progress if the change is to remove major (expensive to implement) requirements being implemented in a Build-in-progress. The milestone reviews within a Build are identical to those in the waterfall methodology, since the Build is implemented using a waterfall. All change requests received during any Build are processed and if accepted are allocated to subsequent Builds. Freezing of the requirements for each Build at the Build SRR means that when the Build is delivered, it is a representation of the customer's needs at the time of the Build SRR. It may not meet the needs of the customer at the time of delivery, but the gap should be small depending on the time taken to implement the Build since the SRR, thus achieving convergence between the needs of the customer and the capability of the as-delivered system.

5.5 THE PROJECT PLAN

The PP is:

- A guide to project execution by providing a reference.
- A communications tool for the present and future.
- The controlling document to manage a project.
- A description of the:
  - Interim and final deliverables the project will deliver.
  - Activities in the managerial and technical processes necessary to develop the project deliverables in the three streams of activities (Section 2.5.2).
  - Resources required in the process of creating, verifying and delivering the project deliverables.
  - Additional plans required to support the project.
  - Change management process (Section 11.3.2).
- Consists of two parts:
  1. The generic part applicable to all projects discussed in Section 5.6.
  2. The specific part applicable to the specific project discussed in Section 5.7.

5.6 GENERIC PLANNING

The generic planning process begins with completing the problem formulation template (Section 3.7.1) for the project, namely:

1. *The undesirable situation:* the need to create a draft PP.
2. *The assumption:* the knowledge of what products will be produced and what will be done (process) to produce the products is available in the planning team or can be accessed when needed. Namely, the planning team has competence in the problem, solution and implementation domains (Section 3.4.6).
3. *The feasible conceptual future desirable solution (FCFDS):* (outcome) having successfully completed the draft PP in a timely manner.

Project Planning

4. *The problem:* to figure out how to create and deliver the products and document the information in the draft PP.
5. *The solution:* use the following process to create the PP.

The process for creating a generic PP is as follows:

1. Find samples of PPs for the methodology to be used in the project from previous projects or on the Internet.
2. Customize the samples to create a list of section headings.
3. Create an annotated outline (Kasser 2018 Section 14.1).
4. Create a list of the products that have to be delivered before acceptance. These will be contract-specific and may include the final product, spare parts, documentation, training, etc.
5. Assuming the waterfall methodology is going to be used, the systems approach to planning starts with the timeline template shown in Table 5.6 for the top level of activities in the project and the states, which begin and end at the major milestones in Table 5.4. Should other methodologies be chosen, an appropriate template might be developed and used and reused. The state numbers identify the state, and when used on a number of projects the same number will always identify the same state. Organizations may add project identification numbers before the three digits in Table 5.6.
6. Using the WP template shown in Table 5.1, create an outline WP for each state listed in Table 5.6 using a spreadsheet. At this time, only fill in items 1, 2, 4, 8, 9, 13, 15, 20 and 22 of the template as shown in Table 5.7 for State 200.
7. Split each activity into three streams of activities (Section 2.5.2) – management, development and test/quality.
8. Working backwards from the final major milestone, use PAM charts (Kasser 2018: Section 2.14) to identify the activities and resources that produce the products at each major milestone.
9. Expand the top-level state WPs in each stream of work into a set of three lower-level activity WPs using PAM charts using a hierarchical structure such as the one shown for WP 210D[6] in Table 5.8. Lower-level activities begin with the high-level number where each of the three streams of activities is identified by either the appropriate letter or a number. For example, WP 210 is split into WP210M, WP210D and WP210T. Table 5.8 contains part of the initial draft for WP 210D.
10. Ensure every management planning activity WP is linked to a development activity WP.
11. Ensure every development activity WP is linked to a testing activity WP.
12. Ensure that every development and testing activity WP is linked to a management activity WP.
13. Continue with the rest of the document production process (Kasser 2018: Section 11.4.5).

---

[6] There would be corresponding WP210M and WP210T; the remaining two parts of WP 210.

TABLE 5.6
SDP State and State Numbers

ID Number	Project/SDP States
000	Project initialization
100	Needs identification
200	System requirements
300	System design
400	Subsystem construction
500	Subsystem test
600	System integration and system test
700	Delivery and handover
800	Project closeout

TABLE 5.7
Typical Initial Draft WP for State 200

Item		
1	WP ID	200
2	Name of activity	System requirements state
4	Narrative description of the activity	Elicit and elucidate requirements
		Gain stakeholder consensus that systems requirements document represents a system that will meet the customer's needs,
		Produce document (printout from database)
		Produce updated PP
		Produce SRR presentation
		Hold SRR
8	Products	210 System requirements document
		220 Updated PP
		230 SRR presentation
		240 SRR
9	Acceptance criteria for products	Customer accepts and signs off on systems requirements document and Updated PP
13	Reason activity is being done	Ensure product to be produced meets customer's needs at the SRR
14	Traceability	Project
15	Prerequisite	Authorization to proceed after completion of OCR
20	Lower-level WP ID's	210 Produce system requirements document
		220 Update PP
		230 Produce SRR Presentation
		240 Produce SRR
22	Internal key milestones	To be determined

Project Planning

TABLE 5.8
Typical Initial Draft of WP 210D

Item		
1	WP ID	210D
2	Name of activity	Producing system requirements document
4	Narrative description of the activity	Elicit and elucidate requirements
		Gain stakeholder consensus that systems requirements document represents a system that will meet the customer's needs
		Produce document (printout from database)
8	Products	210 System requirements document
9	Acceptance criteria for products	Customer accepts and signs off on systems requirements document
13	Reason activity is being done	Ensure product to be produced meets customer's needs at the SRR
14	Traceability	200
15	Prerequisite	Authorization to proceed after completion of OCR
17	Person responsible	TBS (Task leader)
20	Lower-level WPs ID's	211D Elicit and elucidate requirements
		212D Gain stakeholder consensus
		213D Produce system requirements document

At this point, the template for the PP is complete. What remains is to customize the plan and add the specifics for the particular project (Section 5.7).

5.6.1 BUILD A LITTLE TEST A LITTLE

A project's most important resource is time. Projects will be completed faster and at lower cost if waste is eliminated (Malotaux 2010). There are two basic types of waste in the project process. Doing something that:

1. Does not yield value.
2. Doing something poorly.

A common vision of the project goals and well-written requirements prevent creating some items that do not yield value; hence the focus writing the right requirements and creating the common vision in the literature.

The 'build a little test a little' approach:

- Can keep the project focused and swiftly reprioritize work in the event of problems.
- Has been used by electrical engineers to prototype circuits for as long as there have been electrical engineers.

TABLE 5.9
To Do Project Control Spreadsheet

Priority	WP Item or Activity	Date Due	Estimated Time Do Activity	Latest Start Date	Date Completed	Notes

- Organizes the project timeline in chunks of work with specific goals for each chunk.[7] Each activity in the chunk is extracted from the relevant WPs and summarized in a spreadsheet format such as the one shown in Table 5.9. The contents of the spreadsheet are:
  1. *Priority:* the priority of the activity from the WP as modified by the chunking process.
  2. *WP item or activity:* the reference to the WP activity for further details.
  3. *Date due:* the latest date for completion of the activity.
  4. *Estimated time to do activity:* from the WP.
  5. *Latest start date:* the date due minus the estimated time to do the activity.
  6. *Date completed:* once completed may be hidden from view. This item is used to improve the accuracy of future estimates by seeing how accurate this one was.
  7. *Notes:* as appropriate.

Each chunk begins and ends in a short meeting. Each item is examined at the meeting. The project is in trouble if the current date is later than the:

1. Latest start date for any activity.
2. Due date for any uncompleted activity.
3. If people are working on activities not listed in the spreadsheet, then either
   1. They are wasting time,[8] or
   2. The spreadsheet is in error and needs to be updated.

In such a situation, some of the quadruple constraints (Section 4.4.2) need to be adjusted to recover the schedule and the priorities of the activities adjusted. These changes often take the form of completing the activities while delaying lower-priority activities. If the project ends before the lower-priority requirements are completed, then there will have to be a negotiated agreement about how to proceed.

Information from the 'build a little test a little' meetings should end up in the CRIP charts (Section 11.5) and ETL charts (Section 11.6.2) in the formal meetings with the customer.

---

[7] The chunks can be days or weeks depending on the length of the schedule.
[8] And those activities should cease immediately.

Project Planning

5.7 SPECIFIC PLANNING

Specific planning for a project modifies the generic templates to customize them for the scope and type of project as well as adding or deleting WPs as required. This activity requires knowledge of the problem, solution and implementation domains, so if the project planner does not have that knowledge, he or she needs to team with people who have that knowledge.

5.7.1 THE PROCESS FOR CREATING A SPECIFIC PP

The process for creating a specific PP is as follows.

1. Working backwards from the final major milestone customize the PP by deleting unnecessary WPs, combining WPs if necessary and adding additional WPs if necessary.
2. Working back from the final major milestone customize the remaining WPs by determining the:
   1. Scope of work involved in producing the product.
   2. Competency level of the persons who will be doing the work.
   3. Estimated time to perform the work.
   4. Staff who will perform the WPs (Chapter 6).
   5. Estimate the cost to complete the work (Chapter 8).
   6. Estimate the nature of risks associated with the WPs and risk mitigation approaches (Section 10.3).
3. Estimate the number of iterations of the SDP the project will require. In general, the more complex the project or the greater the level of uncertainty associated with the project, the more iterations of the SDP will be needed to realize the system (Kasser and Zhao 2016).
4. Create the project schedule (Chapter 7) by combining the activities described in the WP in series and parallel into the three streams of activities (Section 2.5.2) that come together at each major milestone.
5. Share the draft cost and schedule information with the customer.

If the customer determines the cost is too expensive or the schedule is unacceptable adjust the project until the cost is affordable and/or the schedule is acceptable by:

1. *Reducing the schedule:* by shortening the critical path although, this might increase the cost of the project (Section 9.1.3)
2. *Reducing the cost:* by removing some of the low-priority requirements and/or requirements with a high estimated cost from the final product (Section 9.2.6).

5.8 THE PLANNING PROCESS

Systemic and systematic planning is an interdependent process combining aspects of the process, products and documentation.

5.8.1 The Numbering System

The systems approach to project planning uses the numbering system in a systemic manner. The numbers represent elements of the process and the product to allow missing elements to be easily identified as described in this section. The numbers contain a string of digits, which contain information about:

1. The specific project.
2. The state in the SDP.
3. The WP in the state.
4. The product produced by the WP.
5. The stream of work: management, development or test.

The first part of the identification is the project. The remaining letters and digits provide information about the activity and the product. Consider some examples using a number whose digits can be represented by the letters ABCDEFGHJKL.

The states in the SDP can be numbered using a single digit. Let the letter A define the state in the SDP. Let the first state start with a number 0 and increment by 1 for each subsequent state. The assignment to the A is as shown in Table 5.10. Let the letter B represent the level of the WP in that state where zero is the top level, one is one level down and so on.

Let the letters CDE identify specific WPs in those levels as suggested in Table 5.11 where the first digit identifies the type of activity and the remaining two digits the specific WP performing that activity. For example, WP 00000 represents the needs identification state; WP 10000 represents the system requirements state. WP 32122 is a designing activity (CDE = 122), WP two levels down from the top (B = 2) in the subsystem construction state (A = 3) and WP 32123 is another designing activity in the same state and same level as WP 32122. Projects that don't use the waterfall SDP will have a similar way of distinguishing states. Then WP 11000 represents a WP one level down within WP 10000.

Let the letter F represent the stream of activities, where M is management, D is development and T is test. Accordingly, since every activity is managed and produces a product that needs to be tested, there should always be three WPs with the same number differing in the letter assigned to the position represented by the F.

TABLE 5.10
The First Digit Represents the State

A	State
0	Needs identification state
1	System requirements state
2	System design state
3	Subsystem construction state
4	Subsystem testing state
5	System integration and system test states

TABLE 5.11
Activity Numbering

CDE	Activity
100	Designing
200	Planning
300	Programming
400	Writing
500	Requirements elicitation and elucidation

Project Planning

For example, consider WP 32123. WP 32123M is the WP for the management activities, WP 32123D is the WP for the development activities and WP 32123T is the WP for the test. Any WP that does not come as a set of M, D and T WPs is incomplete.

Let the letters GHJK represent the product being produced in the WP as suggested in Table 5.12 where the first digit represents the type of product, the second and third digits the family of products and the remaining digits represent the specific product. For example, in the product category 2000 for software, the digit represented by H could be allocated to distinguish between source code and object code. Thus, the number 321232D-4040[9] represents a design produced by WP 321232D. This system easily identifies:

- Which activities produce which products.
- If there is an activity that does not produce a product (the digits GHJK are not allocated).
- A product without a planned activity to produce it (the digits ABCDE are not allocated).

In summary, in the systems approach, product and process numbers correlate to make it easy to identify missing elements in the SDP. The allocation is shown in Table 5.13.

5.8.2 Process Architecting

Selecting which process to follow in a project does not mean adopting an existing process in use somewhere else or described in a standard. The process to be used in the project must be architected just as carefully as the product will be architected. Process architecting (Kasser 2005) is a vital part of planning the project yet has been neglected in the literature. Now there are a range of methodologies for use in the development of systems (Avison and Fitzgerald 2003), the traditional waterfall methodology being only one of them. Each methodology fits specific scenarios; however,

TABLE 5.12
Product Numbering

GHJK	Type of Product
10000	Architecture
20000	Software
30000	Requirement
40000	Design
50000	Test procedure

TABLE 5.13
Numbering Representation

Digit position	Represents
A	State in the SDP (Table 5.10)
B	Level in the WP
C	Type of activity (Table 5.11)
DE	Specific activity
F	Stream of work
G	Type of product (Table 5.12)
HJK	Specific product

---

[9] The hyphen makes it easy for the human to read the number.

the real world tends to be more complex than the scenarios taught in the classroom. 'Real world problems do not respect the boundaries of established academic disciplines, nor indeed the traditional boundaries of engineering' (O'Reilly 2004). Accordingly, the optimal project process (Kasser 2005):

- Will probably not be a straightforward unmodified out-of-the text book methodology.
- Is as important to the success of a product development as is the optimal architecture of the product.
- Is a multi-phased time-ordered sequence of activities with constraints on start dates for each activity.

The WBS for the process looks like a hierarchical system-subsystem view of the product. However, little attention seems to have been paid to architecting optimal development processes and consequently, an untailored process may not be optimal in a specific situation. This organizational function of process architecting is to identify the best methodology for the situation and then tailor that methodology to the situation. To make the function more complex, the optimal methodology may be different at different phases in the SDP or the situation may be such that there is no one optimal methodology and parts of several methodologies may be to be assembled into the methodology for the project (Kasser 2002a). The methodology must be tailored to the situation, not the other way around. Once the methodology is chosen, the process for implementing the methodology must be developed. The choices faced by the process architecting function include:

- *Choice of lifecycle:* such as the traditional requirements-driven methodology or a capability-driven methodology.
- *Choice of methodology:* such as (which) soft systems, functional, object-oriented, agile, waterfall, rapid, spiral, etc. one of the factors in this choice is the state of the technology will be incorporated in whatever the project produces (Section 10.4).
- *Choice of process for implementing the methodology, milestone process-products and the checkpoints within the process:* The process must be scaled to the size of the project. Sometimes this may require combining activities or products, e.g. combining the operations concept with the systems requirements documents for small projects, or even choosing to produce milestone documents in the form of PowerPoint presentations instead of text-mode documents.
- *Build-buy decisions:* The decision to build or buy components of the product affects the development process as well as the product architecture. This decision must be made after considering its implications on both the product system and development process.
- *Time to market:* a decision based on holistic thinking. This decision is generally made in a new product development situation. There may be a situation where purchasing some equipment may shorten the schedule, but at the same time increase the cost of the development process. Should this be the

situation, the cost of not shortening the schedule in terms of lost revenues (Section 4.4.1.11) should be determined and used in making this decision (Malotaux 2010).

5.8.2.1 The WBS

The traditional approach to planning a project is to start with a WBS (Kasser 2018: Section 8.18) based on the physical components of the system or product being produced. At the top level is the system; at the next level in the hierarchy are the subsystems and the WBS then breaks out the subsystems into lower-level subsystems which are in turn broken down into even lower-level subsystems until the basic component level. For example, the WBS for a project producing a cup of coffee with milk and sugar could look like the one shown in Figure 5.13.

The traditional approach recognizes that each element in the WBS needs time to produce it and a cost estimate for what it will take to produce it. With that information you can plan, schedule and budget the project. The WBS also serves as a framework for tracking cost and work performance. As the WBS is developed, organizational units and individuals are assigned responsibility for executing the work that produces the elements in the WBS integrating the work with the organization (Larson and Gray 2011: p. 109).

However, by starting with the physical component, the system architecture needs to be designed before the WBS can be created. For example, the WBS for a cup of coffee with milk and sugar shown in Figure 5.13 is only one possible architecture arrangements; and alternative architecture arrangement is shown in Figure 5.14.

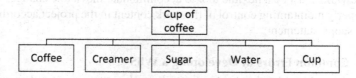

FIGURE 5.13 A traditional example of a WBS for a cup of instant coffee.

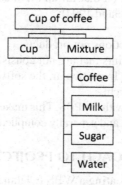

FIGURE 5.14 An alternative example of a WBS.

Both architectures represent the cup of coffee; however, they may lead to different production processes:

- Figure 5.13 might lead to a process where the coffee, water, cream and sugar are all poured into the cup at the same time and then mixed together.
- Figure 5.14 might lead to two different processes where:
  1. The coffee, water, cream and sugar are mixed together and then poured into the cup.
  2. The mixture of coffee, cream and sugar is purchased in a 3 in 1 sachet and then poured into the cup.

The basic principles for creating a WBS are (Cleland 1994):

1. A unit of work should appear at only one place in the WBS.
2. The work content of a WBS item is the sum of the WBS items below it.
3. A WBS item is the responsibility of only one individual, even though many people may be working on it.
4. The WBS must be consistent with the way in which work is actually going to be performed; it should serve the project team first and other purposes only if practical.
5. Project team members should be involved in developing the WBS to ensure consistency and buy-in.
6. Each WBS item must be documented to ensure accurate understanding of the scope of work included and not included in that item.
7. The WBS must be a flexible tool to accommodate inevitable changes while properly maintaining control of the work content in the project according to the scope statement.

5.8.2.2 Common Errors in Developing a WBS

In traditional systems development project, the architecture is created as part of the project. Accordingly, if the final architecture approved in the system design state of the SDP is not the initial architecture upon which the WBS is based, the work can diverge from the PP based on that initial architecture WBS. Common errors in developing a WBS include (Cleland 1994):

- The WBS describes functions not products.
- The WBS has branch points that are not consistent with how the WBS elements will be integrated. In particular, the software and hardware for a unit must be in the same branch.
- The WBS is inconsistent with the PBS. This makes it possible that the PBS will not be fully implemented, and generally complicates the management process.

5.9 THE SYSTEMS APPROACH TO PROJECT PLANNING

The traditional approach to creating a WBS (Cleland 1994) is to plan the activities performed in the project using the WBS as an input tool. The systems approach to project planning does not use the WBS as an input tool; in the systems approach, the WBS is a hierarchical view of some of the information in the project database. The systems approach to project planning is a multi-pass process based on using the products delivered or handed over at each milestone in the SDP (Section 5.4.2) working back from the DRR to the start of the project using the waterfall view as a planning template (Kasser 2018: Section 14.9). The further into the future, the less feasible it is to plan for (Churchman 1968), and so the lower the amount of detail needed. Another reason is changes might affect the work and the time spent on detailed planning in this early state of the project would be wasted. However, the initial draft WPs should be planned in sufficient detail to provide a rough estimate of cost, schedule and risk and ensure that there is no risk that could kill the project should it happen. As the project progresses and the plan is updated for each milestone the level of detail is increased as shown in Figure 5.15.

The steps in the first pass are:

1. Identify the products produced in the system integration and system test states to be delivered at the DRR.
2. Identify the activities performed to produce the products between the IRR and the DRR in the three streams of activities (Section 2.5.2).
3. Construct WPs for the activities using Table 5.1 as a template. Focus on the products, process and risks at this time ignoring cost and schedule. You cannot work out cost and schedule until you have an idea of the staffing, which will be done in step 12.
4. Elaborate WPs into lower-level WPs as necessary using the 'Do Statement' (Chacko 1989, Kasser 2018: Section 3.2.1) or a similar tool.
5. Construct a process out of the management, development and test/quality activities such that the three streams in the system integration and system test states start at the IRR and come together at the DRR.
6. Repeat steps 1–5 for the activities performed in the subsystem testing state between the TRR and the IRR.

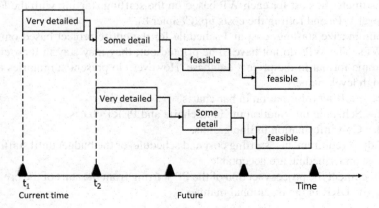

FIGURE 5.15 Planning—level of detail.

7. Repeat steps 1–5 for the activities performed in the subsystem construction state between the CDR and the TRR.
8. Repeat steps 1–5 for the activities performed in the system design state between the SRR and the CDR.
9. Repeat steps 1–5 for the activities performed in the system requirements state between the OCR and the SRR.
10. Repeat steps 1–5 for the activities performed in the needs identification state between the start of project and the OCR.
11. Repeat steps 1–5 for the activities performed after the DRR to deliver and hand over the product to the customer.
12. Examine the risks identified in the WPs for the system integration and test states of the SDP, determine the risk mitigation strategy, and assign appropriate WPs in the earlier states of the SDP to perform the risk mitigation activities[10] as discussed in Section 5.10. Should new risks show up in the risk mitigation WPs, mitigate them in the same way in earlier states of the SDP.
13. Repeat step 12 for the risks identified in the WPs for the subsystem testing state.
14. Repeat step 12 for the risks identified in the WPs for the subsystem construction state.
15. Repeat step 12 for the risks identified in the WPs for the system design state.
16. Repeat step 12 for the risks identified in the WPs for the system requirements state.
17. Repeat step 12 for the risks identified in the WPs for the needs identification state.
18. Repeat step 12 for the risks identified in the post DRR WPs.
19. Determine the staffing for every WP (Section 6.2).
20. Estimate the time for each WP based on the staffing starting with the low-level WPs and level the staffing to create a schedule for the project (Chapter 7).
21. Estimate the cost for each WP based on the staffing starting with the low-level WPs and rolling the costs up (Chapter 8).
22. Summarize staffing, cost and schedule for the entire project based on the WPs. The WPs do not have to be printed out; they may stay in their electronic format in the project database. However, do present summaries at a high level, namely:
    1. Staffing information in bar charts.
    2. Schedule information in Gantt charts and PERT charts.
    3. Cost information in line graphs.
23. Adjust requirements, staffing cost and schedule, or the budget until staffing, cost and schedule are acceptable.
24. Document the process section of the PP starting from the start of the project to the DRR in the traditional manner.

---

[10] Alternatively, modify existing WPs in those early states to add the risk mitigation activities.

Project Planning

The presentation exercise provides examples of the planning process. For example:

- Section 5.11 provides a detailed practical example of steps 1–11.
- Section 6.3 provides a detailed practical example of the staffing step 19.
- Section 7.7 provides a detailed practical example of the scheduling step 20.
- Section 8.8 provides a detailed practical example of the costing step 21.

5.10 USING 'PREVENTION' TO LOWER PROJECT COMPLETION RISK

Preventing risks from turning into events, namely occurring, is reasonably simple when planning is done starting at the end in working back to the beginning. For example, look at the skeleton timeline shown in Figure 5.16. The first risk identified will be in the last state of the SDP, in this case between IRR and DRR. After identifying the risk, and working out what needs to be done to mitigate or prevent that risk, the WP for the risk mitigation activities earlier in the planned SDP can be created and positioned appropriately. In the figure, that would be a WP between CDR and TRR. If the risk needs to be monitored following TRR, appropriate WPs can be inserted into the plan.

5.11 THE PRESENTATION EXERCISE

In the classroom environment, every team exercise during the semester is a project. This section provides an example of applying project planning to a classroom exercise. The knowledge associated with each session in the semester is a mixture of lecture, up to five readings from papers or textbook chapters which the students present (knowledge readings) (Kasser 2018: Section 3.4); the interaction during the knowledge reading presentations and additional knowledge the students bring to the session. The exercise is for student teams to present the knowledge readings for the specific session in the semester.

The level of detail in the following description is overkill for such a simple project. However, the principles apply to larger projects and the knowledge reading exercise makes a good example, especially if the students are doing the knowledge reading exercise in each session. This is because it will help them understand by

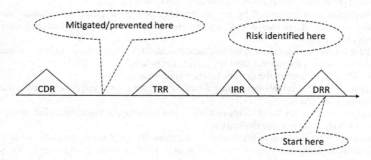

FIGURE 5.16 An example of planned prevention.

TABLE 5.14
Grading Criteria for the Presentation Exercise

1	Conformance to requirements in Section 5.11.2	20%
2	Logic flow of content based on cognitive skills in Table 5.15	20%
3	Correctness of content	20%
4	Originality of additional content	20%
5	Quality of presentation format (viewer friendliness, clarity, ease of reading, colours, font size, appropriate graphics, etc.)	20%

doing, something known as activity-based learning. The WPs developed in this exercise will be used in subsequent chapters to illustrate staffing, scheduling and costing.

5.11.1 The Grading Criteria for the Presentation Exercise

The grading criteria for each session exercise are shown in Table 5.14. The cognitive skills, based on updated Blooms Taxonomy (Overbaugh and Schultz 2013), are shown in Table 5.15. The total grade for the team presentation over the semester is the sum of each individual presentation grade. The passing grade is 50%.[11]

5.11.2 The Requirements for the Presentation Exercise

The requirements for the knowledge reading presentations have evolved to the following. For each reading,[12,13,14] the student team shall:

1. Summarize content of the reading (<1 min).[15]
2. List the main points (<1 min).[16]
3. Prepare a brief on two main points.[17]

---

[11] Experience has shown that the initial presentations in the first session are generally poor and will receive a low grade. Accordingly, the passing grade is set to 50%, which ensures the student team does not fail, and as the grade increases as the semester progresses, the students can look at their grades and see how much they've learned.

[12] Some students' positive feedback indicated that the exercises were a lot of work, but they learnt more doing the exercise than they learned in other classes on similar topics.

[13] These exercises were performed by my postgraduate and professional development students in Singapore. Many students received 'A's on the exercises. If you think the exercises are too overwhelming, remember:
  1. The learning curve applies and the feedback provided to the students let them see it in action.
  2. People behave in the manner they are expected to behave (Section 2.1.4.6).

[14] The students were also given the reasons for the requirements.

[15] The requirement is to facilitate developing the skills to condense the information in the reading and hide (abstract out) details.

[16] This requirement requires the students to analyse and evaluate the knowledge (Taxonomy Levels 4 and 5) to identify and prioritize the main points.

[17] Requirement 4 is for the team to brief on one main point. However, once a team has made a briefing, there is the probability that another team will want to brief the same main point. This requirement requires the students to read two main points in the text and allows the option for the instructor to bypass some of the potential repetition.

TABLE 5.15
Grading Based on Cognitive skills

Grade	Taxonomy Level	Ability Being Tested
A+	6 Creating	Can the student create a new product or point of view?
A	5 Evaluating	Can the student justify a stand, position or decision?
B+/B	4 Analysing	Can the student distinguish between the different parts?
B−	3 Applying	Can the student use the information in a new way?
C+	2 Understanding	Can the student explain ideas or concepts?
C	1 Remembering	Can the student recall or remember the information?

4. Brief on one main point (<1 min per point).[18]
5. Reflect and comment on each reading (<2 min).[19]
6. Compare content with other readings and external knowledge.[20]
7. State why you think the reading was assigned to the session.[21]
8. Summarize lessons learned from the session and indicate source of learning, e.g. readings, exercise, experience, etc. (<2 min).[22]
9. Use a different team leader for each session.[23]
10. Presentation to be less than 15 min.[24]

---

[18] This requirement helps to limit the time for the presentation and minimizes repetition.
[19] This requirement invokes the higher-level cognitive skills by requiring the students to apply, analyse, evaluate and create knowledge (Taxonomy Levels 3–6).
[20] This requirement:
 - Invokes the higher-level cognitive skills by requiring the students to apply, analyse, evaluate and create knowledge (Taxonomy Levels 3–6).
 - Encourages students to research similar material to the assigned readings and compare and contrast the material.
 - Encourages students to make connections between the various readings allocated to a session, developing their ability to see similarities and differences in the assigned and external readings (*Generic* and *Continuum* HTPs).
 - Helps to identify and develop students with problem-solving and problem-formulating skills by requiring the student to apply holistic thinking to the content especially the *Generic* and *Continuum* HTPs. This is where the students can develop and apply their ability to find (Gordon et al., 1974, Kasser 2018: Section 3.3.5):
   - Differences among objects which seem to be similar.
   - Similarities among objects which seem to be different.
[21] This requirement also invokes the higher-level cognitive skills by requiring the students to apply, analyse, evaluate and create knowledge (Taxonomy Levels 3–6). The similarities and differences in this part of the presentation illustrate to the students that different people can draw similar and different conclusions from the same data. In a number of instances, students have drawn innovative applicable conclusions.
[22] This requirement also invokes the higher-level skills by requiring the students to apply, analyse and evaluate knowledge (Taxonomy Levels 3–5).
[23] This requirement tries to minimize the workload on students who tend to be perfectionists and undertake to do most of the team work themselves to compensate for poor performance by individuals.
[24] This requirement puts an upper limit on the length of the entire presentation.

5.11.3 THE PRESENTATION EXERCISE PROJECT PLANNING STATE

The state begins with completing the problem formulation template (Section 3.7.1) for the exercise/project, namely:

1. *The undesirable situation:* the need to successfully[25] complete the exercise in a timely manner.
2. *The assumption:* the exercise has been sized by the instructor so that it can be completed successfully within the allotted time.
3. *The FCFDS:* (outcome) is having successfully completed the exercise in a timely manner with an acceptable grade.
4. *The problem:* is to figure out how to create and deliver a product that meets the requirements of the exercise.
5. *The solution:* is creating and delivering a presentation that meets the requirements of the exercise.

5.11.4 THE PROJECT NUMBERING SYSTEM

The activities performed and products produced in such an exercise can be inherited from previous exercises in other classes based on experience (*Generic* HTP). The numbering system almost in accordance with Section 5.8.1 is as shown in Tables 5.16 and 5.17. The students also felt that they didn't need to separate out the three streams of activities (Section 2.5.2) specifically for such a simple project, so they allocated the management activity to a coordinating WP.

5.11.4.1 The Activity Categories

The students categorized the activities they thought they would be doing and assigned numbers to those categories. They produced Table 5.16, which lists the category sorted by name and number to facilitate the numbering of the WPs.

5.11.4.2 The Product Categories

This project produces two different types of products, draft or initial version and final version. The versions are numbered as shown in Table 5.17.

5.11.4.3 The Exercise Activities, Products and WPs

Very little formality is used in performing a team exercise in a class. If the exercise is performed and documented as a project in some detail, the documentation to create the presentation might look something like the WPs and charts contained in this section. Of course, this is overkill for a class project. However, since you are participating in a class project or have probably participated in such a project, you will be familiar with the context and can use perceptions from the *Generic* HTP to compare what was done for the class project with what needs to be done for (or what was done in) your project.

---

[25] Success is defined as achieving the desired grade.

Project Planning

TABLE 5.16
Activity Numbering

By Number	Activity	By Activity	Number
5	Coordinating	Adding	81
10	Planning	Assigning	12
12	Assigning	Communicating	6n
20	Discussing	Coordinating	5
22	Voting	Creating	4n
24	Reading	Designing	3n
3n	Designing	Discussing	20
4n	Creating	Downloading	63
50	Testing	Instructing	92
6n	Communicating	Integrating	80
61	Speaking	Modifying or Updating	82
62	Uploading	Planning	10
63	Downloading	Presenting	70
70	Presenting	Reading	24
80	Integrating	Reviewing	90
81	Adding	Speaking	61
82	Modifying or Updating	Testing	50
90	Reviewing	Uploading	62
92	Instructing	Voting	22

TABLE 5.17
Versions of Products

Number	Version
100	Draft
200	Final

The WP numbering is assigned according to the states of the SDP. Since the exercise takes place in a context in which the requirements don't change and is concluded within a week, the waterfall methodology (Royce 1987, Kasser 2018: Section 14.9) for the presentation development process is appropriate. So, using the systems approach of working back from the answer and then documenting the plan as a forward process (Kasser 2018: Section 11.8), the sequence of products working from the final delivery was determined to be:

1. The presentation to be presented.
2. The templates for each section of the presentation so that all sections would have the same look and feel.
3. The theme for the presentation to uniquely distinguish the presentation from the others and provide brand recognition.

TABLE 5.18
The Top-Level Sequence of Events for the Activities in Each State of the Exercise

WP	State	Ends After	Section
00000	Presentation planning state	Session presentation planning completed review (SPPCR)	5.11.5
10000	Presentation requirements state	Presentation requirements review (PRR)	5.11.6
20000	Presentation design state	Presentation design review (PDR)	5.11.7
30000	Presentation construction state	Presentation integration review (PIR)	5.11.8
40000	Presentation integration and testing states	Presentation delivery readiness review	5.11.9
50000	Presentation delivery and grading state	Presentation post-mortem review (PPMR)	5.11.10

Accordingly, the top-level sequence of events for the activities in each state of the exercise and the WP number assigned to that state are shown in Table 5.18.

Note:

- The activities in setting up the team are not included in the project. The assumption being they were done in a previous project.
- The final milestone of the exercise is not making the presentation, is not receiving the grade, but it is holding the presentation post-mortem review (PPMR) to understand why a particular grade was received, and to work out what to do to improve the grade in the exercise for the following session.
- Milestones are treated as activities in the planning because they are an activity in the exercise. To be pedantic, the milestone is actually the decision to proceed to the next state of the exercise.
- The numbering of all milestone reviews is WP X9000 where X is the first digit in the state. For example, WP 29000 represents the exercise design review WP in WP 20000.

The WPs for the milestone reviews contain all the activities that have to be performed to prepare for the review as well as actually participating in the review. The WPs in each state are designed so that the three streams of activity (Section 2.5.2) come together at the milestone reviews.

5.11.5 THE PRESENTATION PLANNING STATE (WP 00000)

The presentation planning state (WP 00000) is the initial state in the presentation development process. In this state, the team makes sure that they understand the grading criteria and the instructions for the exercise, and appoint a team leader in accordance with the requirement to do so. In the first session exercise, the team will prepare a PP template; basically, a schedule of what to do based on the instructions, and the assumption that all the presentations will have to meet the same requirements and consequently, the activities will be the same. What is actually

Project Planning

done in each activity will depend on the knowledge reading. The team also creates the presentation theme in this state. At the end of state, and at the end of all subsequent states, the team holds a milestone review to make sure that what they have done is compliant to the requirements and grading criteria to help them obtain the maximum grade.

The products, activities and milestones in each WP are determined using a PAM chart (Section 5.2.1) and inserted into a blank WP template shown in Table 5.1. WP 00000 is split into a number of WPs. The relevant sections of the WPs for the presentation planning state are shown in Table 5.19. WP 00000 contains the managing and top-level activities in the presentation planning state. At the start of the state, it is the set of coordinating activities that ensures that the prerequisites for each of the WPs in the state have been met, the staff and other resources will be available as scheduled, and does the coordinating between the WPs activities during the rest of the state. In a similar manner, the n0000 activities on the remaining WPs perform the same set of functions in their states of the presentation development process.

When developing item 4 in a WP based on Table 5.1, the 'narrative of what the WP actually does', additional prerequisites may show up and need to be included in the prerequisite as, for example, in WP 09000. For example, students might think that all they need to look at is the PP (04142-050) and the presentation theme (04141-205). However, in order to generate that consensus (09000-298) the plan and theme have to be compared against some kind of standard, and that means that the understanding of the grading criteria (04120-295) and instructions for the exercise (00000-295) are also prerequisites for that WP because they will be applied to the plan and theme. Each of the WPs by virtue of the 'narrative' and 'reason for activity' columns should be self-explanatory.

Notice that while WP 00000 is the top-level WP, it also contains some activities of its own (WP 00500). These are generally the management activities that ensure:

- The state can begin.
- The prerequisites have been completed.
- Suitable resources are available for each of the WPs in the state.

Once the WPs have been prepared, the WBS (Section 5.8.2.1) can be viewed. The presentation planning state terminates after the presentation planning completed review (PPCR) (WP 09000).

5.11.6 THE PRESENTATION REQUIREMENTS STATE (WP 10000)

The presentation requirements state (WP 10000) is the state in which the team determines the requirements for the specific exercise and adjusts the PP created in the previous state. The team produces a prioritized requirements specification and the Requirements Traceability Matrix (RTM).

WP 10000 is also split into a number of WPs. The relevant sections of the WPs for the presentation requirements state are shown in Table 5.20. WP 14100 creates the RTM which might take the form of the example in Table 5.21. While creating the RTM, the team also updates the requirements to reflect any changes

TABLE 5.19
The Relevant Sections of the WPs for the Presentation Planning State (WP 00000)

WP	Activity	Prerequisite	Products	Used in WP	Narrative	Reason for Activity
00500	Coordinating 00000	Team	00500-092 ready to start	04120	The exercise initialization activity. The exercise cannot begin without a team, the instructions for the exercise and the grading criteria. This activity also contains the appointment of the team leader for the specific session exercise.	The exercise can't begin until these activities are done.
		Grading criteria	0000-295 Understood Instructions for exercise	14000		
		Instructions for specific exercise	Team leader for exercise.			
04120	Understanding the grading criteria	00500-092 ready to start	04120-295 Better Understanding of the grading criteria	09000	The team receives the instructions for the exercise from the instructor.	Without receiving the instructions, the team won't know what to present.
				04142		
				30500		
				29000		
				39000		
		Grading criteria		34500		
				34600		
04141	Creating PP	04120-295 Better understanding of the grading criteria	04141-050 Initial draft of PP at start of exercise	09000	Using the knowledge learned in class. One or more members of the team prepare a PP for the exercise.	To demonstrate the importance of planning and how having a plan will make the project easier to perform.
				04142		

(*Continued*)

TABLE 5.19 (*Continued*)
The Relevant Sections of the WPs for the Presentation Planning State (WP 00000)

WP	Activity	Prerequisite	Products	Used in WP	Narrative	Reason for Activity
04142	Creating presentation theme	04120-295 Better Understanding of the grading criteria	04142-205 presentation theme	09000	One or more members of the team create a unique presentation theme for their presentations by either choosing one of the PowerPoint samples or creating their own. The team make sure that the contrast is easy to see in the ambient light in the classroom in accordance with the grading criteria.	To provide their presentation with a unique brand.
		04141-050 Initial draft of PP at start of exercise		23200		
				23100		
				23300		
				23400		
				23405		
				23431		
				23432		
				23433		
				23434		
				23435		
				23436		
				23500		

(*Continued*)

TABLE 5.19 (*Continued*)
The Relevant Sections of the WPs for the Presentation Planning State (WP 00000)

WP	Activity	Prerequisite	Products	Used in WP	Narrative	Reason for Activity
09000		04120-295 Better understanding of the grading criteria	09000-298 Consensus to proceed with exercise	10000		
	Holding the SPPCR	04141-050 Initial draft of PP at start of exercise		04142	The team compare the PP and the presentation theme with the instructions for the exercise according to the understanding of the grading criteria and agree they have enough information to proceed with the exercise.	To maximize the knowledge gained from the exercise and the grade received.
		04142-205 Presentation theme				
		04120-295 Better understanding of the grading criteria				
		04141-050 Initial draft of PP at start of exercise	09000-050 Updated PP at PPCR	19100	The coordinating activity that (1) makes sure all the prerequisites have been met and confirms resources are available as planned, and (2) Coordinates all the other activities in the state. This is generally a management task.	To confirm the team and required resources are ready for the state.

Project Planning 165

TABLE 5.20
The Relevant Sections of the WPs for the Presentation Requirements State (10000)

WP	Activity	Prerequisite	Products	Used in WP	Narrative	Reason for Activity
19000	Holding the requirements review	19100-260 Updated PP at requirements review	19000-298 Consensus that the list of requirements for exercise extracted from the instructions is correct	20500	The team reviews the list of requirements for the exercise and the RTM to make sure that they are compliant to the instructions in the grading criteria and the updated PP.	To maximize the probability of getting a high grade. And to make sure they will be learning the right knowledge for the specific session.
		14100-150 RTM		20500		
		14000-151 Prioritized requirements specification				
14100	Creating the RTM	14000-151 Prioritized requirements specification	14100-150 RTM	19000	The team creates the RTM, Updating the requirements document to include changes in the requirements in this state of the exercise.	The RTM will be used to make sure that the presentation meets the requirements.
				20500		
				28200		
14000	Determining the requirements for the exercise	0000-295 Understood instructions for exercise	14000-151 Prioritized requirements specification	14100	The team go through the instructions for the exercise and extract the requirements. The team can do this individually and then pair their findings or they can do it as a group. They also prioritize the requirements.	The team cannot expect to get a good grade without knowing the requirements for the exercise even though they may learn something. In this situation, the priority is based on the grading criteria so that the team can choose to do less work and settle for a lower grade if circumstances are appropriate.
				20500		
				19000		

(Continued)

TABLE 5.20 (Continued)
The Relevant Sections of the WPs for the Presentation Requirements State (10000)

WP	Activity	Prerequisite	Products	Used in WP	Narrative	Reason for Activity
19100	Reviewing the PP	09000-050 Updated PP at PPCR	19100-260 Updated PP at requirements review	19000 20500	The team takes a quick look at the PP to make sure it's still current and makes any necessary changes.	Sometimes the instructor gives additional requirements verbally during the course of a lecture to simulate real-world customer changes during the project. These changes might reflect into different assignments or people or changes in the schedule.
10500	Managing and top-level activities in the requirements state	04141-050 Initial draft of PP at start of exercise 04120-295 Better understanding of the grading criteria 04142-205 Presentation theme	10000-092 Ready to start	19100 14000	The coordinating activity that (1) makes sure all the prerequisites have been met and confirms resources are available as planned, and (2) Coordinates all the other activities in the state. This is generally a management task.	To confirm the team and required resources are ready for the state.

Project Planning

TABLE 5.21
Typical RTM Template for Presentations

ID	Priority	Requirement	Time Limit	Actual Time	Theme Completed	Presentation Page(s)	Pages Tested
1		Summary of presentation	1				
		Reading 1					
1.2		List of main points	1				
1.3		Main point 1	1				
1.4		Main point 2	1				
1.5		Reflect and comment on readings	2				
1.6		Compare content					
1.7		Why the reading was assigned					
1.8		Lessons Learned	2				
1.9		Source of learning					
2		Reading 2					
2.2–3.9		Repeat rows 2–9					
3		Reading 3					
3.2-3.9		Repeat rows 2–9					
N		Repeat for rest of readings (n)					
4		Summary of presentation					
5		different team leader					
6		Total time of presentation	15				

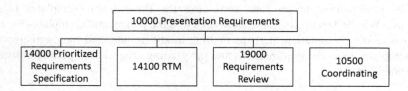

FIGURE 5.17 A partial WBS view of the WPs in the presentation requirements state (10000).

that have taken place in this state. The narrative section of each WP describes what is being done and when more than one thing is being done, the WP might be broken out into sub-WPs. Part of the WBS for the presentation requirements state activities is shown in Figure 5.17.

The presentation requirements state terminates after the Presentation Requirements Review (PRR) (WP 19000).

5.11.7 The Presentation Design State (WP 20000)

The presentation design state (WP 20000) is a state in which each member of the team reads enough of each assigned reading to decide if they want to present that reading. The team then discusses the readings, and if there are more than two, votes on which of the readings are to be presented. The team has a choice of preparing each presentation about the reading together or assigning that preparation to specific members of the team. The WPs are developed showing that the work has been assigned to specific members of the team (member A and member B). The team then produces the templates for each section of that specific presentation. Towards the end of the state, the team updates the RTM and the prioritized requirements document. The updated requirements document will contain any changes in the requirements or priorities for that specific exercise.

The presentation design state (WP 20000) is split into a number of WPs. Each sub-WP activity starts and ends at a key internal milestone, see item 22 in Table 5.1. The relevant sections of the sub-WPs for the presentation design state are shown in Table 5.22. WPs 22200, 23400 and 23500 have lower-level WPs themselves, making a three-level WBS. The relevant sections of WP 22000 are shown in Table 5.23. WP 22000 contains the general management and coordination WP as well as the activities shown in the narrative section in Table 5.22.

WP23400 is expanded in Table 5.24. By requiring WP 23300 as a prerequisite for WP 23100, the intent of the design can be inferred. Instead of creating an introduction and a summary as two separate products, the summary will be created once all the information is in place, and the relevant information in the summary will be used to create the introduction.[26] This approach minimizes differences between summary and the introduction. The relevant sections of the WPs in WP 23400 are shown in Table 5.24. Since both WP 23400 and WP 23500 produce templates for the reading parts of the presentation, the structure will be identical with the same activities. Accordingly, there is no need to show them both.

The PERT chart is shown in Figure 5.18 in which each sub-WP activity starts and ends at a key internal milestone (i.e. A, B and C), see item 22 in Table 5.1.

The presentation design state ends after the Presentation Templates Design Review (WP 29000) in which the team reviewed the presentation templates for each section of the presentation, perhaps making modifications, and then agreeing that the templates meet the requirements and the grading criteria to hopefully ensure an acceptable grade.

5.11.8 The Presentation Construction State (WP 30000)

The presentation construction state (WP 30000) is the state in which the team creates the sections of the presentation using the templates created in the presentation design state, and is split into a number of WPs. Each sub-WP activity starts and ends at a key internal milestone, see item 22 in Table 5.1. The relevant sections of the sub-WPs for the presentation construction state are shown in Table 5.25.

---

[26] This points out the need to have an architecture in place before planning the process can begin.

Project Planning

TABLE 5.22
The Relevant Sections of the WPs for the Presentation Design State (WP 20000)

WP	Activity	Prerequisite	Products	Used in WP	Narrative	Reason for Activity
29000	Holding the presentation templates design review	23100-110 Introduction template	29000-298 Consensus that design of presentation components meets requirements and grading criteria for an acceptable grade	30500	The team reviews the presentation templates for each section of the presentation, perhaps making modifications, and then agreeing that the templates meet the requirements and the grading criteria hopefully ensure an acceptable grade.	It's easy to fix any defects in the presentation template at this time because there is no content. This is also a chance for the team to see the entire template and make sure that the parts are consistent.
		28200-151 Updated prioritized requirements specification at design review				
		04120-295 Better understanding of the grading criteria				
		20000-260 Updated PP at design review	29000-260 Updated PP at design review			
28200	Updating RTM and requirements specification	14100-150 RTM	28200-150 RTM at Design review	48200	The team or an assigned team member updates the RTM to show which elements of the design have been completed. The team also updates the requirements specification to incorporate the effect of any changes during this state of the exercise.	To maximize the probability of producing a presentation compliant to all requirements and grading criteria, accordingly minimizing the probability of leaving out a required section of the presentation.
				38200		
				30500		
		14000-151 Prioritized requirements specification	28200-151 Updated prioritized requirements specification at design review	29000		
				30500		
				38200		
23100	Designing introduction template	04142-205 Presentation theme	23100-110 Introduction template	29000	The team or an assigned member designs the introduction template for the entire presentation.	To ensure a common look and feel for the entire presentation.
		21200-241 Reading 1 assigned to team member A		29000		
		21200-242 Reading 2 assigned to team member B				

(Continued)

TABLE 5.22 (Continued)
The Relevant Sections of the WPs for the Presentation Design State (WP 20000)

WP	Activity	Prerequisite	Products	Used in WP	Narrative	Reason for Activity
23200	Designing lessons learned template	04142-205 Presentation theme	23200-174 Lessons learned template	29000 34300	The team or an assigned member designs the lessons learned template for the entire presentation.	To ensure a common look and feel for the entire presentation.
23300	Designing summary template	04142-205 Presentation theme	23300-180 Summary template	29000	The team or an assigned team member designs the summary template for the entire presentation.	To ensure a common look and feel for the entire presentation.
23400	Designing Reading 1 component templates	04142-205 Presentation theme 21200-241 Reading 1 assigned to team member A	23400-131 Briefing 1 component templates	29000 34505	The assigned member for Reading 1 designs the templates that will be used for each section of briefing 1.	To ensure a common look and feel for the entire presentation.
23500	Designing Reading 2 component templates	04142-205 Presentation theme 21200-242 Reading 2 assigned to team member B	23500-132 Briefing 2 component templates	29000	The assigned member for Reading 2 designs the templates that will be used for each section of Briefing 2.	To ensure a common look and feel for the entire presentation.

(Continued)

TABLE 5.22 (Continued)
The Relevant Sections of the WPs for the Presentation Design State (WP 20000)

WP	Activity	Prerequisite	Products	Used in WP	Narrative	Reason for Activity
21200	Assigning readings	22122-290 Consensus on the two readings to present	21200-241 Reading 1 assigned to team member A	34500	The team leader assigns the readings to volunteers. If nobody volunteers, the team leader finds a way to delegate the work to one of the team members unless it is the team leader's turn to do a reading. Team leaders should not delegate the work to themselves in the event they can't find volunteers.	To make sure that everybody on the team does the same amount of work throughout the semester. This does not necessarily mean that everybody does the same amount work each session.
				34541		
				34542		
				34543		
				34544		
				34545		
				34546		
			21200-242 Reading 2 assigned to team member B	34600		
				34641		
				34642		
				34643		
				34644		
				34645		
				34646		

(Continued)

TABLE 5.22 (Continued)
The Relevant Sections of the WPs for the Presentation Design State (WP 20000)

WP	Activity	Prerequisite	Products	Used in WP	Narrative	Reason for Activity
22000	Discussing readings	14000-151 Prioritized requirements specification 20000-092 Ready to start	22122-290 Consensus on the two readings to present	21200	Each member of the team reads enough of each of the assigned readings to decide if they want to present that reading. The team then discusses the readings as a group. Finally, the team votes on which of the readings to present to get consensus on two of them.	Since the requirement was for two readings, the team needs to get informed consensus on which readings to present.
20500	Coordinating 20000	19000-298 Consensus that the list of requirements for exercise extracted from instructions is correct 14100-150 RTM 14000-151 Prioritized requirements specification	20000-092 Ready to start	22000	The coordinating activity that (1) makes sure all the prerequisites have been met and confirms resources are available as planned, and (2) coordinates all the other activities in the state. This is generally a management task.	To confirm the team and required resources are ready for the state.
		19100-260 Updated PP at Requirements Review	20000-260 Updated PP at Design Review	29000 30500	Verifies that nothing that affects the PP has changed, or if it has, updates the PP to reflect those changes.	To make sure that the delivered presentation is compliant to any changed requirements and that the scope of the work is still feasible.

Project Planning

TABLE 5.23
The Relevant Sections of WP 22000 (Discussing Readings)

WP	Activity	Prerequisite	Products	Used in WP	Narrative	Reason for Activity
22322	Voting on which readings to present	22220-299 List of understood readings	22322-290 consensus on the two readings to present	21200	The team votes on which of the readings to present to get consensus on two of them.	To actually decide which two of the readings to present.
22220	Discussing readings as a group	22124-199 Understanding of what readings are about	22220-299 List of understood readings	22322	The team discusses the readings as a group sharing opinions about the readings.	The team cannot make an informed decision to which readings to present until they've discussed the readings.
22124	Reading enough of each reading	14000-151 Prioritized requirements specification / 22000-092 Ready to start	22124-199 Understanding of what readings are about	22220	Each member of the team reads enough of each of the assigned readings to decide if they want to present that reading.	They need to know enough about the reading to decide if they want to present it.
22500	Coordinating 22000	19000-298 Consensus that the list of requirements for exercise extracted from is correct	22000-092 Ready to start	22124	The coordinating activity that (1) makes sure all the prerequisites have been met and confirms resources are available as planned, and (2) Coordinates all the other activities in the state. This is generally a management task.	To confirm the team and required resources are ready for the state.

TABLE 5.24
The Relevant Sections of WP 23400 (Designing Reading 1 Component Templates)

WP	Activity	Prerequisite	Products	Used in WP	Narrative	Reason for Activity
23405	Coordinating 23430	04142-205 Presentation theme 14000-151 Prioritized requirements specification	23430-092 Ready to start	23431 23432 23433 23434 23435 23436	The coordinating activity that (1) makes sure all the prerequisites have been met and confirms resources are available as planned, and (2) Coordinates all the other activities in the state. This is generally a management task.	To confirm the team and required resources are ready for the state.
23431	Designing list of main points template	14000-151 Prioritized requirements specification 23430-092 Ready to start 04142-205 Presentation theme	23431-120 List of main points template	29000 30500 34580	The team or an assigned member designs the templates that will be used in the section of the briefing that provides the list of the main points.	To ensure a common look and feel for the entire presentation.
23432	Designing main point 1 template	14000-151 Prioritized requirements specification 23430-092 Ready to start 04142-205 Presentation theme	23432-130 Main point 1 template	29000 30500 34500	The team or an assigned member designs the templates that will be used in the section of the briefing that discusses main point 1.	To ensure a common look and feel for the entire presentation.
23433	Designing main point 2 template	04142-205 Presentation theme 14000-151 Prioritized requirements specification 23430-092 Ready to start	23433-140 Main point 2 template	29000 30500 34500	The team or an assigned member designs the templates that will be used in the section of the briefing that discusses main point 2.	To ensure a common look and feel for the entire presentation.

(Continued)

Project Planning 175

TABLE 5.24 (Continued)
The Relevant Sections of WP 23400 (Designing Reading 1 Component Templates)

WP	Activity	Prerequisite	Products	Used in WP	Narrative	Reason for Activity
23434	Designing reflection and comments template	04142-205 Presentation theme 14000-151 Prioritized requirements specification	23434-150 Reflection and comments template	29000 34500 30500	The team or an assigned member designs the templates that will be used in the section of the briefing that discusses the reflection and comments.	To ensure a common look and feel for the entire presentation.
23435	Designing compare content template	04142-205 Presentation theme 23430-092 Ready to start 14000-151 Prioritized requirements specification 23430-092 Ready to start	23435-160 Compare content template	29000 30500 34500	The team or an assigned member designs the templates that will be used in the section of the briefing that compares the content with other readings or/and other team member-located readings.	To ensure a common look and feel for the entire presentation.
23436	Designing why the reading was assigned template	04142-205 Presentation theme 14000-151 Prioritized requirements specification 23430-092 Ready to start	23436-170 Why the reading was assigned template	29000 30500 34500	The team or an assigned member designs the templates that will be used in the section of the briefing that discusses why the reading was assigned.	To ensure a common look and feel for the entire presentation.

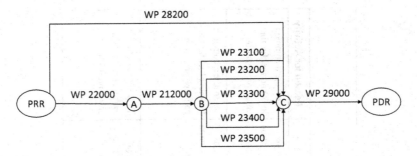

FIGURE 5.18 The PERT chart for the presentation design state (WP 20000).

When creating the WPs, the initial step is to work back from the final deliverable. This produces a number of prerequisites for the different sub-WP's that may sometimes be simplified in the PERT chart corresponding to WP 30000. For example, the initial set of prerequisites for WP 39000 is products 34200-110, 34500-131, 34600-132, 34300-174 and 34400-180. However, when working through the remaining sub-WPs, you might notice that the prerequisite for WP 34200 is 34400-180. You might think that since WP 39990 cannot be started until 34400-180 has been completed, 34400-180 does not need to be a prerequisite for WP 390, because it was a prerequisite for WP 34200. Similarly, since 34500-131 and 34600-1326 are prerequisites for WP 34300, they also do not need to be prerequisites for WP 39000. In this instance, this is NOT the case, since the products are inputs to the review in WP 39000. In general, in other instances it is also preferable not to delete the redundant required products, since some of the subsequent WPs may be delayed during the remaining states of the project, and in knowing which WPs are prerequisites, it might be possible to adjust the work to recover some of the delay.

The activities performed in WP 34500 and WP34600 are identical, so only WP 34500 is expanded in Table 5.26. Since all the activities are creating activities, the tens digit is always a '4' as in 41, 42, etc.

The PERT chart is shown in Figure 5.19 in which each sub-WP activity starts and ends at a key internal milestone, see item 22 in Table 5.1.

5.11.9 The Presentation Integration and Testing States (WP 40000)

The presentation integration and testing states (WP 40000) are the states in which the individual sections of the presentation are integrated into a combined presentation and tested against the requirements and grading criteria. The relevant sections of the sub-WPs for the presentation integration and testing state are shown in Table 5.27.

Part of the WBS view of the WPs in the presentation integration and test state level integration and test state is shown in Figure 5.20.

The presentation integration and testing state terminates after the Presentation Delivery Readiness Review (WP 49000).

Project Planning

TABLE 5.25
The Relevant Sections of the WPs for the Presentation Construction State (WP 30000)

WP	Activity	Prerequisite	Products	Used in WP	Narrative	Reason for Activity
39000	Holding the PIR (presentation ready for integration)	34200-110 Draft Introduction	39000-110 Introduction	40500	The team reviews the presentation templates for each section of the presentation, perhaps making modifications, and then agreeing that the templates meet the requirements and the grading criteria hopefully ensure an acceptable grade.	To produce consensus that all the sections of the presentation are compliant to requirements and to proceed to the integration state
				48000		
		34580-131 Completed individual sections of Reading 1	39000-131 Briefing 1	48000		
				40500		If the team has agreed not to meet any of the requirements for some specific reason, to affirm that decision is still valid.
		34680-132 Completed individual sections of Reading 2	39000-132 Briefing 2	48000		
				40500		
		34300-174 Draft Lessons Learned	39000-174 Lessons learned	48000		
		04120-295 Better Understanding of the grading criteria		40500		
		34400-180 Draft Summary	390000-180 Summary	48000		
		38200-150 RTM at integration review		40500		
		28200-151 Updated prioritized requirements specification at design review	39000-298 Consensus to proceed to the integration state	40500		

(Continued)

TABLE 5.25 (Continued)
The Relevant Sections of the WPs for the Presentation Construction State (WP 30000)

WP	Activity	Prerequisite	Products	Used in WP	Narrative	Reason for Activity
38200	Updating RTM and requirements specification	28200-150 RTM at design review	38200-150 RTM at Integration Review	39000 40500 48200	The team or an assigned team member updates the RTM to show which elements of the design and implement specific requirements. The team also update the requirements specification to incorporate the effect of any changes during this state of the exercise.	To maximize the probability of producing a presentation compliant to all requirements and grading criteria, accordingly minimizing the probability of leaving out a required section of the presentation.
		28200-151 Updated prioritized requirements specification at design review	38200-151 Updated prioritized requirements specification at integration review	39000 40500		
34200	Creating Introduction	34400-180 Draft Summary	34200-110 Draft Introduction	39000 34300	The team or an assigned team member uses the summary as the basis for the introduction or table of contents of the presentation. By using the summary to create the presentation rather than creating them into separate WPs, the summary and introduction will not get out of sync.	Required by the requirements for the exercise.

(Continued)

Project Planning

TABLE 5.25 (Continued)
The Relevant Sections of the WPs for the Presentation Construction State (WP 30000)

WP	Activity	Prerequisite	Products	Used in WP	Narrative	Reason for Activity
34300	Creating draft lessons learned	23200-174 Lessons learned template 34200-110 Draft Introduction 34580-131 Completed individual sections of Reading 1 34680-132 Completed individual sections of Reading 2 34400-180 Draft Summary	34300-174 Draft lessons learned	39000	The team members discuss the lessons they learned participating in the exercise, and the classroom session and create a list of the lessons they learned.	Required by the requirements for the exercise.
34400	Creating Summary	34580-131 Completed individual sections of Reading 1	34400-180 Draft Summary	34300	The team or an assigned team member creates the draft summary from the set of completed individual sections. The summary should be an expanded list of the main points of reading 1 focusing on the aspects of the reading presented.	Required by the requirements for the exercise.

(Continued)

TABLE 5.25 (*Continued*)
The Relevant Sections of the WPs for the Presentation Construction State (WP 30000)

WP	Activity	Prerequisite	Products	Used in WP	Narrative	Reason for Activity
		34680-132 Completed individual sections of Reading 2		34200	The team or an assigned team member creates the draft summary from the set of completed individual sections. The summary should be an expanded list of the main points of reading 2 focusing on aspects of the reading presented.	
				39000		
34500	Creating Briefing 1	21200-241 Reading 1 assigned to team member A	34580-131 Completed individual sections of Reading 1	34300	Team member A abstracts reading to provide the list of main points, creates sections of a presentation on the content associated with each main point, creates the reflections and comments on the reading section, creates the compare content section, creates the why the reading was assigned section. The team leader then accepts completed sections and stores them ready for integration. This is not an integration activity; it's a quality control activity.	Required by the requirements for the exercise.
		23400-131 Briefing 1 component templates				
		14000-151 Prioritized requirements specification		34400		
		04120-295 Better understanding of the grading criteria		39000		
				40500		
				48000		

(*Continued*)

Project Planning

TABLE 5.25 (Continued)
The Relevant Sections of the WPs for the Presentation Construction State (WP 30000)

WP	Activity	Prerequisite	Products	Used in WP	Narrative	Reason for Activity
34600	Creating Briefing 2	21200-242 Reading 2 assigned to team member B	34680-132 Completed individual sections of Reading 2	34400	Team member B abstracts reading to provide the list of main points, creates sections of a presentation on the content associated with each main point, creates the reflections and comments on the reading section, creates the compare content section, creates the why the reading was assigned section. The team leader then accepts completed sections and stores them ready for integration. This is not an integration activity, it's a quality control activity.	Required by the requirements for the exercise.
		23500-132 Briefing 2 component templates		34600		
		14000-151 Prioritized requirements specification		39000		
		04120-295 Better understanding of the grading criteria		40500		
				48000		

(Continued)

TABLE 5.25 (*Continued*)
The Relevant Sections of the WPs for the Presentation Construction State (WP 30000)

WP	Activity	Prerequisite	Products	Used in WP	Narrative	Reason for Activity
30500	Coordinating 30000	23431-120 List of main points template 23432-130 Main point 1 template 23433-140 Main point 2 template 23434-150 Reflection and comments template 23435-160 Compare content template 23436-170 Why the reading was assigned template 23531-120 List of main points 23532-130 Main point 1 23533-140 Main point 2 23534-150 Reflection and comments 235435-160 Compare content 23536-170 Why the reading was assigned 28200-150 RTM at Design review	30000-092 Ready to start	34500 34600	The coordinating activity that (1) makes sure all the prerequisites have been met and confirms resources are available as planned, and (2) Coordinates all the other activities in the state. This is generally a management task.	To confirm the team and required resources are ready for the state.
		20000-260 Updated PP at design review	30000-260 Updated PP at exercise integration review	39000 40500	Verifies that nothing that affects the PP has changed, or if it has, updates the PP to reflect those changes.	To make sure that the delivered presentation is compliant to any changed requirements and that the scope of the work is still feasible.

Project Planning

TABLE 5.26
The Relevant Data Elements in the 34500 Create Briefing 1 WP

WP	Activity	Prerequisite	Products	Used in WP	Narrative	Reason for Activity
34505	Coordinating 34500	23400-131 Briefing 1 component templates	34500-092 Ready to start	34541	The coordinating activity that (1) makes sure all the prerequisites have been met and confirms resources are available as planned, and (2) coordinates all the other activities in the state. This is generally a management task.	To confirm the team and required resources are ready for the state.
		21200-241 Reading 1 assigned to team member A		34542		
		30000-260 Updated PP at exercise integration review		34543		
		04120-295 Better understanding of the grading criteria		34544		
				34546		
34541	Creating List of main points	23431-120 List of main points template	34541-120 List of main points	34580	Team member A abstracts reading to provide the list of main points.	Required by the requirements for the exercise.
		21200-241 Reading 1 assigned to team member A				
		34500-092 Ready to start				
34542	Creating main point 1	23432-130 Main point 1 template	34542-121 Main point 1	34580	Team member A creates a presentation on the content associated with each main point.	Required by the requirements for the exercise.
		21200-241 Reading 1 assigned to team member A				
		34500-092 Ready to start				
34543	Creating main point 2	23433-140 Main point 2 template	34543-122 Main point 2	34580	Team member A creates a presentation on the content associated with each main point.	Required by the requirements for the exercise.
		21200-241 Reading 1 assigned to team member A				
		34500-092 Ready to start				

(Continued)

TABLE 5.26 (Continued)
The Relevant Data Elements in the 34500 Create Briefing 1 WP

WP	Activity	Prerequisite	Products	Used in WP	Narrative	Reason for Activity
34544	Creating reflection and comments	23434-150 Reflection and comments template 21200-241 Reading 1 assigned to team member A 34500-092 Ready to start	34544-150 Reflection and comments	34580	Team member A creates the section on reflections and comments on the reading.	Required by the requirements for the exercise.
34545	Creating compare content	23435-160 Compare content template 21200-241 Reading 1 assigned to team member A 34500-092 Ready to start	34545-160 Compare content	34580	Team member A creates the section on compare content.	Required by the requirements for the exercise.
34546	Creating why the reading was assigned	23436-170 Why the reading was assigned template 21200-241 Reading 1 assigned to team member A 34500-092 Ready to start	34546-170 Why the reading was assigned	34580	Team member A creates the section covering why the reading was assigned.	Required by the requirements for the exercise.
34580	Creating the sections of Briefing 1	34541-120 List of main points 34542-121 Main point 1 34543-122 Main point 2 34544-150 Reflection and comments 34545-160 Compare content 34546-170 Why the reading was assigned	34580-131 Completed individual sections of Reading 1		The team leader accepts completed sections and stores them ready for integration. This is not an integration activity; it's a quality control activity.	To make sure that all the sections that were supposed to be completed have indeed been completed.

Project Planning

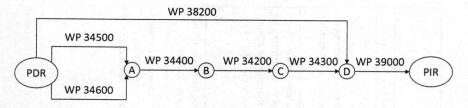

FIGURE 5.19 The PERT chart for the presentation construction state (WP 30000).

5.11.10 THE PRESENTATION DELIVERY AND GRADING STATE (WP 5000)

The presentation delivery and grading state (WP 50000) is the state in which the presentation is uploaded to the class website, the presentation is presented, graded and the grade is discussed by the team and opportunities for improvement discovered. The relevant sections of the sub-WPs for the presentation integration and testing state are shown in Table 5.28. The hierarchical or WBS view of the 50000 level delivery state is shown in Figure 5.21.

The presentation delivery and grading state terminates after the PPMR (WP 59000).

5.11.11 ACHIEVEMENTS AT THIS POINT IN TIME

At this point in time the team has identified:

1. All the activities that need to be performed.
2. All the products that have to be produced.
3. The prerequisites for each activity.

5.12 THE ENGAPOREAN MCSSRP EXERCISE

The background and context for the progressive session exercises is in Appendix 1. The purpose of the exercise is[27,28,29]:

1. To create framework part of the Engaporian[30] Multi-satellite Communications Switching System Replacement Project (MCSSRP) plan.
2. To be used in subsequent session exercises.

---

[27] Some students' positive feedback indicated that the exercises were a lot of work but they learnt more doing the exercise than they learned in other classes on similar topics.
[28] These exercises were performed by my postgraduate and professional development students in Singapore. Many students received 'A's on the exercises. If you think the exercises are too overwhelming, remember:
  - The learning curve applies and the feedback provided to the students let them see it in action.
  - People behave in the manner they are expected to behave.
[29] The students were also given the reasons for the requirements.
[30] A fictitious country.

TABLE 5.27
The Relevant Sections of the WPs for the System Integration and Test State (WP 40000 Level)

WP	Activity	Prerequisite	Products	Used in WP	Narrative	Reason for Activity
49000	Holding the delivery readiness review		49000-298 Consensus that final version of presentation is best one that could be done within constraints	50500		To produce consensus that all the sections of the presentation are compliant to requirements and to proceed to the integration state
		48200-151 Updated prioritized requirements specification at delivery readiness review				If the team has agreed not to meet any of the requirements for some specific reason, to affirm that decision is still valid.
		45000-290 Verified Final presentation				
48200	Updating RTM and requirements specification	38200-150 RTM at Integration Review	48200-150 RTM at delivery readiness review	49000	The team or an assigned team member updates the RTM to show which elements of the design and implement specific requirements. The team also update the requirements specification to incorporate the effect of any changes during this state of the exercise.	To maximize the probability of producing a presentation compliant to all requirements and grading criteria, accordingly minimizing the probability of leaving out a required section of the presentation.
				50500		
		38200-151 Updated prioritized requirements specification at integration review	48200-151 Updated prioritized requirements specification at delivery readiness review	49000		
				50500		

(Continued)

Project Planning   187

TABLE 5.27 (Continued)
The Relevant Sections of the WPs for the System Integration and Test State (WP 40000 Level)

WP	Activity	Prerequisite	Products	Used in WP	Narrative	Reason for Activity
45000	Verifying presentation compliance	44000-190 Final presentation 38200-151 Updated prioritized requirements specification at integration review	45000-290 Verified final presentation	49000 56200 50500	The team or an assigned member checks the quality of the presentation and verifies that it is ready to be uploaded for presentation in class.	To make a last quality check on the presentation document.
44000	Completing final version	48000-190 Complete set of reading presentation sections	44000-190 Final presentation	45000	The team makes any final modifications and holds a run through or rehearsal of the actual presentation.	To make sure that the presentation document is the best one the team can put together within the constraints and rehearse the presentation and comment on what they saw at the rehearsal so that when the presentation is delivered it would be the best one they can do.
48000	Integrating complete team draft	40000-092 Ready to start 39000-110 Introduction 39000-131 Briefing 1 39000-132 Briefing 2 39000-174 Lessons learned 390000-180 Summary	48000-190 Complete set of reading presentation sections	44000	The team or an assigned member integrates all the sections of the presentation into a single presentation paying particular attention to consistency between the sections.	To maximize the probability of producing a consistent presentation to client to all requirements and grading criteria.

(Continued)

TABLE 5.27 (Continued)
The Relevant Sections of the WPs for the System Integration and Test State (WP 40000 Level)

WP	Activity	Prerequisite	Products	Used in WP	Narrative	Reason for Activity
40500	Coordinating 40000	39000-298 Consensus to proceed to the integration state	40000-092 Ready to start	48000	The coordinating activity that (1) makes sure all the prerequisites have been met and confirms resources are available as planned, and (2) coordinates all the other activities in the state. This is generally a management task.	To confirm the team and required resources are ready for the state.
		39000-110 Introduction				
		39000-131 Briefing 1				
		34580-131 Completed individual sections of Reading 1				
		34680-132 Completed individual sections of Reading 2				
		39000-174 Lessons learned				
		38200-151 Updated prioritized requirements specification at integration review				
		38200-150 RTM at integration review				
		390000-180 Summary				
		30000-260 Updated PP at exercise integration review	40000-260 Updated PP at delivery readiness review	49000		
				50500		

Project Planning

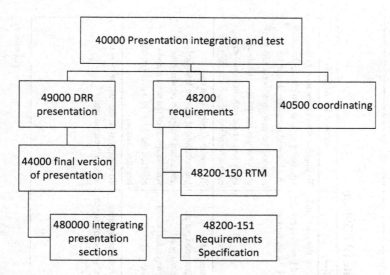

FIGURE 5.20 A partial WBS view of the WPs in the presentation integration and test state (WP 40000).

3. To create and partially populate a set of WPs based on knowledge from this and previous chapters.
4. To practice using the template for a student exercise presentation (Kasser 2018: Section 14.7).

5.12.1 THE REQUIREMENTS FOR THE EXERCISE

The requirements for the exercise are to:

1. Formulate the problem posed by the exercise using the problem formulation template (Section 3.7.1).
2. Identify at least five formal milestones in the MCSSRP.
3. Identify at least two products completed at each formal milestone in the MCSSRP in the development or test/quality stream of work. These may be hardware, software, document, completed system, etc.
4. Draw the Gantt chart/network that links the products (Kasser 2018: Section 2.14).
5. Create partial WPs of at least eight products using Table 5.1 as a template for the following elements:
    1. Identification number.
    2. Name of activity.
    3. Priority.
    4. Narrative description of activity.
    5. Products.
    6. Acceptance criteria for products.

TABLE 5.28
The Relevant Sections of the WPs for the Presentation Delivery and Grading State (WP 50000)

WP	Activity	Prerequisite	Products	Used in WP	Narrative	Reason for Activity
59000	Holding the PPMR	56100-295 Grade (customer sign off)	59000-279 Increased understanding (non-tangible)	Outcome	The team needs to discuss the grade and instructs comments and how to apply them to in state zero of the upcoming session.	To improve the presentation in the upcoming session. No blame for issues that lowered the grade should be awarded, because there were plenty of opportunities for the team to check every part of the presentation during each state of the session exercise.
56100	Receiving grade	57000-290 Completed making presentation (delivered the product)	56100-295 Grade (customer sign off)	59000	The instructor comments on the presentation pointing out good points and where things were not good suggesting alternative ways that would be good, and awards the grade.	To provide feedback on the presentation, of session knowledge, and award the grade pointing out which aspects of the presentation affected the grade in a positive or negative manner.
57000	Making presentation	56200-190 Uploaded presentation	57000-290 Completed making presentation (delivered the product)	56100	The team makes the presentation using one or more team members according to their preferences.	To comply with the requirement to make a presentation.

(*Continued*)

TABLE 5.28 (Continued)
The Relevant Sections of the WPs for the Presentation Delivery and Grading State (WP 50000)

WP	Activity	Prerequisite	Products	Used in WP	Narrative	Reason for Activity
56200	Uploading presentation	NNN-298 Go ahead to make presentation in class	56200-190 Uploaded presentation	57000	The team leader or an assigned team member volunteer uploads the presentation to the class website and notifies the rest of the team that the presentation has been uploaded.	To comply with the requirement that each session exercise be uploaded, a meta-requirement for the session exercise and accordingly not listed in the exercise requirements.
		50000-092 Ready to start				
		45000-290 Verified final presentation				
50500	Coordinating 50000	49000-298 Consensus that final version of presentation is best one that could be done within constraints	50000-092 Ready to start	56200	The coordinating activity that (1) makes sure all the prerequisites have been met and confirms resources are available as planned, and (2) coordinates all the other activities in the state. This is generally a management task.	To confirm the team and required resources are ready for the state.
		45000-290 Verified final presentation				

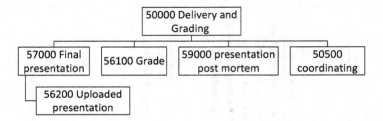

FIGURE 5.21 The WBS view of the 50000 presentation delivery and grading state.

    7. Reason activity is being done.
    8. Prerequisites.
    9. Lower-level WPs (if any).
   10. Decision points (if any).
   11. Internal key milestones (if any).
   12. Assumptions not stated elsewhere.
6. Use spreadsheet then export (copy/paste) to PowerPoint.
7. Create WBS for activities (Section 5.8.2.1).
8. Create Gantt charts linking the activities (Section 5.2.1).
9. Leave out times (no data yet).
10. Plan and prioritize the tasks.
11. Format presentation according to the applicable parts of the sample four-part template for a management review presentation (Section 5.4.2.2.1).
12. Number all slides.
13. Prepare <5 min presentation that includes:
    1. The problem posed by the exercise formulated according to the problem formulation template (Section 3.7.1):
    2. The MCSSRP formal milestones.
    3. Summary of WPs.
    4. A Gantt chart (Section 5.2.1).
    5. Where at least one identified risk will be mitigated.
    6. Lessons learned.
    7. A compliance matrix (Kasser 2018: Section 9.5.2) for meeting the requirements for the exercise.
14. Save presentation as file '5-teamname'.pptx.

5.13 SUMMARY

Successful projects are planned. This chapter helped you to understand planning by focusing on product-based planning, the project and SLCs and planning methodologies. The chapter discussed plans, the difference between generic and project-specific planning, how to incorporate prevention into the planning process to lower the completion risk and shows how to apply WPs instead of WBSs. The chapter ended by showing how to use product-based planning in an exercise.

REFERENCES

Arthur, Lowell Jay. 1992. *Rapid Evolutionary Development*. Hoboken, NJ: John Wiley & Sons.
Avison, David, and Guy Fitzgerald. 2003. *Information Systems Development: Methodologies, Techniques and Tools*. Maidenhead: McGraw-Hill Education (UK).
Boyd, John. 2017. *The essence of winning and losing* 1995 [cited 13 November 2017]. Available from www.pogoarchives.com.
Chacko, George K. 1989. *The Systems Approach to Problem Solving*. New York: Prager.
Churchman, C. West. 1968. *The Systems Approach*. New York: Dell Publishing Co.
Clark, Wallace. 1922. *The Gantt Chart a Working Tool of Management*. New york: The Ronald Press Company.
Cleland, David I. 1994. *Project Management: Strategic Design and Implementation*. New york: Mcgraw-Hill
DERA. 1997. *DERA Systems Engineering Practices Reference Model*. Farnborough, Hampshire: Defence Evaluation and Research Agency (DERA).
Gordon Grace, Ann E. MacEachron and G. Lawrence Fisher. 1974. A contingency model for the design of problem solving research program. *Milbank Memorial Fund Quarterly* no. 52 (2):184–220.
Kasser, Joseph Eli. 2000. A framework for requirements engineering in a digital integrated environment (FREDIE). In *the Systems Engineering, Test and Evaluation Conference (SETE)*, Brisbane, Australia.
Kasser, Joseph Eli. 2002a. The cataract methodology for systems and software acquisition. In *SETE 2002*. Sydney, Australia.
Kasser, Joseph Eli. 2002b. The cataract methodology for systems and software acquisition. In *SETE 2002*, Sydney, Australia.
Kasser, Joseph Eli. 2005. Introducing the role of process architecting. In *the 15th International Symposium of the International Council on Systems Engineering (INCOSE)*, Rochester, NY.
Kasser, Joseph Eli. 2008. Luz: From light to darkness: Lessons learned from the solar system. In *the 18th INCOSE International Symposium*, Utrecht, Holland.
Kasser, Joseph Eli. 2012. Getting the right requirements right. In *the 22nd INCOSE International Symposium*, Rome, Italy.
Kasser, Joseph Eli. 2017. Enabling complex developments through systems thinking. In *SWISSED* 2017, Zurich, Switzerland.
Kasser, Joseph Eli. 2018. *Systems Thinker's Toolbox: Tools for Managing Complexity*. Boca Raton, FL: CRC Press.
Kasser, Joseph Eli. 2019. *Systemic and Systematic Systems Engineering*. Boca Raton, FL: CRC Press.
Kasser, Joseph Eli, and Robin Schermerhorn. 1994. Gaining the competitive edge through effective systems engineering. In *the 4th Annual International Symposium of the NCOSE*, San Jose, CA.
Kasser, Joseph Eli, and Yang-Yang Zhao. 2016. Simplifying solving complex problems. In *the 11th International Conference on System of Systems Engineering*, Kongsberg, Norway.
Lano, Roberto. 1977. The N^2 chart. In *TRW Software Series*. Redondo Beach, CA.
Larson, Erik W., and Clifford F. Gray. 2011. *Project Management the Managerial Process*. 5th edition, The McGraw-Hill/Erwin series operations and decision sciences. New York: McGraw-Hill.
Malotaux, Neils. 2010. *Predictable Projects Delivering the Right Result at the Right Time*. Keo, Japan: N R Malotaux Consultancy.
Miller, George. 1956. The magical number seven, plus or minus two: Some limits on our capacity for processing information. *The Psychological Review* no. 63:81–97.

O'Reilly, John E. 2004. Message from the president of the IEE. In *the International Engineering Management Conference*, Singapore.

Osborn, Alex F. 1963. *Applied Imagination Principles and Procedures of Creative Problem Solving*. 3rd revised edition. New York: Charles Scribner's Sons.

Overbaugh, Richard C., and Lynn Schultz. 2013. *Bloom's taxonomy*. Old Dominion University 2013 [cited 13 March 2013]. Available from ww2.odu.edu/educ/roverbau/Bloom/blooms_taxonomy.htm.

Royce, Winston W. 1987. Managing the development of large software systems: Concepts and techniques. In *ICSE '87 the 9th International Conference on Software Engineering*, Los Angeles

Stauber, B. Ralph, Harry M. Douty, Willard Fazar, Richard H. Jordan, William Weinfeld, and Allen D. Manvel. 1959. Federal statistical activities. *The American Statistician* no. 13 (2):9–12.

6 Successful Project Staffing

Because projects are staffed by people, this chapter discusses aspects of staffing projects, teams and distributing assignments to members of the project team. The chapter explains:

1. The need for high-performance teams.
2. That people are not interchangeable; namely, one person does not necessarily equal another.
3. How to staff a project.

The chapter discusses:

1. People in Section 6.1.
2. The systems approach to staffing a team in Section 6.2.

6.1 PEOPLE

People are one of the seven P's of project management (Section 2.1). The people issues pertaining to project staffing discussed in the section are:

- Availability.
- Competencies and skills.
- Compatibility.
- Permanent or temporary.
- Costs.
- Teams.

6.1.1 AVAILABILITY

People are generally a scarce resource; good people are even scarcer. This means that project managers (PMs) might not always get the people they want to staff a project. Sometimes acquiring the desired person will take some politicking and negotiation with other managers. Moreover, small organizations tend not to have a large pool of possible project members. In such situations, outsourcing may be a solution to finding suitable personnel.

6.1.2 COMPETENCIES AND SKILLS

People are not equal or interchangeable. They have different levels of experience, knowledge, competence and motivation. A traditional way of measuring a person's

TABLE 6.1
Traditional Competency Levels Based on Years of Experience

Labour Category	Years of Professional Experience
Senior	Eight or more
Mid-Level	Three to seven
Junior	Less than three

skill level is to base it on the number of years of experience such as in the example shown in Table 6.1. However, while this is easy to do, it does not give any indication of the effectiveness of the person. It does not show if the person has improved continuously over those years or is repeating the same behaviour in each of those years. See the resumes in Appendix 5 for examples. A better approach is some kind of competency assessment such as one based on the competency maturity model (CMM) (Section 2.1.3) which can be applied across all disciplines, because it covers cognitive skills in problem-solving and can be defined for different disciplines and domains.

6.1.3 COMPATIBILITY

Compatibility is often overlooked. People working in a team must be compatible and able to get along with each other. This means they should not have habits or behaviour traits that irritate other members of the team.[1]

6.1.4 PERMANENT OR TEMPORARY

Some people are assigned to the project for the duration of the project; others can be assigned for specific tasks. For example, during the system design state of the SDP personnel with designing skills need to be assigned to the project, similarly during the testing states, personnel with testing skills need to be assigned to the project. Sometimes these can be the same people if they have both sets of skills. Sometimes people need to be assigned temporarily to replace permanent project staff who are absent for some reason.

6.1.5 COSTS

Different people have different associated costs. In general, the more senior or more experienced a person is, the more expensive they are. This means that in general

---

[1] Many years ago, the team I was working in needed an additional person. When we interviewed the candidates, one of the candidates met all the qualifications but did not seem to have a sense of humour. Since the team did a lot of kidding around while they were working, having a team member without a sense of humour would have been disruptive. Since other candidates were equally qualified and seemed to have a sense of humour, that candidate who did not seem to have a sense of humour was not considered for the position.

Successful Project Staffing

expensive people should not be assigned to do tasks that less expensive people perform equally well. The more experienced and knowledgeable people might be better utilized guiding and mentoring the less experienced and knowledgeable personnel as well as performing some of the tasks.

6.1.6 Teams

Most projects are done by a group of people namely more than one person. This means they work in teams. Teams consist of organizational units whose members report to the same superior in the structural hierarchy (Howe 1981). So being able to manage teams is a required skill for a PM. The best teams are made up of people with different areas of expertise working together. The team leaders can themselves be members of a higher-level team. Teams can:

- *Produce something:* products, services, reports, advice, etc.
- *Manage something:* organizations, groups, contractors, etc.
- *Test something:* products, concepts, etc.

6.1.6.1 The Lifecycle of a Team

The five-phase lifecycle of a team has been stated as (Tuckman 1965):

1. *Forming:* The team meets and learns about the opportunity and challenges, and agrees on goals and begins to tackle the tasks. Team members:
   - Tend to behave quite independently.
   - May be motivated but are usually relatively uninformed of the issues and objectives of the team.
   - Are usually on their best behaviour.
   - Are very focused on self.
2. *Storming:* The team addresses issues such as what problems they are really supposed to solve, how they will function independently and together and what leadership model they will accept. Team members:
   - Open out to each other and confront each other's ideas and perspectives.
3. *Norming:* Team members:
   - Adjust their behaviour to each other as they develop work habits that make teamwork seem more natural and fluid.
   - Often work through this stage by agreeing on rules, values, professional behaviour, shared methods, working tools and even taboos.
   - Begin to trust each other.
   - Get more acquainted with the project and motivation increases.
   - May lose their creativity if the norming behaviours become too strong and begin to stifle healthy dissent and the team begins to exhibit groupthink.
4. *Performing:* Team members
   - Have become interdependent.
   - Are motivated and knowledgeable.

- Are now competent, autonomous and able to handle the decision-making process without supervision. Dissent is expected and allowed as long as it is channelled through means acceptable to the team.
5. *Adjourning:* This phase involves completing the task and breaking up the team.

Tuckman's well-known five-phase team lifecycle is generally applied to all types of teams. However, perceptions from the *Scientific* HTP infer that it seems applicable to Theory Y teams (Section 2.1.4.2) in environments where the team is knowledgeable, but the goals of the project and the roles and responsibilities have not been clearly stated. Accordingly:

- If the goals of the project and the roles and responsibilities are clearly stated, there should be hardly any storming.
- The leadership style needs to be adjusted to match the team lifecycle phase, the motivation (Section 2.1.4) and the competence of the team members (Section 2.1.3).

6.1.6.2 Characteristics of Effective Teams

Effective teams usually have high morale and can be recognized by the following attributes (Kasser 1995):

- *Mutual trust amongst the members:* a willingness to communicate openly without fear of reprisal or that someone else will take undue credit. Team members have overlapping skills and may cross train to provide a degree of backup in the event a member is absent for a while.
- *Mutual respect amongst the members:* respect for the leader, customer and each other's specialty, and are willing to listen, i.e. hear something from another perspective.
- *Mutual sharing of resources:* anything from information sources to computer equipment.
- *Well-defined roles and responsibilities:* everyone knows what they are supposed to do, and what everyone else is responsible for.
- *High esprit de corps:* the team knows what they are supposed to do and know they are doing it well. Morale is high.

6.1.6.3 Creating and Staffing Effective Teams

This section describes one way of managing the project skill pool in an optimal manner across all work packages (WPs) with reference to the project organization (Section 4.4.3). The PM uses the WP to develop the staffing profile which defines skill levels, skill types and competencies by position description. Once the task leader (TL) completes the staffing profile, the TL will:

- Tentatively assign personnel to the task.
- Ensure the appropriate skills and expertise of each person have been exercised at the required level of competence within the previous 5 years, and then provisionally assign specific personnel to the task.

Successful Project Staffing

- Give priority to those who have demonstrated the required level of competence within the previous 2 years.
- The sequential steps to identify needed personnel are:
  1. Within the project from other activities where the identified skill need is being reduced or eliminated.
  2. External to the project but within the existing personnel resources of the project organization.
  3. Temporary personnel from the specialty skills bank.
  4. Temporary personnel from external sources.
  5. New hires from external sources.

Contention for specialized skills should normally only arise where there are very few personnel available for assignment. Consequently, while the TL will provide a list of requested staff to the PM, it is the PM who has the ultimate authority for staffing the task. When the PM and TLs meet to evaluate a WP, they will take a broad view of activities and schedules across the project. There is often some degree of flexibility in making minor schedule adjustments to permit use of existing contract resources at the appropriate time. This approach, which results in a stable staffing profile level that avoids the need for temporary staff build-up and decline and thereby retains the "institutional knowledge" on the task, is a key to successful WP management.

The activities in a WP may be performed by an individual or by two or more persons in the form of an interdependent or cross-functional team. Anecdotal evidence notes that when a WP is performed by a single person:

- Something usually gets left out because it is difficult if not impossible for a single person to be accomplished in all the functional skills necessary to perform the task.
- The undocumented background knowledge which the person gains, is lost to the project when that person leaves the project.

Each team consists of interdependent people each with specialized knowledge in different areas yet having some overlap of core knowledge. Each TL will have specialized knowledge in the WP their team will be performing. This approach to staffing projects makes use of:

- Interdependent teams of key personnel and other personnel.
- A skill bank of consultants (external or other resources within the organization) available on an as-needed basis.
- Participative management techniques.
- A key person.

The synergism resulting from this approach is a win–win scenario for both the project organization and the customer, namely:

- The key person:
  - Provides the expert skills, other personnel do the actual work, resulting in lower labour costs.

- Provides the guidelines and monitors the schedule. Since the necessary functional skills may be different for different stages of the work, the make-up of the team may change accordingly.
- Expands the capabilities of the junior personnel adding to the corporate capabilities.
- Can work on several projects at the same time, working with and guiding a number of other persons.
- The project knowledge will remain if any one person leaves. This allows for:
  - Subsequent reassignment or promotion of personnel without impacting the project.
  - Other people to be rotated through the team.
- The team members tend to:
  - Pick up the tasks they most enjoy doing, so those tasks are done well.
  - Act like a quality circle[2] and continuously lower the cost of doing the work.

Conventional wisdom staffs a project based on the activities that need to be carried out. The seniority level for the job is decided, and a requisition posted. The manager interviews candidates, makes a selection and then introduces the candidate to the existing employees, and puts the now new employee to work. Some managers never achieve a team, just a collection of individuals, reporting to that manager. Some people are not prepared or willing to participate in a team, and sometimes personalities clash. Other approaches include:

- Choosing people based only on their knowledge, experience and competency as documented in their resumes.
- Asking candidates within the organization who they wish to work with/for, rather than just assigning them to slots on an organization chart (Kasser 1995).
- Posting (internal) the job openings in new projects as team positions, and asking employees to provide proposals as a team rather than just submitting individual resumes (DeMarco and Lister 1987). DeMarco and Lister also suggested that if management can find a team of successful people who want to work on a project, they should let them. Of course, the manager still has to guide or coordinate the team so that synergism takes place, to ensure the team is an effective one.
- Using teams of interdependent people each with specialized knowledge in different areas yet having some overlap of core knowledge but who have never worked together.

6.2 THE SYSTEMS APPROACH TO STAFFING A TEAM

The systems approach to staffing a team is as follows:

1. Breakout what needs to be done into specific activities in the three streams of activities (Section 2.5.2) as discussed in Section 6.2.1.

---

[2] A participative technique to manage and improve quality.

Successful Project Staffing

2. Determine the skills and competencies needed to perform those activities as discussed in Section 6.2.2.
3. Partition the activities and the necessary skills and competencies to perform the activities into job position descriptions as discussed in Section 6.2.3.
4. Staff each of the job positions using the methods as discussed in Sections 6.1.6.3 and 6.2.4.

If the staffing need is to add people to an existing team, managers using the systems approach allow the members of the team who will be working with the new employee to interview the candidates and make their recommendations as to the suitability of the candidate for joining the team.

6.2.1 Breaking Out What Needs to Be Done into Specific Activities in the Three Streams of Work

Determining what needs to be done on a project in the three streams of activities (Section 2.5.2) was performed as part of the specific planning (Section 5.7.1). The specific planning process documents the products that each WP is going to produce and the activities that produce the products.

6.2.2 Determining the Necessary Skills and Competencies to Do the Activities

Once the activities have been determined, the skills and competencies necessary to produce the product have to be identified. For example, if the activity is going to produce:

- Upgraded legacy software written in COBOL,[3] the programmers will need to understand COBOL as well as knowing how to design software.
- A test plan for this hardware or software, the person writing the test plan will need to know how to write test plans and understand the item being tested and the requirements it is supposed to meet.

Take the list of WPs and activities created in Section 6.2.1 and expand it into a matrix where the columns are associated with competence (Section 2.1.3) domain and discipline knowledge and skills and the rows are associated with the WPs, activities and estimated time to perform the activities as shown in the example Table 6.2. The perceived complexity of Table 6.2 is one reason why the simpler years of experience in Table 6.1 is frequently used.

6.2.3 Partitioning the Activities into Jobs

It seems that over the years every profession has organized itself into interdependent jobs. Many of these jobs are actually functions. For example, a baker bakes,

---

[3] Common Business Oriented Language.

TABLE 6.2
Partial Competences, Knowledge and Skills Matrix

WP	Activity	Competencies (CMM)	Knowledge – Java	Knowledge – Programming	Knowledge – Testing	Skills	Estimated Time
210M	Manage software creation and testing						2
310D	Create software components		X	X			20
320T	Test software components		X		X		5
410D	Integrate software components		X	X	X		20
420D	Integration test		X	X	X		5

a computer programmer programs, a tester tests, and so on. A typical job description includes much of the following:

- *The title of the position:* e.g. supervisor, engineer, tester, programmer, etc.
- *The duties:* the functions or the tasks performed by the person in the job.
- *The supervisor:* to which the person in the job reports.
- *The requirements:* the competencies, knowledge and skills required to do the job.
- *The salary range:* although once the position is filled the specific salary of the person in the job is not generally published.
- *The timeline:* the starting date and if applicable, the ending date of the position. If it's a permanent position there won't be an ending date. If it's a temporary position there will.

6.2.4 STAFFING EACH OF THE JOB POSITIONS

The activities in the WPs for the project now need to be assigned to jobs. One way of doing this is to use a spreadsheet such as the one shown in Table 6.3. The estimated time for the person or persons to perform an activity must be at least equal to the time allocated to the activity in Table 6.2. The activities are assigned to jobs according to the following rules.

1. The total number of hours assigned to job positions must equal the total number of hours estimated for the set of activities in the project.
2. The number of hours estimated in a week must be less than the project's approved number of hours for that week (Malotaux 2010).
3. A job can contain more than one activity as long as the number of hours is not maximized.

TABLE 6.3
Partial Activity Job Position Matrix

WP	Activity	Job Position	Estimated Time	Person for the Job
210M	Manage software creation and testing	TL	2	
310D	Create software components	Programmer 1	10	
310D		Programmer 2	10	
320T	Test software components	Tester 1	5	
410D	Integrate software components	Tester 1	5	
420D	Integration test	Tester 1	5	

4. The title of the job does not determine the activities performed in that job. For example, there is no reason why the job position of programmer could not contain testing activities as well as programming activities. Moreover, while the job title generally reflects what the person in the job does, there is no reason that it has to.[4]

An alternative way of assigning people to the positions is to work out the number of jobs based on the activities and let the team self-select which person does which activities and what their job position titles (Section 6.1.6.3).

6.3 THE PRESENTATION EXERCISE

The presentation exercise WPs (Section 5.11) provides an example of staffing a project. It was chosen as an example because students generally do team exercises in postgraduate classes and should be able to relate the examples to the team activity and apply them in the Multi-satellite Communications Switching System Replacement Project (MCSSRP) Exercise. Assume the team doing the presentation exercise consists of five people with the following skills:

1. *Andy:* who is good at testing and likes finding problems.
2. *Brian:* who is musical which is a skill that is totally irrelevant to this project.
3. *Lisa:* who is artistic and likes painting and math.
4. *Michael:* who is good at using computer software.
5. *Susan:* who is logical and good at planning.

The team gets together at the beginning of session 1 to decide on which of them will work in the different WPs. They planned the first draft sequentially from WP 00000 to WP 50000. The relevant information for WP 00000 is shown in Table 6.4. Since they

---
[4] For example, when I was working at Ford Aerospace, the job I was doing was systems engineering, but my job title was 'Engineering Specialist'.

TABLE 6.4
Staffing Profile for WP 00000

WP	Activity	Time	Andy	Brian	Lisa	Michael	Susan
00500	Coordinating WP 00000	6	1	2	1	1	1
04120	Understanding the grading criteria	5	1	1	1	1	1
04141	Creating the PP	1	0	0	0	0	1
04142	Creating presentation theme	1	0	0	1	0	0
09000	Holding the SPPCR	5	1	1	1	1	1
Total	WP 00000	18	3	4	4	3	4

didn't have any experience on how much time would actually take, they estimated the relative amount of time for each activity and insert that estimate into the WPs. They:

- Chose Brian as the team leader. Since Brian is the team leader, he will have to spend some time coordinating the others and that shows up in the coordinating WP (WP 00500). The team recognizes that Brian will have a heavier workload for the session as team leader, but each of them will take their turns as team leaders so over the semester the workload should level out.
- Realized that they have to work together to understand the grading criteria (WP 04120) and would all be participating in the session presentation planning completed review (WP 09000).
- Delegated to Susan the task of creating the PP (WP 04141) which includes the staffing, cost and schedule elements, because one of her skills was planning.
- Delegated the task of creating the presentation theme (WP 04142) to Lisa because she is artistic.
- Assigned Brian as the person responsible for each WP that the team performed as a team.

The team then moved on to WP 10000, the relevant information is shown in Table 6.5. They:

TABLE 6.5
Staffing Profile for WP 10000

WP	Activity	Time	Andy	Brian	Lisa	Michael	Susan
10500	Managing and top-level activities in the requirements state	3	1	2	0	0	0
14000	Determining the requirements for the exercise	10	2	2	2	2	2
14100	Creating the RTM	5	1	1	1	1	1
19000	Holding the PRR	5	1	1	1	1	1
19100	Reviewing the PP	5	1	1	1	1	1
Total	WP 10000	28	6	7	5	5	5

Successful Project Staffing

- Estimated that the management and top-level activities could be shared between Andy and Brian to level the workload since Andy had less work to do in WP 00000.
- Decided to spend twice as much time determining the requirements for the exercise to make sure they will complete and correct as they spent on any of the other activities.

The team then moved on to WP 20000. They started by staffing the low-level WPs and then continued with higher-level WPs. The relevant information is shown in Table 6.6. They:

- Delegated assigning the readings to Andy.
- Delegated designing the templates for each section of both presentations to Lisa to follow-on her presentation theme design.
- Noted that Lisa's workload was away above everybody's else's and decided to make changes later when they had completed staffing all the WPs.

TABLE 6.6
Staffing Profile for WP 20000

WP	Activity	Time	Andy	Brian	Lisa	Michael	Susan
20500	Coordinating WP 20000	3	0	3	0	0	0
21200	Assigning readings	4	1	0	1	1	1
22124	Reading enough of each reading	10	2	2	2	2	2
22220	Discussing readings as a group	10	2	2	2	2	2
22322	Voting on which readings to present	5	1	1	1	1	1
23100	Designing introduction template	1	0	0	1	0	0
23200	Designing lessons learned template	1	0	0	1	0	0
23300	Designing summary template	1	0	0	1	0	0
23405	Coordinating 23430	1	0	1	0	0	0
23431	Designing list of main points template	1	0	0	1	0	0
23432	Designing main point 1 template	1	0	0	1	0	0
23433	Designing main point 2 template	1	0	0	1	0	0
23434	Designing reflection and comments template	1	0	0	1	0	0
23436	Designing the why the reading was assigned template	1	0	0	1	0	0
23500	Designing reading 2 component templates	6	0	1	6	0	0
28200	Updating RTM and requirements specification	1	0	0	0	1	0
29000	Holding the presentation templates design review	5	1	1	1	1	1
Total	WP 20000	54	7	11	22	8	7

- Delegated updating the requirements traceability matrix (RTM) and requirements specification to Michael since he had less work assigned in WP 10000.

The team then moved on to WP 30000. They started by staffing the low-level WPs and then continued with higher-level WPs. The relevant information is shown in Table 6.7. They:

- Rolled up the coordinating activities for all the lower-level WPs into WP 30500.
- Cut down on Lisa's workload because of her heavy workload in the previous state of the project.

The team then moved on to WP 40000. The relevant information is shown in Table 6.8. They:

- Lessened Lisa's workload because of her heavy workload earlier in the project.

TABLE 6.7
Staffing Profile for WP 30000

WP	Activity	Time	Andy	Brian	Lisa	Michael	Susan
30500	Coordinating WP 30000	2	0	2	0	0	0
34200	Creating introduction	1	0	0	0	1	0
34300	Creating draft lessons learned	1	0	0	0	0	1
34400	Creating summary	1	0	0	0	1	0
34500	Creating Briefing 1	1	1	0	0	0	0
34541	Creating list of main points	1	1	0	0	0	0
34542	Creating main point 1	1	1	0	0	0	0
34543	Creating main point 2	1	1	0	0	0	0
34544	Creating reflection and comments	1	0	0	0	0	1
34545	Creating Compare content	1	0	0	0	0	1
34546	Creating why the reading was assigned	1	0	0	0	0	1
34580	Creating the sections of Briefing 1	1	1	0	0	0	0
34600	34600	8	0	5	0	0	3
38200	Updating RTM and requirements specification	1	0	0	0	1	0
39000	Holding the PIR	5	1	1	1	1	1
Total	WP 30000	27	6	8	1	4	8

Successful Project Staffing

TABLE 6.8
Staffing Profile for WP 40000

WP	Activity	Time	Andy	Brian	Lisa	Michael	Susan
40500	Coordinating WP 40000	2	0	2	0	0	0
44000	Completing final version	2		0	0	2	0
45000	Verifying presentation compliance	1	1	0	0	0	0
48000	Integrating complete team draft	2	0	0	0	0	2
48200	Updating RTM and requirements specification	1	0	0	1	0	0
49000	Holding the DRR	5	1	1	1	1	1
Total	WP 40000	13	2	3	2	3	3

The team then moved on to WP 50000. The relevant information is shown in Table 6.9.

Now that they had the staffing information and the time spent on the activities, they were ready to add cost and schedule information into the WPs.

They looked at the staffing summary in Table 6.10 and noticed that Lisa had the heaviest workload. Lisa felt that that was unfair, because she would also have a heavy workload when it was her turn to be the team leader. She pointed out that her heavy

TABLE 6.9
Staffing Profile for WP 50000

WP	Activity	Time	Andy	Brian	Lisa	Michael	Susan
50500	Coordinating WP 50000	1	0	1	0	0	0
56100	Receiving grade	5	1	1	1	1	1
56200	Uploading presentation	1	0	0	0	1	0
57000	Making presentation	1	1	0	0	0	0
59000	Holding the PPMR	5	1	1	1	1	1
Total	WP 50000	13	3	3	2	3	2

TABLE 6.10
Staffing Summary for Session 1 Presentation Exercise

Activity	Time	Andy	Brian	Lisa	Michael	Susan
WP 00000	18	3	4	4	3	4
WP 10000	28	6	7	5	5	5
WP 20000	54	7	11	22	8	7
WP 30000	27	6	8	1	4	8
WP 40000	13	2	3	2	3	3
WP 50000	13	3	3	2	3	2
Total	153	27	36	36	26	29

workload was a result of being allocated all the template design activities in WP 20000. Accordingly, the team went back to the spreadsheet parts of WP 20000 and reallocated some of the activities as shown in Table 6.11. This produced the revised staffing summary shown in Table 6.12 and made Lisa happy because she had the lowest workload in the exercise.

TABLE 6.11
Revised Staffing Profile for WP 20000

WP	Activity	Time	Andy	Brian	Lisa	Michael	Susan
20500	Coordinating 20000	3	0	3	0	0	0
21200	Assigning readings	4	1	0	1	1	1
22124	Reading enough of each reading	10	2	2	2	2	2
22220	Discussing readings as a group	10	2	2	2	2	2
22322	Voting on which readings to present	5	1	1	1	1	1
23100	Designing introduction template	1	0	0	1	0	0
23200	Designing lessons learned template	1	0	0	1	0	0
23300	Designing summary template	1	0	0	1	0	0
23405	Coordinating 23430	1	0	1	0	0	0
23431	Designing list of main points template	1	1	0	0	0	0
23432	Designing main point 1 template	1	0	0	1	0	0
23433	Designing main point 2 template	1	0	0	0	1	0
23434	Designing reflection and comments template	1	0	0	0	1	0
23436	Designing the why the reading was assigned template	1	1	0	0	0	0
23500	Designing Reading 2 component templates	7	2	1	2	2	0
28200	Updating RTM and requirements specification	1	0	0	0	1	0
29000	Holding the presentation templates design review	5	1	1	1	1	1
Total	WP 20000	55	11	11	14	12	7

TABLE 6.12
Revised Staffing Summary for Session 1 Presentation Exercise

Activity	Time	Andy	Brian	Lisa	Michael	Susan
WP 00000	18	3	4	4	3	4
WP 10000	28	6	7	5	5	5
WP 20000	55	11	11	14	12	7
WP 30000	27	6	8	1	4	8
WP 40000	13	2	3	2	3	3
WP 50000	13	3	3	2	3	2
Total	154	31	36	28	30	29

Successful Project Staffing

6.3.1 Achievements at This Point in Time

At this point in time the team has identified:

1. All the activities that need to be performed.
2. All the products that have to be produced.
3. The prerequisites for each activity.
4. The people who will perform the activities.

6.4 THE ENGAPOREAN MCSSRP EXERCISE

The background and context for the progressive session exercises is in Appendix 1. The purpose of the exercise is:

1. To practice using the template for a student exercise presentation (Kasser 2018: Section 14.7).
2. To add staffing information to the set of WPs developed in Exercise 5.12.

6.4.1 Requirements for the Exercise

The requirements for the exercise are to:

1. Formulate the problem posed by the exercise using the problem formulation template (Section 3.7.1).
2. Group work (activities) in WPs from previous session into staff positions.
3. Examine the resumes in Appendix 5.
4. Pick out people who fit staff positions.[5]
5. Add people resources to the relevant WPs.
6. Format presentation according to the applicable parts of the sample four-part template for a management review presentation (Section 5.4.2.2.1).
7. Prepare <5 min presentation that includes:
   1. The problem posed by the exercise formulated according to the problem formulation template (Section 3.7.1).
   2. Organization chart.
   3. People and why they were picked.
   4. List of positions by project lifecycle state.
   5. Staffing summary:
      1. By project state.
      2. By type of person.
      3. Explaining any non-level loading.
   6. Lessons learned, including what was noticed about the resumes.
   7. A compliance matrix (Kasser 2018: Section 9.5.2) for meeting the requirements for the exercise.

---

[5] While reviewing this manuscript Dr Xuan-Linh Tran suggested adding 'and providing reasons for selecting them'.

8. Number the slides.
9. Save presentation as file '6-teamname'.pptx.

6.5 SUMMARY

Projects are staffed by people, so this chapter discussed aspects of staffing projects, teams and distributing assignments to members of the project team. The chapter explained the need for high-performance teams, that people are not interchangeable; namely, one engineer does not necessarily equal another and ended with a discussion and example of how to staff a project.

REFERENCES

DeMarco, Tom, and Timothy Lister. 1987. *Peopleware.* New York: Dorset House Publishing Company.

Howe, Roger J. 1981. *Building Profits Through Organizational Change.* New York: AMACOM.

Kasser, Joseph Eli. 1995. *Applying Total Quality Management to Systems Engineering.* Boston, MA: Artech House.

Kasser, Joseph Eli. 2018. *Systems Thinker's Toolbox: Tools for Managing Complexity.* Boca Raton, FL: CRC Press.

Malotaux, Neils. 2010. Predictable Projects delivering the right result at the right time. EuSEC 2010 Stockholm: N R Malotaux Consultancy.

Tuckman, Bruce W. 1965. Developmental sequence in small groups. *Psychological Bulletin* no. 63 (6):384–399. doi:10.1037/h002210.

7 Successful Project Scheduling

This chapter explains scheduling and adhering to schedules, an important function of project management since exceeding schedule is one of the characteristics of failed projects. Consequently, it is important to get the right schedule for the project. Accordingly, this chapter explains:

1. Scheduling in Section 7.1.
2. Estimating the correct amount of time for activities in Section 7.2.
3. Levelling the workload in Section 7.3.
4. The critical path in Section 7.4.
5. The theory of constraints in Section 7.5.
6. The critical chain in Section 7.6.

7.1 SCHEDULING

The basic principles behind scheduling are:

- The total work is sum of all work.
- The total activities are sum of all activities.
- The total activities form a chain of activities.
- The activities may be performed in series or in parallel depending on the availability (or lack of availability) of resources and the prerequisites for the activity.

Schedules:

- Are visualizations of the Work Packages (WPs) from the *Temporal* HTP.
- Are shown in Gantt charts (Section 5.2.2).

Scheduling should begin once:

1. The products to be produced have been specified and the WPs that will produce those products have been determined.
2. The personnel have been assigned to the WPs (Section 6.2.4). The activities in the WPs may be performed in series or sometimes in parallel if the personnel are available. That is why scheduling doesn't begin until after the personnel have been assigned to all the activities.

211

WP		5	10	15	20	25
210M	Manage software creation and testing					
310D	Create software components					
320T	Test software components					
410D	Integrate software components					
420D	Integration test					

FIGURE 7.1 Corresponding initial schedule based on Tables 6.2 and 6.3.

One of the key points in scheduling is that generally the first estimate of the time to perform at an activity will be lower than the actual time it will take. Scotty, the chief engineer in the star ship *Enterprise* in the *Star Trek* series, always used to double his completion time estimate before he gave it to Captain Kirk or any other officer requesting such an estimate. Since he always finished ahead of schedule, his estimating approach got him a good reputation. Captain Kirk realized this and generally asked Scotty to complete the task in half the estimated time. The relationship between estimated completion times and actual completion times for U.S. Defense projects resulted in a universal fantasy factor of 33%, namely a schedule overrun of one-third (Augustine 1986).

When the information in Tables 6.2 and 6.3 are combined into a schedule, the corresponding Gantt chart is shown in Figure 7.1. The figure assumes that programmer 1 and programmer 2 are available and so the calendar time for the task will equal the actual time. If only one programmer is available, then WP 310D will take twice as long namely 20 time units. What this partial staffing plan and schedule does not show is what activities tester 1 will be performing in WP 310T. Presumably, tester 1 will be creating the software to test plans and software test procedures. Similarly, the management stream will be making sure that the activities are done according to the schedule and budget and that any special resources needed for those activities will be available as and when needed.

7.2 ESTIMATING THE CORRECT AMOUNT OF TIME FOR ACTIVITIES

While traditional schedule estimating is based on people working for a full week of 40h on the project, in reality, the net available time is about 26h or two-thirds due to people spending time on other activities (Malotaux 2010). If this factor is not considered at planning time, the project will take one-third longer than the planned schedule. Perceptions from the *Generic* HTP compare this statement to Augustine's fantasy factor that also results in underestimating project schedules by a third (Augustine 1986).

When estimating the time for activities, the estimate will generally have a tolerance of plus or minus some time. For example, the activity might be estimated as taking 16 days plus of minus 2 days. Accordingly, the shortest time is 14 days, the nominal time is 16 days and the longest time is 18 days. When showing the estimated time in schedules, always show the longest time. Similarly, when costing the activity always use the cost for the longest time.

Successful Project Scheduling

WP 10000	░░	░░					
WP 20000			░░	░░			
WP 30000			░░	░░	░░	░░	
WP 40000							░░

FIGURE 7.2   Initial schedule estimate.

7.3   LEVELLING THE WORKLOAD

The estimated scheduled calendar time for the project developed in Section 7.1 depended on the availability of programmers. If one programmer would be available, the programming task would take twice as long as it would take if two programmers were available. That was a simple example of levelling the workload. Levelling the workload is basically making sure that the number of people on a project remains the same for the parts of the project in which their skills, competencies and expertise are needed. Sometimes this can cover the entire project and sometimes it only covers part of the project depending on the type of project. For example, during the programming state of the software project, levelling the workload means that the number of programmers remain constant for the programming state.

The workload is levelled to keep a constant staffing profile and avoid conflicting use of scarce resources. Members of the project team, in general, cannot be hired and fired at will. Even if they could, they may not be available when needed. Temporary help from outside can only do so much work. Moreover, in large organizations it's not always possible to transfer people to other projects and in small organizations it generally is not possible. And over time can only be used up to a point.

Levelling begins once the first estimate of the schedule has been produced. Consider a typical planned project to develop a new product. The initial schedule is outlined in Figure 7.2 and the corresponding staffing profile is shown in Figure 7.3. The large increase in staff in WP 30000 is deemed to be unacceptable so the project manager has to level the workload.

When levelling a project workload, you adjust the start and finish dates of the activities so that fewer specific people work on WPs in series and schedule WPs in

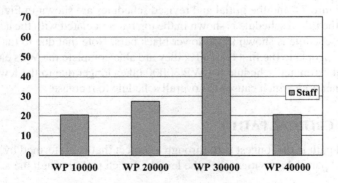

FIGURE 7.3   The project first estimated staffing profile.

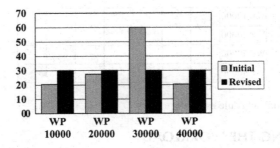

FIGURE 7.4 The project initial and revised estimated staffing profiles for the first four WPs.

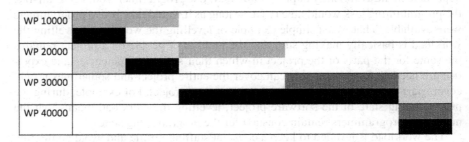

FIGURE 7.5 The project initial and revised estimated schedules for the first four WPs.

parallel. For example, in Section 7.1 the initial staffing profile was levelled by using one programmer to perform both programming tasks in series rather than using two programmers to perform the tasks in parallel. For example, if a WP is estimated to be completed in 2 weeks when staffed with six people, then when staffed with three people they should be able to complete the same WP in 4 weeks.[1] The result will be fewer people but a longer schedule.

Levelling the workload is an iterative activity of adjusting the schedule and the staffing to reach an acceptable staffing level and at the same time an acceptable schedule. For example, when the staffing profile in Figure 7.3 was levelled, the resulting staffing profile is 30 people. The initial and revised staffing profile estimates are shown in Figure 7.4 and the initial and revised schedules are shown in Figure 7.5. In Figure 7.5, the initial schedule is shown in the top bar associated with the activity and the revised schedule is shown in the lower black bars. Note that due to the ability to assign more people to the first two WPs, they are able complete the WPs earlier than in the initial estimated schedule but WP 30000 takes longer due to the fewer people being assigned to it which causes the overall schedule to increase.

7.4 THE CRITICAL PATH

The critical path is the longest path through a project that is performed by a number of serial and parallel activities when the length of each path through the activities is

---

[1] Up to a point (Brooks 1972).

Successful Project Scheduling

added up. There can be more than one critical path if different path lengths add up to identical longest path times. The critical path is usually seen on a program evaluation review technique (PERT) chart (Kasser 2018: Section 8.10) which shows the dependencies between the activities. A PERT chart consists of a number of circles[2] (milestones) joined by lines with an arrow at the destination end of the line as shown in Figures 5.6 and 7.6. To keep the drawings simple, the milestones represented by the circles are labelled as numbers and the activities are labelled by their starting and ending milestone numbers. Each activity is also labelled with its estimated time, for example et: 6 for activity 1-2. The time for each path through the network is shown in Table 7.1, which identifies path 1-2-4-6-7-8 as the shortest path taking 14 time units and path 1-3-4-5-8 is the longest path taking 15.5 time units.

The critical path is usually marked identified in some way often by colouring the critical path lines in red or using dashes in grey-scale drawings. It is more useful to colour the milestones along the critical path as well to really point out the critical path as shown in Figure 7.7.

Figure 7.7 and Table 7.1 provide two views of the same information.

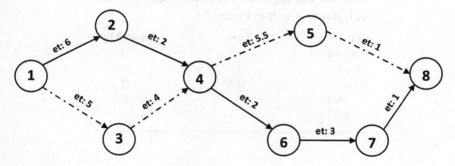

FIGURE 7.6 A partial PERT chart.

TABLE 7.1
Time for Each Path from Milestones 1 to 8 in Figure 7.6

Path	Times
1-2-4-5-8	14.5
1-3-4-5-8	15
1-3-4-5-8	15.5
1-3-4-6-7-8	15
1-2-4-6-7-8	14

---

[2] Or rectangles.

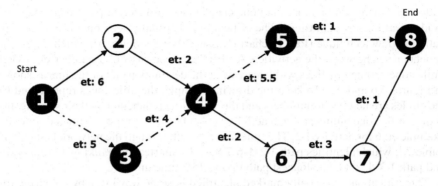

FIGURE 7.7 Highlighting the critical path.

TABLE 7.2
Original and Adjusted Times for Each Path from Milestones 1 to 8 in Figure 7.6

Path	Times	Adjusted Times
1-2-4-5-8	14.5	14
1-3-4-5-8	15	14.5
1-3-4-5-8	15.5	15
1-3-4-6-7-8	15	15
1-2-4-6-7-8	14	14

- The graphic view in the PERT chart in Figures 7.6 and 7.7 identifies the critical path.
- The tabular view in Table 7.1 can show you what happens if you decide to shorten the critical path, for example, if you decide to shorten the critical path by shortening either activity 4-5 or activity 5-8 by 0.5 time units, the resulting times for each pass will be as shown in Table 7.2 creating two critical paths.

7.4.1 Slack Time, Early and Late Finishes

Slack time is defined as the unused time in a path where there is more than one path between two milestones. For example, in Figure 7.6, there are two paths between milestones 1 and 4 and there are two paths between milestones 4 and 8. Table 7.3 shows:

- The critical path and the slack times for each path between milestones 1 and 4.
- Slack time as the difference (1) between the time to complete a path (8) and the time to complete the critical path (9).

Successful Project Scheduling

TABLE 7.3
Slack Times in Figure 7.6

	Path 1-2-4		Path 1-3-4	
Path	Activity	Time	Activity	Time
	1-2	6	1-3	5
	2-4	2	3-4	4
Total		8		9
Slack time		1		0

7.4.2 The Fallacy in the Use of Slack Time in Fixed Resource Situations

Conventional wisdom is that when there is slack time between milestones, the starting time for an activity can be adjusted as long as the ending time does not exceed the ending time of the critical path between those two milestones. For example, the schedule for the activities listed in Table 7.3 is shown in Figure 7.8. The schedule is planned with the assumption that activity 1-2 starts at the same time as activity 1-3 resulting in the slack time after activity 2-4. The overall schedule would still be met if:

- Activity 1-2 was delayed by 1 time unit. Since activity 2-4 depends on the completion of activity 1-2, the delay in the start of activity 1-2 will result in a delay in the start of activity 2-4.
- Activity 1-2 starts on time but activity 2-4 is delayed by 1 time unit.

However, there is no contingency time in the event of a delay in either or both of these activities due to the late start. Starting the activities at the planned starting time is when the benefits of the slack time are seen. Because should there be a delay of less than 1 time unit in that path, the overall schedule will not be affected.

The fallacy in the use of slack time is that if there is a resource that is needed by the critical path, the slack time cannot make use of it and so using slack time as a contingency is a good idea but care has to be taken (risk management) to make sure there is no conflict of resources by the activities in the parallel paths.

	1	2	3	4	5	6	7	8	9
1-3	■	■	■	■	■				
3-4						■	■	■	■
1-2	■	■	■	■	■	■			
2-4							■	■	

FIGURE 7.8 The corresponding schedule for Table 7.3.

7.4.3 Accuracies of the Estimated Schedules

The schedules on the WPs during the planning stage are only estimates of the time that the activity would take. Accordingly, there is a tolerance on that estimated time which means there is a shorter estimated time and a longer estimated time as well as the nominal estimated time. The accuracy of the estimate is stored in the WP (for example, item 6 in Table 5.1). The three estimates are often shown in PERT charts as labels on the activity lines, for example, activity 1-2 in Figure 7.6 would be labelled et: 4, 6, 9 to show that the shortest estimated time is 4 time units, the nominal estimated time is 6 time units and the longest estimated time is 9 time units. The task durations can also be presented in tabular format such as the one shown in Table 7.4. Use of this information:

1. Provides three different estimated times for a project, namely:
    1. *A success-oriented time estimate:* based on the assumption that every activity will complete at the shortest estimated times.
    2. *A nominal time estimate:* based on the assumption that every activity will complete at the nominal estimated times.
    3. *A worst-case time estimate:* based on the assumption that every activity will complete at the longest estimated times.

    Each of the project networks based on the three different estimated times for a project might have different critical paths which complicates shortening the critical path.
2. Identifies opportunities resulting from early finishes. In many instances, an early finish in one activity cannot be taken advantage because the subsequent activity is waiting for an input from a parallel activity. However, if the subsequent activity is not dependent on any other input and the resources are available then that subsequent process could have an early start. Accordingly, contingency planning needs to make sure that any necessary resources for the subsequent activity would be available, or could be made available, in the event of the opportunity for an early start.

TABLE 7.4
Task Duration Accuracies

Task	Expected	Early	Late	Uncertainty
1-2	6	4	9	5 (−2, +3)
1-3	5	3	7	4
2-4	2	2	3	1
3-4	4	4	5	1
4-5	5.5	4	5.5	1.5
4-6	2	2	2	0
5-8	1	0.5	2	1.5
6-7	3	2	5	3
7-8	1	1	2	1

Successful Project Scheduling

3. Assists in risk management because the factors causing the estimated late finish are generally anticipated identified risks. If these factors are considered as risks and prevented or mitigated during the planning process, the estimated late finish will be reduced.

7.5 THE THEORY OF CONSTRAINTS

The theory of constraints (Goldratt 1990) was originally conceptualized in a general management production situation. Given a production process that contains a number of processes such as the situation shown in Figure 7.9, management using the theory of constraints places inventorying buffers between the processes and monitors the throughput of the entire system taking corrective action in two situations to maintain the buffer levels in a steady-state condition:

1. *Excessive inventory build-up:* in which case the corrective action is to reduce the production to lower the inventory level to the predefined steady-state condition.
2. *Excessive inventory depletion:* in which case the corrective action is to increase production to increase the inventory level to the predefined steady-state condition.

Perceptions from the *Generic* HTP indicate that this approach seems to be just in time (JIT) production with a small excessive inventory.

7.6 THE CRITICAL CHAIN

The critical chain is a concept that comes out of the theory of constraints (Goldratt 1990). Given a production process that contains a number of processes such as the situation shown in Figure 7.9, *the critical chain is the path through the network of processes that has the lowest throughput.* Perceptions from the from the *Generic* HTP note that concept is similar to the critical path hence the name. Each process shown in Figure 7.9 is identified by a letter followed by a number which represents the maximum throughput of that process. Since there are relatively few well-organized process elements, it is easy to see that the lowest number is associated with a process;

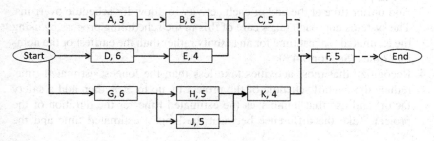

FIGURE 7.9 The critical chain.

hence, the critical chain is A-B-C-F and is identified by the dashed line. The critical chain can be used in several ways including:

- Deciding which process to upgrade to improve the overall throughput. You look along the critical chain, find the process with the lowest throughput and focus on improving that process.
- If the overall process does not need a throughput of more than three units (in this example), the fact that the other processes in the system have a higher throughput represent spare capacity. In the event that one of those processes needs to be changed, for example, machinery needs to be replaced, then acquiring production capacity with a throughput is in excess of three units does not contribute to increasing the overall throughput.
- If the overall process does need a throughput of more than three units (in this example), then it is obvious that the throughput of process A needs to be increased. If the throughput is increased to 4, there will then be three critical chains A-B-C-F due to A having an increased throughput of four, D-E-C-F due to E having a throughput of four and G-H-K-F due to K having a throughput of four. Any further improvement would mean increasing the throughput of A, E, G and K (currently at four) to the next limiting throughput value which is five (C, D, F, H and J).

7.6.1 Use of the Critical Chain in Project Management

The resources are people and the inventory buffers store time. The systems approach to project management incorporates the critical chain according to the following principles:

1. Make sure a person is not assigned to more than one activity at the same point in time. The systems approach takes care of this in the scheduling process.
2. Set due dates for completion of tasks not the amount of time allocated to an activity. The systems approach takes care of this in the planning process by focusing on products and working from completion milestone to starting milestone (the due date is the date of the milestone), starting each task as late as possible.
3. Add buffer time at the end of each activity to allow for schedule overruns. The systems approach takes care of this in the scheduling process by using the estimated longest time for an activity rather than the earliest or the nominal times (Section 7.4.3).
4. Recognize that most activities take less than the longest estimated time, reduce the overall schedule of the project by up to 50% then add a safety factor[3] and use that number as the estimated time for the duration of the project. Take the difference between the longest estimated time and the

---

[3] Sometimes known as a 'fudge' factor.

Successful Project Scheduling

revised number and add that time as a buffer at the end of the project. This means that if the project does overrun in the middle it still has a chance to complete on time. For example, if the longest time for an activity is 10 time units, reduce the activity by 50% to 5 time units and add a safety factor of 2 time units. The revised estimated time for the activity is now 7 time units. Accordingly, schedule the activity for 7 time units and add the remaining 3 time units into the end of project schedule buffer. The justification for using the 50% estimates is that half of the tasks will finish early and half will finish late, so that the variance over the course of the project should be zero.

The systems approach does not incorporate this principle in development projects for the following reasons:
1. The justification for the 50% ignores the flaw of averages (Savage 2009).
2. A visible buffer at the end of the project is an easy target for removal by upper management who don't understand critical chain principle.
3. The principle only seems to work if there is one due date for the project namely the final delivery date. For example, a project using the waterfall method will have a set of milestones with their own due dates. If the buffer is not built in to each state in the waterfall process, the milestones will slip and the project will appear to be delayed, even though the final milestone due date may be met.[4] It is better to build buffers in to the project before each milestone. This is done using the longest time estimate during the systems approach to scheduling (Section 7.2).
5. In the execution state of a project, personnel should focus on completing each activity as soon as possible. This requires motivation (Section 2.1.4).
6. In the execution state of a project, knowing that some activities will finish before and other activities will finish after their estimated due dates, there is no need to strictly hold each activity to its due date. The systems approach does this for all the activities within a state in the SDP (Section 5.4) by only comparing the estimated due dates for a milestone with the actual accomplished date namely the top-level activity in that state of the project. There is no need to formally compare the estimated due dates and actual dates of the lower-level activities at intermediate milestones.

7.7 THE PRESENTATION EXERCISE

Having decided which team members are going to staff which of the WPs (Section 5.12), the team now has to assign the schedule. They went back to the staffing estimate for WP 00000 shown in Table 6.4 and noted that each activity was allocated 1 time unit. They looked at the prerequisites for each of the activities in Table 5.19 and worked back from WP 09000 and realized that each task had to be performed in sequence as shown in the Gantt chart in Figure 7.10 in which each WP except WP 00500 takes the same amount of time namely 1 time unit. WP 00500 is the coordinating and management activity for WP 00000 and accordingly is active for the whole of

---

[4] Mind you, the project manager in the last state of the project will get a reputation for being a miracle worker when that late project actually completes on time.

Time units	1	2	3	4
WP 00500	■			
WP 04120	■			
WP 04141		■		
WP 04142		■	■	
WP 09000				■

FIGURE 7.10 The presentation exercise schedule for WP 00000.

WP 00000. When they added up the time to perform each of the WPs in series, they determined that the total calendar time for WP 00000 was 4 time units even though the team would be putting in 18 time units of work.

They moved on to the staffing estimate for WP 10000 shown in Table 6.5. Once more, starting from the last activity WP 19000 they looked at the prerequisites for each activity in Table 5.20 and realized that each activity except WP 19100 had to be performed in sequence. When they looked at the prerequisite for WP 19100 they noted that it could be performed at any time during WP 10000. Since it was an activity that they would all be involved in, and they were all working on other activities, they decided to place it as the last activity before WP 19000 so that they would be able to get the benefits of any knowledge gained from working on the other WPs. Accordingly, the schedule for WP 10000 is as shown in the Gantt chart in Figure 7.11. When they added up the time to perform each of the WPs in series, they determined that the total calendar time for WP 10000 was 5 time units even though the team would be putting in 28 time units of work.

They moved on to the revised staffing estimate for WP 20000 shown in Table 6.11. Once more, starting from the last activity WP 29000 they looked at the prerequisites for each activity in Table 5.22 and realized that WP 23100, WP 23200, WP 23300, WP 23400 and WP 23500 could be performed in parallel, and WP 28200 could be performed in parallel with other WPs as shown in the PERT chart in Figure 7.12. They looked at the staffing assignments, and created the schedule for WP 20000 shown in the Gantt chart in Figure 7.13. They realized that while WP 28200 could be performed in parallel with other WPs, it would best be done right before WP 29000. When they added up the calendar time to perform each of the WPs in series,

Time units	1	2	3	4	5
Cumulative	5	6	7	8	9
WP 10500	■				
WP 14000		■	■		
WP 14100			■		
WP 19100				■	
WP 19000					■

FIGURE 7.11 The presentation exercise schedule for WP 10000.

Successful Project Scheduling

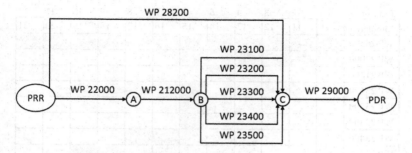

FIGURE 7.12 The PERT chart for WP 20000.

Time units	1	2	3	4	5	6	7	8	9	10	11	12	13	14
Cumulative	10	11	12	13	14	15	16	17	18	19	20	21	22	23
WP 20500														
WP 22000														
WP 21200														
WP 23100														
WP 23200														
WP 23300														
WP 23400														
WP 23500														
WP 28200														
WP 29000														

FIGURE 7.13 The presentation exercise schedule for WP 20000.

they determined that the total time for WP 20000 is 28 time units in 14 calendar time units.

Tables 6.6 and 6.11 do not contain any information about WP 22000 because that information is in the lower-level WP in Table 5.23. So, the assignment of 5 time units for WP 22000 is based on the information in Table 5.23. The information for WP 23400 is in Table 5.24, and they estimated the same amount of effort for WP 23500. When they looked at WP 231000, WP 23200 and WP 23300 they saw that they had to put them in series because they had all been assigned to Lisa. So, they looked at the other people's workload to see if they could assign somebody else to do them in parallel and decided that they could give one WP to Andy and one WP to Brian. This would lower Lisa's workload by 2 time units, raise Andy and Brian's by 1 time unit each, and save 2 calendar time units as shown in the second revised schedule for WP 20000 in Figure 7.14. Lisa noticed that her total workload on the project had dropped by 2 more time units, so she was even happier.[5] The second revised staffing estimate is shown in Table 7.5. These schedule and staffing adjustment activities demonstrate that staffing and scheduling can be an iterative process of levelling the workload (Section 7.3) and adjusting the length of the schedule (Section 9.2.2).

---

[5] She also realized that since this wasn't the only presentation they would make in the class, she would probably make up those 2 time units later in the semester, and probably add some more time.

Time Units	1	2	3	4	5	6	7	8	9	10	11	12
Cumulative	10	11	12	13	14	15	16	17	18	19	20	21
WP 20500												
WP 22000												
WP 21200												
WP 23100												
WP 23200												
WP 23300												
WP 23400												
WP 23500												
WP 28200												
WP 29000												

FIGURE 7.14 The second revised schedule for WP 20000.

TABLE 7.5
Second Revised Staffing Summary for Session 1 Presentation Exercise

Activity	Time	Andy	Brian	Lisa	Michael	Susan
WP 00000	18	3	4	4	3	4
WP 10000	28	6	7	5	5	5
WP 20000	55	12	12	12	12	7
WP 30000	27	6	8	1	4	8
WP 40000	13	2	3	2	3	3
WP 50000	13	3	3	2	3	2
Total	154	32	37	26	30	29

They moved on to the staffing estimate for WP 30000 shown in Table 6.7. Once more, starting from the last activity WP 29000 shown in Table 5.25 they looked at the prerequisites for each activity and ended up with the PERT chart shown in Figure 7.15.

The PERT chart shows that WP 34500 and WP 34600 can be performed in parallel. However, the team lacked the resources to perform them in parallel in the minimum time; in the previous state of the project, they chose to perform WP 23400 and WP 23500 in series for the same reason. This time they chose to perform WP 34500 and WP 34600 in parallel but in the same 10 time units. This choice gave them some

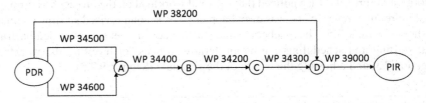

FIGURE 7.15 The PERT chart for WP 30000.

Successful Project Scheduling

Time units	1	2	3	4	5	6	7	8	9	10	11	12	13	14	15
Cumulative	22	23	24	25	26	27	28	29	30	31	32	33	34	35	36
WP 30500															
WP 34500															
WP 34600															
WP 34400															
WP 34200															
WP 34300															
WP 38200															
WP 39000															

FIGURE 7.16 The presentation exercise schedule for WP 30000.

flexibility so that if they took longer on one of the WPs they could make up the time by shortening the time to perform another WP. Alternatively, if they finished early on one of the WPs, they had the flexibility to spend more time on the later WPs. Accordingly, the schedule for WP 30000 is as shown in Figure 7.16.

They moved on to the staffing estimate for WP 40000 shown in Table 6.8. Once more, starting from the last activity WP 49000 shown in Figure 7.17 they looked at the prerequisites for each activity and noticed that each activity had to be performed in sequence except for WP 482000 which had no prerequisites in that state. Accordingly, they chose to put it right before the end of state review WP 49000 in the same way they had allocated WP 38200 and WP 28200. Another benefit of doing it this way is should they run over time on the earlier WPs they can shorten the schedule by doing WP 48200 in parallel with WP 34300 or any other earlier WP. The resulting schedule for WP 40000 is shown in Figure 7.17.

They moved on to the staffing estimate for WP 50000 shown in Table 6.9. Once more, starting from the last activity WP 59000 shown in Table 5.28 they looked at the prerequisites for each activity and noticed that each activity had to be performed in sequence. Accordingly, the schedule for WP 50000 as shown in Figure 7.18 which brought the total schedule for the project to 47 calendar time units even though as Table 7.5 shows they would actually be putting in 154 time units of work. Since they had more than 47 calendar time units to do the exercise, the estimate was feasible and accepted by the members of the team.

Since some of the WPs are assigned to a single person, there will be time when their teammates will be waiting for their input. This is slack time (Section 7.4.1) as far as they are concerned.

Time units	1	2	3	4	5	6	7
Cumulative	37	38	39	40	41	42	43
WP 40500							
WP 48000							
WP 44000							
WP 45000							
WP 48200							
WP 49000							

FIGURE 7.17 The presentation exercise schedule for WP 40000.

Time units	1	2	3	4
Cumulative	44	45	46	47
WP 50500				
WP 56200				
WP 57000				
WP 56100				
WP 59000				

FIGURE 7.18 The presentation exercise schedule for WP 50000.

7.7.1 ACHIEVEMENTS AT THIS POINT IN TIME

At this point in time the team has identified:

1. All the activities that need to be performed.
2. All the products that have to be produced.
3. The prerequisites for each activity.
4. The people who will perform the activities.
5. The estimated period of performance (time to complete).

7.8 THE ENGAPOREAN MCSSRP EXERCISE

The background and context for the progressive session exercises is in Appendix 1. The purpose of the exercise is:

1. To practice using the template for a student exercise presentation (Kasser 2018: Section 14.7).
2. To develop and add scheduling estimates to activities in the WPs produced in Exercise 6.3. The exercise is limited to a nominal schedule for each activity.

7.8.1 THE REQUIREMENTS FOR THE EXERCISE

The requirements for the exercise are to:

1. Formulate the problem posed by the exercise using the problem formulation template (Section 3.7.1).
2. Develop and add scheduling estimates to activities in the WPs produced in Exercise 6.3.
3. Format presentation according to the applicable parts of the sample four-part template for a management review presentation (Section 5.4.2.2.1).
4. Prepare <5 min presentation that includes:
   1. The problem posed by the exercise formulated according to the problem formulation template (Section 3.7.1).
   2. A Gantt chart (Kasser 2018: Section 8.4) for the total project schedule.

3. Gantt charts for the schedules of each state of the SDP between the milestones.
4. At least one PERT chart (Kasser 2018: Section 8.10) showing the critical path(s) where there is more than one path between the milestones.
5. Lessons learnt.
6. A compliance matrix (Kasser 2018: Section 9.5.2) for meeting the requirements for the exercise.
5. Number the slides.
6. Save presentation as file '7-teamname'.pptx.

7.9 SUMMARY

Scheduling and adhering to schedules is an important function of project management, exceeding schedule is one of the characteristics of failed projects. Consequently, it is important to get the right schedule for the project. Accordingly, this chapter explained how to create the project network, how to create a project schedule, the critical path and its importance and the fallacy in slack time in fixed resource situations.

REFERENCES

Augustine, Norman R. 1986. *Augustine's Laws*. New York: Viking Penguin Inc.
Brooks, Fred. 1972. *The Mythical Man-Month*. Boston, MA: Addison-Wesley Publishing Company.
Goldratt, Eliyahu M. 1990. *Theory of Constraints*. Great Barrington, MA: North River Press.
Kasser, Joseph Eli. 2018. *Systems Thinker's Toolbox: Tools for Managing Complexity*. Boca Raton, FL: CRC Press.
Malotaux, Neils. 2010. Predictable Projects delivering the right result at the right time. EuSEC 2010 Stockholm: N R Malotaux Consultancy.
Savage, Sam L. 2009. *The Flaw of Averages*. Hoboken, NJ: John Wiley and Sons.

8 Successful Project Cost Estimating

This chapter explains that there is always a cost to develop a product. Project managers need to estimate costs before a project begins to determine if the project is affordable and/or will deliver the required return on investment (ROI). Accordingly, estimating and adhering to costs is an important function of project management.

Exceeding cost is one of the characteristics of failed projects. Consequently, it is important to get the right cost estimate for the project. Project managers also need to re-estimate the costs once the project commences and more detail is known about the seven interdependent P's of a project. However, before trying to estimate costs the project manager needs to understand the nature of costs, where they come from and what they represent. Accordingly, this chapter explains:

1. The three axes of cost-effective projects in Section 8.1.
2. Why project managers estimate costs in Section 8.2.
3. Methods for estimating project costs in Section 8.3.
4. Accuracies of cost estimates in Section 8.4.
5. Categories of costs in Section 8.5.
6. Ways of controlling costs in Section 8.6.
7. Types of cost contracts in Section 8.7.

8.1 THE THREE AXES OF COST-EFFECTIVE PROJECTS

The interdependent P's affect the cost of the project as shown in Figure 8.1, the chart of a generic process to produce a product. The process or timeline lies on the X-axis and the product lies on the Y-axis, and the organization is represented by the Z-axis which is not shown. The generic simplified process or timeline has a starting milestone (S), a checkpoint or milestone (M)[1] and a terminating milestone (E). The cost of producing a product (C_p) to meet specifications is a function (f) of all three axes and may be represented as:

$$C_p = f(aX) + f(bY) + f(cZ)$$

where a, b and c are parameters for an axis in the specific organization.[2] Note the baseline from S to E on the Y-axis slopes upwards showing the cost of the product on the Y-axis.

---

[1] Which represents all the project milestones.
[2] The difficulty in trying to separate the parameters is that they seem to be very tightly coupled so that, for example, parameter 'a' may contain elements from the Y- and Z-axis.

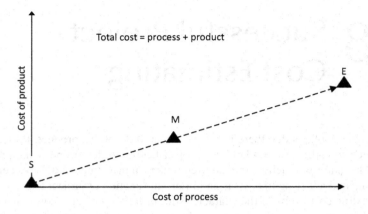

FIGURE 8.1 The generic cost of a process and its product.

When the project begins, the line from S to M to T is estimated so it is represented as a dashed line as shown in the figure. It is a straight line, because the assumption is everything will go according to plan. When the progress of the project is measured, deviations might be seen. One of the ways project progress is measured is by measuring the costs incurred by the project at time intervals. These costs represent work performed, material consumed and supplies procured. When these actual costs are compared against the estimated costs, a variance is often seen. This variance may be due to:

- Errors in estimating the costs before the project begins.
- Some factors in the implementation of the project.
- Both of the above.
- Other causes.

The process contains three categories of costs, the costs of doing the work and preventing defects (risk management and prevention) (C_m), and the costs to complete after the check point (C_c). Each category may contain different elements such as fixed and variable costs depending on the specific situation. The cost of the process (C_p) is the sum of the costs in each category, namely:

$$C_p = C_m + C_c.$$

If the costs of doing the work prior to the milestone escalate above the planned costs, the angle of the line between point S and point M shown in Figure 8.1 will be greater than the angle in shown in Figure 8.1. If that is projected to time M and time E the cost of the product will increase as shown in Figure 8.2 by the cost to produce the defect (C_m). The cost to complete will also be increased by the cost of the activities to fix the defects produced before the milestone. The extra costs incurred can be represented as the difference in the length of the lines in Figures 8.1 and 8.2. From this perspective the purpose of the milestones in the project is to ensure that if the line from S to M is deviating from the baseline to increase the cost, activities are taken to adjust the direction of

Successful Project Cost Estimating

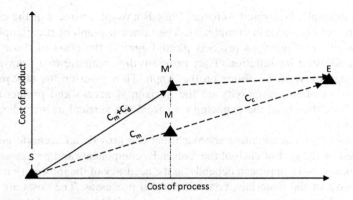

FIGURE 8.2 The cost of a process that is exceeding estimated costs.

the line back to the baseline from each milestone, preventing additional costs to produce defects and repair them. If the deviation can be detected before the milestone and the correction activities initiated at the point of detection of the deviation, the effect of the deviation will be minimized, lowering the additional costs of the deviation.

Estimating the costs of a project is initially performed in the project initialization and project planning states, a time when information is limited and may not be accurate. As the project progresses, the cost estimates are refined based on work that's been done (those costs are known) and more information about the future work. By the end of the project, the cost estimate is extremely accurate because it's based on the work that has been done and those costs are known. Cost estimates are refined during the project to make sure that the project will not exceed the budget.

8.2 THE REASON FOR ESTIMATING PROJECT COSTS

We estimate costs for a number of reasons. For example, we estimate the cost of:

- A project to see if it's affordable, namely there are sufficient funds in the budget to develop or purchase it.
- Developing a new product to see if the ROI is desirable. Namely, we must sell a product at a large enough profit to justify spending the money to acquire[3] the product.
- A change request to determine if it is affordable.

8.3 METHODS FOR ESTIMATING PROJECT COSTS

There are a number of methods for estimating project costs including:

- *Analogy:* a qualitative approach that provides a rough estimate based on perspectives from the *Generic/Temporal* HTPs to make comparisons with similar products developed in the same or in similar organizations.

---
[3] Develop or purchase.

For example, Federated Aerospace needs a rough estimate of the cost of a project of a certain complexity. They draw a graph of the complexity of number of previous projects plotted against the costs of those projects adjusted for inflation. They program the computer to a draw a line through most of the points on the graph. They position the new project in terms of its complexity on the horizontal access and project up to the line, then read the expected cost from the vertical axis as shown in Figure 8.3.

- *Bottom-Up:* a quantitative approach that can provide an accurate number based on the cost of each of the materials, components and processes. The accuracy of the approach depends on the accuracy of the information about the cost of the materials, components and processes. The costs are categorized in various ways such as fixed and variable costs, and the costs of prevention, appraisal and failure. This is the traditional approach used to determine the cost of each of the staffed work packages (WP) developed in the staffing session exercise in Chapter 7.
- *Expert:* a qualitative approach that provides a rough estimate based on accumulated knowledge and experience.
- *Parametric methods:* a quantitative approach similar to the analogy but generally providing a more accurate estimate suitable for use with projects with a high degree of repeatability. The proposed product and process are modelled using computer software to produce an estimate based on information from various sources appropriate to the project including manufacturers and suppliers, industry and professional associations and technology experts and cost analysers.
- *Conceptual design to cost (CDTC):* a systemic and systematic hybrid approach generally for use in new product development situations discussed in Section 8.3.1.

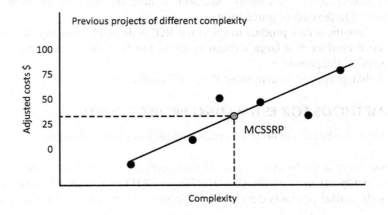

FIGURE 8.3 The *Generic/Temporal* costing approach (analogy method).

Successful Project Cost Estimating

8.3.1 Conceptual Design to Cost

CDTC (Hari, Shoval, and Kasser 2008):

- Helps the project manager and system engineer to quickly evaluate the major manufacturing costs of candidate solution concepts using sketchy information.
- Distinguishes the difference between cost and price where:
  - *Cost* is the amount of money that must be spent to realize (develop or acquire) a product or service.
  - *Price* is the amount of money that a customer is willing to pay for the product or service, namely what that product or service can be sold for.
  - *Profit or ROI* is the difference between the cost and the price.
- Is a systemic and systematic hybrid methodology based on using a mixture of the:
  - Design to cost (DTC) methodology (Michaels and Wood 1989).
  - Pareto principle also known as the 80-20 rule.
  - Knowledge and experience of the organizational experts.
  - Risk associated with the availability of technology (Section 10.4).
- Is not used as a stand-alone process; CDTC is built-in to the project management methodology. However, for educational purposes, this section discusses CDTC as a process containing the following stages (as adapted for project management):
  - Identifying and prioritize the needed capability discussed in Section 8.3.1.2.
  - Benchmarking the capability discussed in Section 8.3.1.3.
  - Determining the price target for the product discussed in Section 8.3.1.4.
  - Estimating the cost of the product discussed in Section 8.3.1.5.
  - Adjusting the cost estimate discussed in Section 8.3.1.6.

8.3.1.1 Benefits of Using CDTC

The contribution of CDTC is to increase the accuracy of cost estimates during the conceptual design stage of a new product. CDTC was developed in design laboratory experiments about 20 years ago. In these experiments, 29 new product development teams (129 members) were instructed to apply the methodology as part of their customized development process known as the integrated customer-driven conceptual design method (ICDM) (Hari and Weiss 1996) which has been evaluated, tested and successfully implemented in the Israeli high-tech industry (Hari, Weiss, and Zonnenshain 2004). Results included:

- The probability of achieving design target cost without using CDTC was relatively low (30% of the teams failed to achieve the target cost). Teams that used CDTC improved their total cost estimates by an average of 23%.
- The type of metrics used for the design quality measurement (Hari, Weiss, and Zonnenshain 2001) and their relative weighting.

- The success rate of each metrics type and of the whole concept, the improvement in the design quality during the conceptual design process.
- A paradigm shift in the use of costs during the design states of the system development process (SDP). The traditional approach estimates costs when design details are known. The CDTC approach estimates costs in the project definition and conceptual design stages of the SDP, and uses estimated costs as selection criteria when deciding between alternative conceptual solutions.

8.3.1.2 Identifying and Prioritizing the Needed Capability

The project design team elicits and elucidates the needs through discussions with the customer perceiving the situation from the *Operational* and *Functional* HTPs and from the *Generic* HTP to inherit capability from similar situations. Once the needs are identified, the team prioritizes the needs. The team discusses each capability with the customer prioritizing and sorting the capability based on the customer's willingness to pay for the capability into the following three willingness to pay/priority groups:

1. *Essential:* high priority; the major reasons the customer is willing to pay for the product.
2. *Beneficial:* medium priority; some needs that the customer may be willing to pay for (a relatively small percentage of the total price of the essential capability).
3. *Nice to have:* low priority; capability that the customer will be happy to have but is not willing to pay for.

The team uses the appropriate prioritization tool depending on the number of needs (Kasser 2018: Chapter 4).

8.3.1.3 Benchmarking the Capability

The team benchmark the price of the capability in similar products to determine an approximate target price for that capability (the analogy method (Section 8.3) is a suitable method). The team totals the estimated price of each of the needs groups.

8.3.1.4 Determining the Target Price for the Product

The target price for the product is:

- The price the customer is willing to pay for the product.
- Determined jointly via discussions between the project team and the customer.

The discussion starts with the 'essential' needs:

- If the customer is not willing to pay the price of all the 'essential' needs, then some of the lower-priority essential needs are removed until the discussion agrees to a set of essential needs the customer will settle for at a price that the customer is willing to pay. This agreement sets the target price.

Successful Project Cost Estimating

- If the customer is willing to pay more than the price for all the 'essential' needs, the discussion moves on to the 'beneficial' needs.

When the discussion reaches the 'beneficial' needs:

- If the customer is not willing to pay the price for all the 'beneficial' needs, then some of the lower-priority 'beneficial' needs are removed until the discussion agrees to a mixture of 'essential' and 'beneficial' needs the customer will settle for at a price that the customer is willing to pay. This agreement sets the target price.
- If the customer is willing to pay more than the price of all the 'beneficial' needs, the discussion moves on to the 'nice to have' needs.

When the discussion reaches the 'nice to have' needs:

- If the customer is not willing to pay the price for all the 'nice to have' needs, then some of the lower-priority 'nice to have' needs are removed until the discussion agrees to a set of 'essential', 'beneficial' and 'nice to have' needs the customer will settle for at a price that customer is willing to pay. This agreement sets the target price.
- If the customer is willing to pay more than the price for all the 'nice to have' needs, the discussion agrees to the whole set of needs at a price the customer is willing to pay.

The project manager then has the price target for the product; namely, the price the customer is willing to pay and a set of prioritized needs which can be converted to a set of prioritized requirements. The rest of the process determines if the product can be realized within the target price.

A benefit of this approach is that it eliminates low-priority high-cost needs/requirements very early in the product lifecycle.

8.3.1.5 Estimating the Cost of the Product

The project manager estimates the cost is using modified bottom-up approach. The traditional 'bottom-up' approach estimates the cost of every WP in the project and then adds them all up to produce the final cost. CDTC recognizes that in the project initialization state of a project, the estimates are approximate and so it makes use of the Pareto principle to simplify and reduce the time taken to create the estimates.

The project manager identifies the nine probable most expensive WPs based on factors such as staff and materiel, and then estimates the costs of those WPs. This number is assumed to be 80% of the total costs of the project. According to the Pareto principle, the remaining WPs will contribute the remaining 20% of the cost. So, when the estimated 80% and the remaining 20% are summed, the result is the estimated cost of the total project.

As the project progresses and more information is known about what each WP will be doing and producing, the cost estimates will be refined as part of the DTC activities in the remaining states of the SDP.

8.3.1.6 Adjusting the Cost Estimate

If the estimated cost added to the expected ROI for the project is less than equal to the target price, there is no need to adjust the cost estimate. If the target price cannot be met, then the organization needs to decide if they really want the project. If the organization does want the project, then the cost estimate or the expected ROI or both need to be lowered.

Perceived from the *Generic* HTP, the approach to adjusting the cost estimate is similar to the approach used to shorten the critical path (Section 7.4). Starting with the most expensive WP, each of the cost elements are examined to see if the cost can be reduced by various approaches such as:

- Reducing the cost of doing the planned work by using lower-paid staff.
- Reducing the time spent on the work (schedule) by using higher-paid staff with more expertise. The trade-off here is that the higher-paid staff will be paid for fewer hours than the lower-paid staff if the schedule wasn't shortened.
- Removing some of the work activities/packages and accepting a greater risk of not meeting the cost and schedule targets.
- Using technology to reduce the cost of doing the work even though there will be an initial expense in acquiring the technology.
- Using cheaper components or materials without sacrificing conformance to specifications. Namely avoiding over specifying unneeded attributes.
- Purchasing commercial-off-the-shelf (COTS) instead of developing needed items in-house (Section 8.6.7).

8.3.2 Lifecycle Costing

A lifecycle cost estimate (LCCE):

- Estimates the total cost of a conceptual solution over the entire system or product lifecycle as part of an analysis of alternatives (AoA) study (OAS 2013) required for U.S. government acquisitions in the needs identification state of the SDP (Section 5.4.1).
- Includes costs incurred for research and development, investment, operations and support and end of life disposal as well as the procurement cost of the system.
- The LCCE can provide the following insights to inform acquisition decisions:
  - Total cost to the system acquirer of developing, procuring, fielding and sustaining operations for each alternative for its expected lifecycle.
  - The annual breakdown of costs expected for the alternative by funding categories.
  - Trade-off analysis/cost as an independent variable (CAIV) (Section 8.3.3) to identify solutions that, given a fixed cost, provide the greatest (may be less than 100% solution) capability (Section 8.3.3).

Successful Project Cost Estimating

- The cost drivers of alternatives (i.e. those items having the greatest impact on the overall costs).
- Cost of enablers and operational support for the capability being evaluated.
- Estimated lifecycle costs that represent what is necessary to deliver the predicted operational effectiveness for each alternative.
- Projected costs associated with various operational, basing, fielding or programmatic decisions expected for each alternative evaluated.
- Uncertainty and risk associated with the cost estimate.

The cost of each alternative (baseline and all proposed alternatives) must be evaluated for the same lifecycle time frame. The time frame should start at the end of the AoA and continue until the end of the lifecycle as defined in the study (e.g. 20-year lifecycle). This allows for a fair comparison of each conceptual alternative and may require service-life extension efforts for other alternatives (including the baseline) with expected shorter useful lives or the calculation of residual values for alternatives that may continue to provide capability past the study cut-off dates. Categories of AoA costs include:

- *Sunk costs:* those that either already occurred or will be incurred before the AoA can inform any decisions on their expenditure.
- *Research and development (R&D) costs:* costs of all R&D phases, including advanced technology demonstration (including concept development), technology development and engineering and manufacturing development, are included in this cost element.
- *Investment costs:* (low-rate initial production, full-rate production and fielding) includes the cost of procuring the prime mission equipment and its support. This includes training, data, initial spares, support equipment, integration, pre-planned product improvement items and construction; the cost of acquisition, construction or modification of facilities (barracks, mess halls, maintenance bays, hangers, training facilities, etc.) necessary to accommodate an alternative, as well as all related procurement (including transportation, training, support equipment, etc.).
- *Operations and maintenance (O&M) Costs:* those programme costs necessary to operate, maintain and support system capability through its operational life. These costs include all direct and indirect elements of a defence programme and encompass costs for personnel, consumable and repairable materiel, and all appropriate levels of maintenance, facilities and sustaining investment.
- *Disposal costs:*
  - Represent the cost of removing excess or surplus property or materiel from the inventory.
  - May include costs of detoxification, divestiture, demolition, redistribution, transfer, donation, sales, salvage, destruction or long-term storage.
  - May also reflect the costs of hazardous waste disposition of storage and environmental cleanup.

- May occur during any phase of the acquisition cycle. If during development or testing some form of environmentally unsafe materials are created, the costs to dispose of those materials are captured here.
- *Pre-fielding costs:* associated with maintaining the existing capabilities being analyzed in the AoA until a specific alternative can be fielded to provide them. There may be costs associated with ramp-up of new alternatives and a corresponding ramp-down of existing baseline capabilities.

8.3.3 Cost as an Independent Variable

In the DOD, CAIV is a technique for varying the expected cost of the alternative(s) and changing performance and schedule to determine the impact of funding limitations. This technique allows the cost team to perform 'what if' analysis with funding levels even before such levels have been determined or included in budgets. It is good practice for the cost team to fluctuate the estimate they have developed by decrements (for example, 10% and 25%) and then, with the alternatives, the development team derives the number of units, performance characteristics and schedules that such reduced funding levels would represent. It is likely this effort will identify a point at which it is not advisable to proceed with one or more alternatives.

In commercial terms, CAIV seems to be a method of matching the amount of functionality that could be achieved for the amount the customer is willing to pay (Denzler and Kasser 1995) and is similar to CDTC (Section 8.3.1).

8.4 ACCURACIES OF COST ESTIMATES

Factor affecting the accuracies of cost estimates include:

- Creating rough estimates as a response to customer's haste.
- Extending the project timeline; the longer the timeline the lower the accuracy.
- Failing to identify risks and prevent/mitigate them will result in additional unplanned work.
- Lacking experience with similar projects.
- Not fully understanding the scope of the work it takes to produce the products in the WPs.
- Not updating cost estimates after change requests affecting the scope of work have been accepted.
- Scheduling activities assuming that personnel will work 100% of the time when a more likely figure is 80%. For example, if an activity is estimated as taking 80 hours, it should be scheduled as 100 person-hours not 80.

Each nominal estimate shall constitute a value between the highest and the lowest expected amount. The highest and lowest amounts may be computed, approximated as ±10% or developed statistically. The accuracy of the estimate is stored in the WP (e.g. in item 6 in Table 5.1).

Successful Project Cost Estimating

8.5 CATEGORIES OF COSTS

While costs can be categorized in many different ways, the most useful ones are as follows:

- *Fixed and variable costs* where:
  - *Fixed costs:* costs of equipment and materiel that is not consumed by the project. A computer is a fixed cost because that cost is incurred as long as the computer remains for the project irrespective of how much use it has. Similarly, a software package is a fixed cost for the same reason.
  - *Variable costs:* costs of items consumed and resources are used by the project. For example, computer paper is a variable cost because the cost depends on how much paper is used.
- *Costs of prevention, production, appraisal and failure* where:
  - *Prevention costs:* costs of activities that are supposed to prevent defects in the design and development of the system (Crosby 1979).
  - *Production costs:* costs of activities that produce the product and the defects in the design and the development of the system.
  - *Appraisal costs:* costs of the activities that determine whether the system being engineered conforms to its requirements (Crosby 1979).
  - *Failure costs:* costs of the activities associated with the products not meeting their specifications (Crosby 1979).

These costs are incurred in the different activities in the different states of the SDP as shown in Table 8.1. Perceived from the *Quantitative* HTP, the lowest overall cost would be achieved from a mix of prevention, production, appraisal and failure costs. However, when making the decision as to allow a failure in a very small percentage of instances, perceptions from the *Big Picture* HTP indicate the cost of failure may not only affect the project or product, but it may also negatively affect the perception of the quality of the product and the organization producing it, something that has to be considered.

TABLE 8.1
Cost Categories

Cost Category	Mainly in State in SDP	Project Activity
Prevention	Needs identification, System requirements and system design	Planning, analysing failure modes, etc.
Production	Subsystem construction	Designing, constructing, coding, etc.
Appraisal	System integration and system testing	Prototyping, inspecting, and testing, etc.
Failure	Operations and maintenance	Redesigning, reworking, scraping, etc.

8.6 WAYS OF CONTROLLING COSTS

The effect of a cost increase was shown graphically in Figure 8.2. Most of the ways of controlling costs in the systemic and systematic approach to project management are based on performing risk management by viewing undesirable outcomes from the *Continuum* HTP, identifying the consequences of those undesirable outcomes in terms of cost increases and other attributes and then preventing or mitigating those outcomes. Consider the following undesirable outcomes of:

1. Failing to communicate the vision discussed in Section 8.6.1.
2. Failing to understand the customer's real requirements discussed in Section 8.6.2.
3. Failing to plan ahead discussed in Section 8.6.3.
4. Failing to control changes discussed in Section 8.6.4.
5. Failing to apply lessons learned from the past discussed in Section 8.6.5.
6. Failing to document the reasons for decisions discussed in Section 8.6.6.
7. Failing to maximize use of COTS discussed in Section 8.6.7.

8.6.1 FAILING TO COMMUNICATE THE VISION

The major consequences of failing to communicate and maintain the vision include overruns in cost and schedule, complete project failure and cancellation or initial failure then recovery. Without a clear vision, different sections of the project proceed in different directions at different rates. Figure 8.4 illustrates the communications

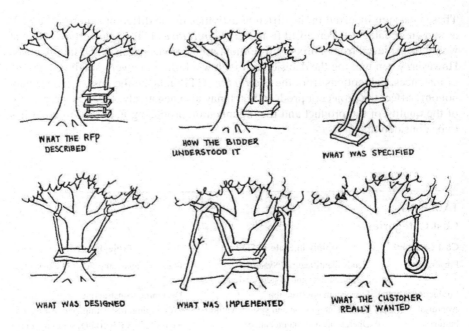

FIGURE 8.4 The effect of failing to communicate the vision.

Successful Project Cost Estimating

aspect of the situation.[4] If there was a clear vision, all the trees would have the same tire hanging from their branches and the cartoon would have no relevance.

The project generally produces a vision of a conceptual system which then has to be documented and communicated to everyone working on the team as well as representatives from the customers, operators and users. The PP and CONOPS are two documents that help communicate the vision. These documents provide the answers to the Kipling questions (Kasser 2018: Section 7.6), who, what, where, when, why and how, are critical to the success of a project and must be tailored as applicable to the complexity of the project. Both documents are initially written early in the system planning state to clarify and organize the systems' objective and constraints. Subsequently as the system evolves through the SDP and requirements and constraints change, each document has to be updated for each major milestone to keep the vision current. Other techniques include informal reviews (Section 5.4.2.1) and newsletters. If the vision is communicated, people will know where their piece fits, and will have a sense of contributing to the system. This technique is one way of implementing Deming's 9th point, 'break down the barriers between staff areas' (Deming 1986: p. 62).

8.6.2 Failing to Understand the Customer's Real Requirements

The major consequences of failing to understand the customer's real requirements are unhappy customers. Unhappy customers go elsewhere for their subsequent needs (no follow-on work), and make difficulties at delivery and payment time (Kasser 1995). This risk is mitigated by activities that include:

- Communicating with the customer as frequently as practical.
- Getting the customer involved in the project.
- Establishing effective personal relationships between the project personnel and the customer.

A serendipitous bonus of getting the customer involved in the project is the availability of the customer's undocumented information pertaining to the product being produced and the environment in which it will be used.

The additional risk incurred is the risk that the project personnel will make changes as a result of informal conversations with the customer. This risk can be mitigated by training project personnel to say 'no' politely, and refer requests for changes to the customer's designated project person in a way that does not harm the relationship with the customer. The schedule and cost impact of the resulting change request must then be evaluated and negotiated.

8.6.3 Failing to Plan Ahead

The major consequences of failing to plan ahead are cost escalation due to reactive management. Reactive management is where the project manager reacts to events that

---

[4] This sketch has been around for at least 50 years and is just as applicable today as it was when I first saw it in 1970.

occur rather than planning ahead and either preventing those events or identifying ways to mitigate them should they occur by applying a combination of both planning and prevention. From the systems perspective, planning ahead incorporates risk management. By using perspectives from the *Operational* and *Functional* HTPs to plan the process assuming nothing goes wrong, and perspectives from the *Continuum* HTP to investigate the effect of something going wrong and incorporating prevention or mitigation into the plan, risk management is incorporated as part of the normal systemic and systematic planning process. These prevention and mitigation planning activities are known as contingency planning in the traditional management paradigm. The systems approach incorporates them into the planning process rather than performing them as a separate add-on activity to the planning process.

8.6.4 Failing to Control Changes

The major consequences of failing to control changes are cost and schedule escalations.

Changes are a way of life during the SDP. The waterfall method (Royce 1970) provided a template for MBO (Mali 1972), (Section 11.7) for systems whose requirements were known. There were well-defined milestones and evaluation criteria. The method broke down, for several reasons including:

- *Poor implementation:* people blamed the methodology for failing to deliver on schedule and within budget, instead of accepting their poor change management.
- *Time:* The project took so long that the needs changed before the product was delivered.
- *Human nature:* When customers had a chance to use early versions of the system, they tended to decide that they really wanted something else because they hadn't thought through what they wanted in the first place.

The common element in these factors is change. A change is a change in the requirements for the system. The effect of the change must be assessed against the PP (Kasser 1995) using the change control process (Section 11.3.2).

8.6.5 Failing to Apply Lessons Learned from the Past

The major consequences of failing to apply lessons learned from the past are cost escalation and schedule delays due to repeating the:

- Experience and learning the lessons in the present (Santayana 1905).
- Mistakes and paying for them on the current project.

The systems approach incorporates applying lessons learned at the beginning of each state in the SDP. The milestone review at the end of each state incorporates a lessons-learned presentation. The management activity at the beginning of each state incorporates a review of the lessons-learned in the previous state[5].

---

[5] From the same and past appropriate similar projects.

8.6.6 FAILING TO DOCUMENT THE REASONS FOR DECISIONS

The major consequence of failing to document the reasons for decisions is that earlier decisions are questioned throughout the design, implementation and test states of the project. Many projects do not document reasons for decisions. Some do, but in project notebooks, so the data is difficult if not impossible to find.

8.6.7 FAILING TO MAXIMIZE USE OF COTS

The major consequence of failing to use COTS is that the costs go up over the lifecycle of the product being produced. Unless a manufacturing company is vertically integrated, it is generally cheaper to integrate a component that has been built and tested than to develop and test a new component that will perform the same functionality. *COTS reduce COSTS if used correctly.*[6] There should be decisions about building or buying material and equipment along the whole system SDP rather than just assuming the material or equipment will be designed and built. That is the major advantage of COTS; however, there is also a negative side to the decision. COTS products are upgraded when the COTS manufacturer decides to do so. Sometimes an upgrade does not provide interoperability with, or the same functionality as, the un-upgraded product. This factor needs to be considered as a risk criterion in the build-buy decision. For example:

- *In software development:* there are commercially available software libraries that perform many of the functions of a software product. It is generally not necessary to reinvent those functions. The risk of incompatibility and future upgrades may be mitigated by purchasing the source code of the library.
- *In hardware development:* the project building the hardware item such as a modem, telephone and compact disc (CD) player does not build its own resistors, capacitors and printed circuit boards. It acquires them and integrates them.
- *In systems integration projects:* the item being integrated is generally purchased rather than produced. For example, in a communications system, the transmitters, receivers and modems are generally purchased and then integrated into communication system.

Note the use of costs does not eliminate the need for testing. COTS products may not be tested under all the conditions that the product that will be using the COTS may encounter. For example, the first Arianne V rocket reused software from the Arianne IV which was not tested under all conditions expected in the Arianne V. The result was that the rocket became a submarine[7] when the software experienced an overrun leading to a launch failure.

---

[6] Notice how COSTS was reduced to COTS.
[7] The range safety officer destroyed the rocket so it became lots of little submarines when the pieces hit the water.

8.7 TYPES OF CONTRACTS

Projects are typically performed within the framework of a contract. A contract:

- Is the overt manifestation of an agreement; it is not the agreement.
- Documents who (supplier) is to supply what (product) to whom (customer), the amount of, and how the payment is to be made.
- Any other legal arrangement between the parties.
- May be:
  1. *Fixed price:* supplier and customer agree on a fixed price for the work to be performed.
  2. *Cost plus:* also known as time and materials (T&M). These contracts come in several variations and provide for the payment of reasonable, allowable and allocatable costs incurred by the supplier in the performance of the contract.

From the customer's perspective, the choice as to which type of contract to use is determined by the element of risk (the probability of successful completion of the contract) and who pays for it (customer or supplier). Contracts in which there is risk, namely there are known unknowns in the performance of the contract tend to be cost plus contracts. Contracts in which there is little or no risk, namely the job is simple or the job has been done before, tend to be fixed price contracts.

A fixed price contract is negotiated based on the customer being willing to pay the supplier being more than the cost to produce the product.

- From the supplier's perspective the price can be based on the:
  - Estimated costs plus a profit factor.
  - Going market rate based on the customer's willingness to pay.
- From the customers' perspective the price is always based on the amount that the customer is willing to pay (Section 8.3.1.6).

In a fixed price contract, cost overruns lead to a smaller profit, excessive cost overruns lead to a loss and really excessive overruns can lead to bankruptcy. In a cost-plus contract, cost overruns are generally covered by the supplier until the supplier cancels the contract. By understanding the contractual background to the project, the project manager can use the rules in cost-effective manner. Each type of contract has its advantages and disadvantages, yet the goal of reducing the costs of the project applies to both of them.

8.8 THE PRESENTATION EXERCISE

In the classroom environment the team would not cost the project; however, for the purpose of providing an example of how costing is done, this section will cost the project based on the information in the previous sections. The labour category and salaries for each of the team members based on Table 6.1 are shown in Table 8.2. The salaries for the two junior and two senior staff members are slightly different

Successful Project Cost Estimating

TABLE 8.2
Team Member Salary Rates and Labour Categories

Member	Labour Category	Rate/Time Unit
Andy	Junior	$20.00
Brian	Senior	$60.00
Lisa	Mid-level	$40.00
Michael	Junior	$25.00
Susan	Senior	$55.00

reflecting their slightly different levels of experience which does happen quite often in a postgraduate class.[8]

The number of time units each person works on each with top-level package is shown in the revised staffing summary shown in Table 7.5. These time units are multiplied by the salary rates in Table 8.2 to estimate the costs of each high-level WP. When the overhead rate of 15% is added, the resulting cost estimate for the session 1 exercise is shown in Table 8.3.

Table 8.3 does not show the information very well. The information is more understandable if shown in a graph such as the ones in Figures 8.5–8.7 in which:

- Figure 8.5 shows the cumulative of cost of the entire project.
- Figure 8.6 shows the cost of each state. Information necessary to ensure the cash flow is positive. In other words, Figure 8.6 shows the amount of funding needed to complete each state of the project.
- Figure 8.7 is a compound line and bar chart (Kasser 2018: Section 2.4) showing a combination of Figures 8.5 and 8.6; traditionally used in management briefings summarizing the cost information of a project.

TABLE 8.3
Presentation Project Costs

	Andy	Brian	Lisa	Michael	Susan	Direct Labour	Overhead	Total	Cumulative
WP 00000	$60.00	$240.00	$160.00	$75.00	$220.00	$755.00	$113.25	$868.25	$868.25
WP 10000	$120.00	$420.00	$200.00	$125.00	$275.00	$1,140.00	$171.00	$1,311.00	$2,179.25
WP 20000	$240.00	$720.00	$480.00	$300.00	$385.00	$2,125.00	$318.75	$2,443.75	$4,623.00
WP 30000	$120.00	$480.00	$40.00	$100.00	$440.00	$1,180.00	$177.00	$1,357.00	$5,980.00
WP 40000	$40.00	$180.00	$80.00	$75.00	$165.00	$540.00	$81.00	$621.00	$6,601.00
WP 50000	$60.00	$180.00	$80.00	$75.00	$110.00	$505.00	$75.75	$580.75	$7,181.75
TOTAL	$640.00	$2,220.00	$1,040.00	$750.00	$1,595.00	$6,245.00	$936.75	$7,181.75	

---

[8] Susan is being paid less than Brian because Brian has 5 more years of experience than Susan; the company is not sexist.

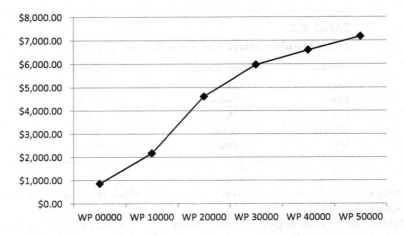

FIGURE 8.5 The estimated costs for the session 1 presentation exercise (line chart).

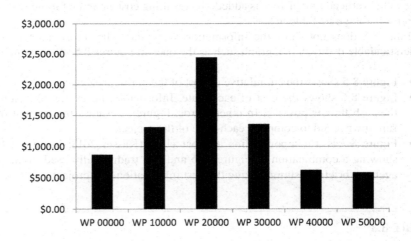

FIGURE 8.6 The estimated costs for the session 1 presentation exercise (bar chart).

8.8.1 Achievements at This Point in Time

At this point in time the team has identified:

1. All the activities that need to be performed.
2. All the products that have to be produced.
3. The prerequisites for each activity.
4. The people who will perform the activities.
5. An estimate of the period of performance (time to complete).
6. An estimate of the cost of the project.

Successful Project Cost Estimating

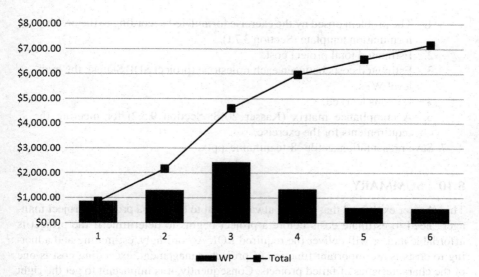

FIGURE 8.7 Graphical summaries of the estimated costs for the session 1 presentation exercise.

8.9 THE ENGAPOREAN MCSSRP EXERCISE

The background and context for the progressive session exercises is in Appendix 1. The purpose of the exercise is:

1. To practice using the template for a student exercise presentation (Kasser 2018: Section 14.7).
2. To develop and add costing estimates to activities in the WPs produced in Exercise 7.7 using the bottom-up estimating methodology. The exercise is limited to estimating the nominal cost for each activity.

8.9.1 THE REQUIREMENTS FOR THE EXERCISE

The requirements for the exercise are to:

1. Formulate the problem posed by the exercise using the problem formulation template (Section 3.7.1).
2. Make realistic assumptions on salary fixed and overhead costs.
3. Update the WPs from Exercise 7.7 with estimated cost information.
4. Format presentation according to the applicable parts of the sample four-part template for a management review presentation (Section 5.4.2.2.1).
5. Number the slides.
6. Prepare <5 min presentation that includes:

1. The problem posed by the exercise formulated according to the problem formulation template (Section 3.7.1).
2. Estimated total project cost.
3. Estimated costs between each milestone (project SDP States); the high-level WPs.
4. Lessons learned.
5. A compliance matrix (Kasser 2018: Section 9.5.2) for meeting the requirements for the exercise.
7. Save presentation as file '8-teamname'.pptx.

8.10 SUMMARY

This chapter explained that there is always a cost to develop a product. Project managers need to estimate costs before a project begins to determine if the project is affordable and/or will deliver the required ROI. Accordingly, estimating and adhering to costs is an important function of project management, exceeding cost is one of the characteristics of failed projects. Consequently, it is important to get the right cost estimate for the project. Project managers also need to re-estimate the costs once the project commences and more detail is known about the seven interdependent P's of a project. However, before trying to estimate costs the project manager needs to understand the nature of costs, where they come from and what they represent. Accordingly, this chapter explained:

1. Factors influencing the quality of estimates.
2. Methods for estimating project costs.
3. The different types of cost contracts.
4. Ways of controlling costs.

REFERENCES

Crosby, Philip B. 1979. *Quality is Free*. New York: McGraw-Hill.
Deming, W. Edwards. 1986. *Out of the Crisis*. Cambridge, MA: MIT Center for Advanced Engineering Study.
Denzler, David W. R., and Joseph Eli Kasser. 1995. Designing budget tolerant systems. In *the First International Symposium on Reducing the Cost of Spacecraft Ground Systems and Operations , Rutherford Appleton Laboratory*, England.
Hari, Amihud, and Menachem P. Weiss. 1996. ICDM - An inclusive method for customer driven conceptual design. In *the Second Annual Total Product Development Symposium*, Pomona, CA.
Hari, Amihud, Menachem P. Weiss, and Avigdor Zonnenshain. 2001. Design quality metrics used as a quantitative tool for the conceptual design of a new product. In *the ICED 01 the International Conference on Engineering Design*, Glasgow, Scotland.
Hari, Amihud, Shraga Shoval, and Joseph Eli Kasser. 2008. Conceptual design to cost: A new systems engineering tool. In *the 18th International Symposium of the INCOSE*, Utrecht, Holland.
Hari, Amihud, Menachem P. Weiss, and Avigdor Zonnenshain. 2004. ICDM - An integrated methodology for the conceptual design of new systems. In *System Engineering Test and Evaluation Conference SETE 2004*, Adelaide, Australia.

Kasser, Joseph Eli. 1995. *Applying Total Quality Management to Systems Engineering.* Boston, MA: Artech House.
Kasser, Joseph Eli. 2018. *Systems Thinker's Toolbox: Tools for Managing Complexity.* Boca Raton, FL: CRC Press.
Mali, Paul. 1972. *Managing by Objectives.* New York: John Wiley & Sons.
Michaels, Jack V., and William P. Wood. 1989. *Design to Cost.* New York: John Wiley & Sons.
OAS. 2013. *Analysis of Alternatives (AoA) Handbook, A Practical Guide to Analyses of Alternatives*, edited by Air Force Materiel Command (AFMC) Office of Aerospace Studies. Kirtland AFB, NM: Air Force Materiel Command (AFMC) OAS/A5.
Royce, Winston W. 1970. Managing the development of large software systems. In *IEEE WESCON*, Los Angeles, CA.
Santayana, George. 1905. Reason in common sense. In *The Life of Reason* (Vol. 1). New York: Charles Scribner and Sons.

9 Successfully Adjusting Project Schedules and Costs

A project spends most of its time in the project performance state (Section 5.3.1.3) when predicted and unpredicted events have the potential to increase costs and cause delays. Accordingly, the chapter explains:

1. Shortening the schedule in Section 9.1.
2. Reducing costs in Section 9.2.
3. Cancelling a project (the ultimate in cost and schedule reduction) in Section 9.3.

9.1 SHORTENING THE SCHEDULE

Schedules should be shortened if the value of the time saved is greater than the cost of saving the time. Costs of saving the time include the costs of the additional resources, such as the cost of adding a new person, an extra person, as well as the costs of performing the crashing activity. Some reasons for shortening the schedule are:

- The project is late; there will be a penalty cost that is greater than the crashing costs.
- The customer has offered a bonus for early completion in the middle of the project. The customer might have developed an urgent need for the product the project is producing.
- The resources used by the project are needed for another higher-priority project.

Schedules can be shortened because they are based on calendar time rather than actual time. In calendar time, a day has 24 h but the actual time spent on a project will depend on the number of people working. If one person is working for the whole 24 h in the day, then then 24 h of work has been done. If two people are working for the whole day then 48 h of work has been done. So, ways of shortening the schedule include:

1. Increasing the number of daily working hours by the use of overtime discussed in Section 9.1.1.
2. Increasing the number of daily working hours by adding people discussed in Section 9.1.2.

3. Crashing the project to reduce the schedule discussed in Section 9.1.3.
4. Improving productivity discussed in Section 9.1.4.
5. Reducing the work by removing some of the work packages (WPs) discussed in Section 9.2.3.
6. Subcontracting some of the work; effectively temporarily adding the subcontractor's personnel to the project (Tran 2019).
7. Redesigning the product to use commercial-off-the-shelf (COTS) rather than developing the unit in-house (Tran 2019).

It is up to the project manager faced with the problem of shortening the schedule to decide which approach to use for the specific type of project.

9.1.1 OVERTIME

Overtime is an obvious method of increasing the number of daily working hours to shorten the schedule and has been used for many years. However, there are downsides to the use of overtime including the effects on:

- *Cost:* generally, people working extra hours are paid for that time often at an agreed a higher hourly rate than the regular hourly rate. This will increase the cost of the project.
- *Personnel:* Overtime works up to a point, but once that point is exceeded people get tired in, make mistakes and their productivity decreases. In addition, the extra hours worked have to be taken from other activities such as spending time with their families, working on hobbies and resting and relaxing. So, these extra hours worked tend to decrease a person's morale and can lead to depression. Accordingly, over time, if used, should only add a small percentage to the person's workload.

However, in the 1980s, Luz in Israel had a 40-h workweek for 4 days a week as compared with the regular 5 or even 5.5 h working day in other companies. Everybody worked Monday to Thursday, and had a 3-day weekend. This allowed the Muslims to have their day of rest on Friday, the Jews to have their day on Saturday and the Christians to have their day on Sunday. However, from time to time, the company did ask employees to work overtime for a few weeks to meet a delivery deadline. However, the company did not pay any employee overtime for the first 8 h of overtime in a week. This provided the company with the reputation of working a 4-day week in a time where most companies worked 5 or more days, and the benefit of not having to pay employees for overtime for their first 8 h of overtime each a week. It also provided the employees with the incentive to focus on the work and complete the project on schedule to minimize the amount of unpaid overtime they would have to put in.

The employees enjoyed a 4-day working week for most of the year, and since lunchtime was included in the 10-h working day, the time they spent at work during those 4 days was not that much greater than they would have spent in an 8 h working day with a lunch on their own time.

Adjusting Project Schedules and Costs

9.1.2 Adding People

Adding people is an obvious way of increasing the number of daily working hours to shorten the schedule and has been used for many years. Some of the literature and experience states the effect is positive while some of the literature and experience states the effect is negative. Perceived from the *Continuum* HTP, the literature and experience quoted are not contradictory they just lack systems thinking because they are often set in a single context and then generalized to all projects. The effect of adding people may improve (shorten) some project schedules but not others, depending on the type of project. For example:

- *The negative effect:* adding people does not shorten the schedule for a number of reasons including (Brooks 1972):
  - *The learning curve or ramp-up time:* the additional people do not become productive immediately.
  - *The communications overhead:* the additional people have to be added into the communications loop, which makes the loop more complex.
- *The positive effect:* the impact new people provide is particularly useful on a challenging software project where they may be able to have an immediate impact in two ways (Needham 2009):
  1. *Driving for simplicity:* new people joining a team may well be best placed to judge the quality of a code base. Since they have very little to no context of the decisions made, they only have the code to tell them the story of the project so far. Consequently, they are the best placed to tell whether or not the code has been written in an obvious way because they will probably have trouble understanding what's going on if that isn't the case.
  2. *Enthusiasm:* new people often have much higher enthusiasm than people who may have been working on a project for an extended period of time and could be feeling a bit jaded. They haven't fought the battles or experienced the pain that current team members may have done so they arrive with few preconceptions and a desire to improve the way that things are being done. An added benefit is that enthusiasm is contagious so some of it is bound to rub off on other members of the team thereby leading to improved morale.
- *The nature of the task:* some tasks can be partitioned so that more than one person can perform them, others cannot be partitioned. For example, it takes a woman 9 months to make a baby, adding more women to the task will not change the schedule.

9.1.3 Crashing a Project

Crashing a project is an iterative loop that gradually reduces the schedule estimate until the target schedule is reached as follows:

1. Identify the critical path (Section 7.4) (the previous iteration may have changed the critical path).

2. Determine if you can split a WP in the critical path into two parallel WPs. This is the reverse of how the staffing was levelled (Section 7.3). If this split reduces the schedule estimate for the project to the target amount, the project schedule has been crashed and the crashing process loop ends.
3. Starting with the WP with the longest schedule in the critical path, reduce the schedule of each WP in the critical path by examining the assumptions in the WP, the resources needed and the materials used until the target schedule for the entire project has been achieved. If the target schedule is not achieved, check if the critical path has changed. If the critical path has changed, go back to step 1 if it has not changed repeat steps 2 and 3.
4. If the target schedule cannot be achieved in this manner, then try to reduce the required capability of the product as discussed in Section 8.3.1.6.
5. If the target schedule still cannot be achieved, it is a doomed project; cancellation is recommended but may not be the right thing to do politically.

9.1.4 Improving Productivity

Improving productivity means getting more things done within the allotted time or completing the tasks within the allotted time (shortening the schedule). Ways of improving productivity include:

- Automating repetitive tasks.
- Dealing with emails twice in batches. Open your email once a day and read all the mail. For each email requiring a response, either just think about it or write an immediate reply but don't send it. Save it for a day, reread the reply to make sure there are no words that you would regret before sending it. Do not deal with emails as they arrive during the day.
- Delegating activities and not doing everything yourself, provided you have people that you can delegate those activities to. If you don't have those people, acquire them.
- Focusing on what needs to be done and avoiding doing any tasks that don't assist in completing your goals.
- Maintaining your health by exercising and keeping fit. You can multitask exercising time by listening to relaxing music or educational material.
- Focusing on a single task for at least 10 min without interruption. If you have a number of tasks to get through in a day schedule them in specific time slots rather than multitasking on an ad hoc basis.
- Planning the day, each morning, adjusting the previous day's plan if necessary. Planning in the morning allows your subconscious mind to think about the issues while you sleep.
- Prioritizing the items on a to-do list and putting them in a schedule so that the item that needs to be completed first is assigned highest priority. Assign the completion dates and times using the just in time-decision-making approach (Kasser 2018: Section 8.6). Table 5.9 is also a useful tool for individual time management as well as project time management.

- Recharging by taking time off. The biblical concept of 1 day of rest in 7 was an idea ahead of its time. Take that day, switch off from work completely and do other things.[1]
- Recognizing the difference between urgent and important (Covey 1989) and focusing on completing the important tasks.
- Sleeping between 7 and 8 h a night. Taking short naps during the day when you feel tired rather than consuming sugar and caffeine to stay awake.
- Working smarter not harder, using the Pareto principle to prioritize the tasks that need to be done and focusing on those that produce 80% of the required results.
- Build lollygagging time into your daily schedule (Section 12.1).

9.2 REDUCING COSTS

Costs are one of the quadruple constraints (Section 4.4.2). Reducing costs can only be achieved by changing at least one of the other three constraints as for example:

1. Replacing the personnel assigned to an activity with lower-cost personnel.
2. Increasing the schedule.
3. Deleting some of the work without changing the requirements.
4. Reducing materiel costs.
5. Redesigning the product.
6. Removing the low-priority requirements.
7. Negotiating alternatives to paying resources by giving them other packages instead of money, e.g. company products and/or services, which have the same retail dollar value, but in fact cost the organization fewer dollars (Tran 2019).

9.2.1 Replacing Personnel Assigned to an Activity with Lower-Cost Personnel

For example, in the presentation exercise, the numbers show that the cost of the project could be reduced if Susan and Brian switched activities in the presentation exercise (Section 9.5). However, lower-cost people generally mean lower levels of competence, which may result in errors and mistakes that would then increase the costs.

9.2.2 Increasing the Schedule

If two WPs are being performed in parallel by two groups of people, it may be possible for the people working on one WP to work on the second package when they finished the first WP. This approach reduces costs but extends the schedule by the time taken for the second WP. As long as the schedule path through those WPs does

---

[1] I have often found that insurmountable problems I was facing on a Friday were solved by my subconscious by the time I got back to work the following week.

not change the critical path there will be no schedule impact. If the changed path becomes the new critical path because it now takes longer than the previous critical path, there will be a schedule impact which may or may not be important.

9.2.3 Deleting Some of the Work without Changing the Requirements

If WPs are deleted, it will be possible to remove the people who were assigned to work on those packages from the project. However, those WPs should have been in the project for a reason and removing them will incur a risk. When budgets are cut, training and testing WPs tend to be deleted from projects, which may lead to cost and schedule overruns later in the system development process (SDP). There is a need to take a systems approach to budget cuts and delete those elements that reduce the costs across the entire project lifecycle as opposed to just picking on WPs without considering their impact of removing them over the SLC.

9.2.4 Reducing Materiel Costs

Reducing materiel costs by using cheaper parts and equipment. This generally results in less reliable parts and equipment, which may or may not be important. For example, if a part is going to be used in a missile, it need not have a very long operational life because the missile will only be operating for a few minutes once it's fired. Accordingly, there is no need to purchase components with long operational lives in that situation. However, as the missile can be expected to have a long standby life before it is deployed, the part should have a long standby life.

In the electronics industry, many gold-plated or silver-plated connectors are used for high reliability. There are situations where the use of gold and silver plating on connectors is not necessary and cheaper connectors could be used instead.

9.2.5 Redesigning the Product

Redesigning the product using a different technology can often result in a cheaper product.

9.2.6 Removing the Low-Priority Requirements

Removing the low-priority requirements generally results in a cheaper product. This is achieved by removing the WPs that produce and test the low-priority requirements.

9.3 CANCELLING A PROJECT

Cancelling the project:

- Is the ultimate adjustment. Projects are cancelled for several reasons including:
  - Being out of time.
  - Being out of funds.

Adjusting Project Schedules and Costs

	1	2	3	4	5	6	7
Needs	■						
Requirements		■					
Design			■				
Realization				■			
Integration					■		
Test						■	
O&M							■

FIGURE 9.1 The widget project: original schedule.

- Being no longer needed.
- Experiencing too many problems.
- The priority of the project has dropped to the point where higher-priority projects will take up the resources required by the project.
- Political reasons.
- Requires an orderly closeout which includes:
  - Providing personnel with a clear vision of their post-project assignments.
  - Storing project data.
  - Allocating unused resources including unspent funds.
- May be a project in itself depending on size of project being cancelled.

9.4 THE WIDGET SYSTEM PROJECT

The widget system project provides an example of how a problem that shows up in the system test state of the SDP can impact the schedule. Federated Aerospace is developing the widget system in which the SDP starts in the needs identification state (Section 5.4.1). The project is the first of its kind: there are no similar systems in existence. The widget system comprises two subsystems, Part A and Part B. The original 7-month widget project schedule shown in Figure 9.1[2] was planned in the traditional manner as a single pass through the waterfall (Section 14.9) assuming there would be no serious problems during the SDP. Note the subsystem construction and subsystem test states have been combined into a realization state in the figure. However, 'The best laid schemes o' mice an' men gang aft agley, [often go awry]' (Burns 1786).

All went well with the widget system as it passed through the sequential states of the SDP until a major problem showed up in the system test state. At this point, the problem impacted the development schedule and the SDP reverted to a second needs identification state as shown in the revised schedule in Figures 9.2 and 9.3. Figure 9.2 clearly shows the iteration by adding the second iteration of the SDP to the first, while Figure 9.3 shows the delay in the traditional manner. The second iteration of the SDP began in Month 7 and the original operations and maintenance (O&M) state was delayed to Month 13.

---

[2] The milestones are omitted for simplifying the chart because the purpose of the figure is to show the task relationships.

	1	2	3	4	5	6	7	8	9	10	11	12	13
Needs	■						■						
Requirements		■						■					
Design			■						■				
Realization				■						■			
Integration					■						■		
Test						■						■	
O&M													■

FIGURE 9.2 The widget project: revised schedule.

Months	1	2	3	4	5	6	7	8	9	10	11	12	13
Needs	■												
Requirements		■											
Design			■										
Realization				■									
Integration					■								
Test						■							
Needs-2							■						
Requirements-2								■					
Design-2									■				
Realization-2										■			
Integration-2											■		
Test-2												■	
O&M													■

FIGURE 9.3 The widget project: alternate view of the revised schedule.

While the solution to the original problem demonstrated that the revised design would meet the functional requirements, at the end of the second system test state in Month 12, the design still needed to be validated for the non-functional and manufacturing requirements. Accordingly, at this point in time, the SDP iterated back to the third system design state as shown in the revised schedule in Figures 9.4 and 9.5 and for an additional delay of 2 months. The third realization and system test states were skipped since the changes did not affect the design and the project was so far behind schedule.

9.4.1 Changes Due to Delays

When a schedule timeline is first drawn, it is drawn as an ideal or notional schedule. However, in the real world, events happen that cause delays (Burns 1786). These delays need to be shown in the corresponding Gantt chart. For example, suppose the original widget project schedule shown in Figure 9.1 suffered a 2-month delay during the design state which caused the critical design review (CDR) to be delayed and must be shown on the modified project schedule. This is a

Adjusting Project Schedules and Costs

	1	2	3	4	5	6	7	8	9	10	11	12	13	14	15
Needs	■						■								
Requirements		■						■							
Design			■						■			■			
Realization				■						■					
Integration					■						■				
Test						■								■	
O&M															■

FIGURE 9.4 The widget project: extended revised schedule.

Months	1	2	3	4	5	6	7	8	9	10	11	12	13	14	15
Needs	■														
Requirements		■													
Design			■												
Realization				■											
Integration					■										
Test						■									
Needs-2							■								
Requirements-2								■							
Design-2									■						
Realization-2										■					
Integration-2											■				
Test-2												■			
Design-3													■		
Test-3														■	
O&M															■

FIGURE 9.5 The widget project: modified schedule with delays due to the unanticipated problems.

delay, not a required iteration as shown in Figure 9.3. One way of showing delays is illustrated in Figure 9.6, which rolls the revised schedule in Figures 9.2 and 9.3 up into the extended test state. The 2-month delay in Months 4 and 5 are shown as a half-height bar in Figure 9.6. Sometimes delays are shown as red bars on the chart whereas green bars represent the notional timeline, and blue bars represent activities being performed ahead of schedule. Whichever convention is in use, the prime directive in the Gantt chart, in fact the innovation that it brought to project management, is to show the original timeline *and* the modified adjusted timeline *on the same* chart.

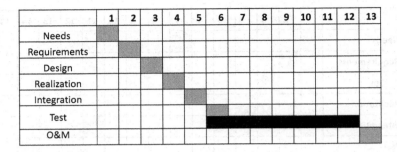

FIGURE 9.6 The widget project: traditional revised schedule.

9.4.2 Comments

The simple example in the widget project has illustrated how:

- The SDP was delayed by the activities in the second and third iterations as can be seen by comparing the original schedule in Figure 9.1 with the adjusted schedule in Figure 9.6. Accordingly, the widget project's original optimistic success-oriented 7-month planned schedule turned into a 15-month project with corresponding cost escalations due to unforeseen problems in the system design.
- First of a kind system development projects which correspond to Shenhar and Bonen's Type D projects with super-high technological uncertainty (Section 10.4) should use a schedule containing two or three passes through the waterfall rather than the single success-oriented approach commonly used. This concept may be generalized as 'the more complex the system, the more iterations of the SDP will be needed to realize the system'.
- The original single pass waterfall iterated back to the needs identification state once the problem occurred. The literature generally illustrates the iteration from the *Functional* HTP by drawing a line from one state to the other in the waterfall view shown in Figure 9.6. This approach tends to gloss over the accompanying schedule delays since activities must be repeated and have to be inserted into the project timeline.
- Solutions gave rise to problems as the SDP progressed.
- The PP should contain some slack time at the end of the system test state after the tests have been completed and before the milestone review to allow for defects to be dealt with. Simple defects may be fixed at that time and not require iteration back to an earlier state of the SDP. If no defects show up, and there are no tasks to complete, then the milestone at the end of the state can be moved forward in time and the project becomes ahead of schedule.
- The degree of iteration in the SDP should a problem arise depends on the nature of the problem.

Adjusting Project Schedules and Costs

9.4.3 Lessons Learned

Lessons learned included the more complex the system, the more iterations of the SDP will be needed to realize the system (Kasser and Zhao 2016).

9.5 THE PRESENTATION EXERCISE

One of the team members (Susan) had to travel out of town for a family emergency and cannot take part in the exercise. To keep things simple, assume Susan had to leave at the start of the Session 1 exercise. The other members of the team agreed to take up Susan's share of the work for that session in return for Susan making it up by reducing their workloads in a future session.

9.5.1 Modifying the Staffing Level

The first task in replanning the session exercise was to modify the staffing to remove Susan as a resource. They examined and changed each WP in turn and if changes were necessary, made them on the basis of trying to equalize the workload across the WP.

The staffing profile for WP 00000 (Table 6.4) was revised to the one shown in Table 9.1 by reassigning Susan's activity in WP 04141 to Andy. This reassignment could change the schedule depending on Andy's assignments. The remaining WPs were not affected by the change since they were joint activities.

The staffing profile for WP 10000 (Table 6.5) was revised to the one shown in Table 9.2 by removing Susan. Since all the WPs were joint activities none of the other team members had to take on additional work, so there would be no need to change the schedule for WP 10000.

The second revised staffing profile for WP 20000[3] (Table 7.5) was then revised by removing Susan. Since all the WPs were joint activities none of the other team members had to take on additional work, so there would be no need to change the schedule for WP 20000.

TABLE 9.1
Staffing Profile for WP 00000 without Susan

WP	Activity	Time	Andy	Brian	Lisa	Michael	Susan
00500	Coordinating 00000	5	1	2	1	1	0
04120	Understanding the grading criteria	4	1	1	1	1	0
04141	Creating the PP	4	1	0	0	0	0
04142	Creating presentation theme	1	0	0	1	0	0
09000	Holding the SPPCR	4	1	1	1	1	0
Total	WP 00000	15	4	4	4	3	0

---

[3] The latest revised version must be used when changing the cost, schedule or staffing.

TABLE 9.2
Staffing WP Profile for 10000 without Susan

WP	Activity	Time	Andy	Brian	Lisa	Michael	Susan
10500	Managing and top-level activities in the requirements state	3	1	2	0	0	0
14000	Determining the requirements for the exercise	8	2	2	2	2	0
14100	Creating the RTM	4	1	1	1	1	0
19000	Holding the requirements review	4	1	1	1	1	0
19100	Reviewing the PP	4	1	1	1	1	0
Total	WP 10000	23	6	7	5	5	0

The staffing profile for WP 30000 (Table 6.7) was revised by reassigning Susan's activities in WP 34543, WP 34544, WP 34545 and WP 34546 to Michael. This reassignment could change the schedule depending on Michael's assignments.

The staffing profile for WP 40000 (Table 6.8) was revised to the one shown in Table 9.3 by reassigning Susan's 2 time units of activities in WP 48000 to split them equally between Andy and Lisa. This reassignment could change the schedule depending on Andy and Lisa's assignments.

The staffing profile for WP 50000 (Table 6.9) was revised by removing Susan. Since all the WPs were joint activities, none of the other team members had to take on additional work, so there would be no need to change the schedule for WP 50000.

The revised staffing summary for Session 1 presentation exercise without Susan is shown in Table 9.3 and the changes are shown in Table 9.4. Susan's 29 time units consisted of 10 time units when she was working on a WP by herself and spending time in joint meetings. The meeting time was deducted from the time budget and the unique activities were assigned to Andy, Lisa and Michael. Had they been costing the effort, the cost of the project would have come down by the 19 time units Susan didn't spend in the meetings and the difference in salary between Andy, Lisa, Michael and Susan for the remaining 10 time units.

TABLE 9.3
Revised Staffing Summary for Session 1 Presentation Exercise without Susan

Activity	Time	Andy	Brian	Lisa	Michael	Susan
WP 00000	15	4	4	4	3	0
WP 10000	23	6	7	5	5	0
WP 20000	48	12	12	12	12	0
WP 30000	26	6	8	1	11	0
WP 40000	12	3	3	3	3	0
WP 50000	11	3	3	2	3	0
Total	135	34	37	27	37	0

Adjusting Project Schedules and Costs

TABLE 9.4
Staffing Assignment Changes

Activity	Time	Andy	Brian	Lisa	Michael	Susan
Original	154	32	37	26	30	29
Revised	135	34	37	27	37	0
Difference	−19	2	0	1	7	−29

9.5.2 Adjusting the Schedule

The only schedules that might need revising at this point in the replanning exercise are:

1. *WP 00000:* depending on Andy's assignments. In fact, the schedule didn't change, because WP 04141 could not be performed until after WP 04120. So, Andy performed WP 04141 instead of Susan without any impact on the schedule.
2. *WP 30000:* depending on Michael's assignments. In fact, the schedule didn't change, because Michael had not been assigned to any WPs for calendar time occupied by WP 34543, WP 34544, WP 34545 and WP 34546.
3. *WP 40000:* depending on Andy and Lisa's assignments. Since Susan's 2 time units of activities in WP 48000 were equally between Andy and Lisa they would be able complete WP 48000 in 1 calendar time unit (Unit 37 in Figure 7.16) instead of the two Susan would have taken to do the work (Unit 37s and 38). Accordingly, the schedule for WP 40000 was shortened by 1 calendar time unit, which provided an opportunity to reschedule WP 50000 to start early and finish early by the same amount of time.

This finding made them seriously consider shortening the entire schedule by splitting the work in the WPs instead of assigning single team members to the different WPs. They realized that they could shorten the schedule by doing that. But then Susan pointed out, that they had plenty of time in the schedule to complete the project even if they went 100% over schedule because they had a whole week to complete it. Moreover, she commented sharing the work in a WP would actually increase the amount of time spent on that WP because not only would the team members be doing their assigned work, they would also be spending extra time coordinating and communicating their efforts inside the WP.

Susan's absence was the only change experienced in the exercise. Her absence reduced the schedule by 1 time unit and correspondingly reduced the cost to $1,541.

9.5.3 Adjusting the Costs

Susan's absence reduced the cost by $1,541. The reduced cost based on the revised staffing summary without Susan (Table 9.3) is shown in Table 9.5 and Figure 9.7.

TABLE 9.5
Actual Costs of the Presentation Project without Susan

WP	Andy	Brian	Lisa	Michael	Susan	Direct Labour	Overhead	Total	Cumulative
WP 00000	$80.00	$240.00	$160.00	$75.00	$0.00	$555.00	$83.25	$638.25	$638.25
WP 10000	$120.00	$420.00	$200.00	$125.00	$0.00	$865.00	$129.75	$994.75	$1,633.00
WP 20000	$240.00	$720.00	$480.00	$300.00	$0.00	$1,740.00	$261.00	$2,001.00	$3,634.00
WP 30000	$120.00	$480.00	$40.00	$275.00	$0.00	$915.00	$137.25	$1,052.25	$4,686.25
WP 40000	$60.00	$180.00	$120.00	$75.00	$0.00	$435.00	$65.25	$500.25	$5,186.50
WP 50000	$60.00	$180.00	$80.00	$75.00	$0.00	$395.00	$59.25	$454.25	$5,640.75
Total	$620.00	$2,040.00	$1,000.00	$850.00	$0.00	$4,905.00	$735.75	$5,640.75	

Adjusting Project Schedules and Costs

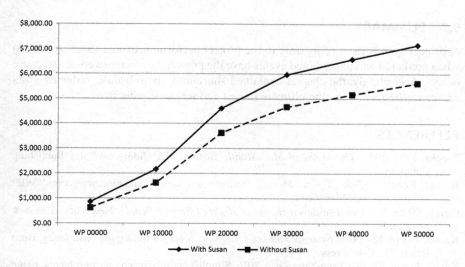

FIGURE 9.7 The actual costs of the presentation project without Susan.

The planned costs are shown as a solid line. The actual costs shown as a dotted line diverging from the plan right from the start.

9.6 THE ENGAPOREAN MCSSRP EXERCISE

The background and context for the progressive session exercises is in Appendix 1. The purpose of the exercise is:

1. To practice using the template for a student exercise presentation (Kasser 2018: Section 14.7).
2. To update the project information from Exercise 5.12 after the occurrence of an unanticipated event.
3. For all teams to deal with the same event so they can learn from each other's efforts. In future exercises, each team will select its own event, and the events will be different.

9.6.1 THE REQUIREMENTS FOR THE EXERCISE

Just before the systems requirements review (SRR), the customer informed the project manager that the project schedule was unacceptable and needed to be speeded up (reduced in time) by 25%. Accordingly:

1. Calculate effect on cost, schedule and staff due to customer's change order.
2. Prepare a presentation according to the instructions in Appendix 3.
3. Save presentation as file '9-teamname'.pptx.

9.7 SUMMARY

A project spends most of its time in the project performance state (Section 5.3.1.3) when predicted and unpredicted events have the potential to increase costs and cause delays. Accordingly, the chapter explained shortening the schedule, reducing costs and cancelling a project (the ultimate in cost and schedule reduction).

REFERENCES

Brooks, Fred. 1972. *The Mythical Man-Month*. Boston, MA: Addison-Wesley Publishing Company.

Burns, Robert. 1786. To a Mouse, https://www.poetryfoundation.org/poems/43816/to-a-mouse-56d222ab36e33.

Covey, Steven R. 1989. *The Seven Habits of Highly Effective People*. New York: Simon & Schuster.

Kasser, Joseph Eli. 2018. *Systems Thinker's Toolbox: tools for managing complexity*. Boca Raton, FL: CRC Press.

Kasser, Joseph Eli, and Yang-Yang Zhao. 2016. Simplifying solving complex problems. In *the 11th International Conference on System of Systems Engineering*, Kongsberg, Norway.

Needham, Mark 2009. The effect of adding new people to project teams. markhneedham.com. https://markhneedham.com/blog/2009/10/21/the-effect-of-adding-new-people-to-project-teams/ access date 13 Dec 2018.

Tran, Xuan-Linh. 2019. Personal communication, 30 January 2019.

10 An Introduction to Managing Risk and Uncertainty over the Project Lifecyle

This chapter explains risk management, focusing on risk prevention as well as mitigation. The chapter explains:

1. The definitions of the terminology used in risk management in Section 10.1.
2. Risks and opportunities in Section 10.2.
3. Risk management in Section 10.3.
4. Risks based on the availability of technology in Section 10.4.
5. Risk profiles in Section 10.5.
6. Risk mitigation or risk prevention in Section 10.6.
7. Cascading risks in Section 10.7.
8. Contingencies and contingency plans in Section 10.8.

10.1 DEFINITIONS OF THE TERMINOLOGY

- *Uncertainty:* The lack of complete certainty, that is, the existence of more than one possibility. The 'true' outcome/state/result/value is not known (Hubbard 2009).
- *Measurement of uncertainty:* A probability assigned to an uncertain outcome or consequence. For example, 'there is a 60% chance of rain this afternoon'.
- *Risk:* A state of uncertainty where some of the possibilities involve a loss, catastrophe or other undesirable outcome or consequence. Risks:
  - Are concerned with the future; the past cannot be changed.
  - Occur as a result of changes.
  - Are the result of decisions and actions.
  - Are unavoidable, but can often be prevented or mitigated.
- *Opportunity:* a state of uncertainty where some of the possibilities involve a desirable outcome or consequence.
- *Event:* a state of certainty associated with a risk. Something expected or unexpected has occurred; what was a risk, opportunity or consequence has occurred and turned into an event. For example, it rained this afternoon.
- *Accident:* an unforeseen event that results in a loss, catastrophe or other undesirable outcome or consequence.

10.2 RISKS AND OPPORTUNITIES

Risks and opportunities occur as a result of an action, a decision or something totally unexpected happening. For example:

- Taking an action means doing something. For example, turning left at the traffic light may result in finding a faster route (an opportunity) or a slower route due to congestion (a risk).
- Making a decision means deciding to do something. For example, deciding what food to have for dinner may result in food poisoning (a risk).
- Standing on a log may result in falling down and breaking an elbow (a risk). When this event occurs, it is known as an accident.

Uncertainties in outcomes of actions can produce both risks and opportunities. For example, you can look at a bet as an opportunity to make money or as a risk of losing money. Considering risks and opportunities, they have a:

- *Probability of occurrence:* ranging from very unlikely to highly likely.
- *Severity of impact:* ranging from hardly noticeable to catastrophic: it must be prevented or mitigated in the event of a risk, or too good an opportunity to miss.

There is no single uniform standard metric for the probability of the occurrence of the event or the severity of the impact of that event. For a project, the levels can be defined as:

- *Probability:* Almost certain (5), Likely (4), Possible (3), Unlikely (2) and Rare (1).
- *Severity:* Extreme (5), Major (4), Moderate (3), Minor (2) and Insignificant (1).

10.2.1 THE RISK RECTANGLE

The traditional approach of summing up the risk and the severity of an event is to multiply the probability by the severity and show the results in a matrix known as a risk rectangle. A risk rectangle based on these values of severity and probability is shown in Table 10.1 where, in a coloured version:

- *E: extremely high:* 25 (5*5) is coloured red.
- *H: high risk:* greater or equal to 20 (4*4) is coloured orange.
- *M: moderate risk:* between 5 and 20 is coloured yellow.
- *L: low risk:* less than or equal to 4 (2*2) is coloured green.

In order for the risk rectangle to be a practical tool, the qualitative descriptive terms must be quantified into specific numbers which is not an easy thing to do. For example, numbers need to be assigned to the terms significant and severe. These ought to be specified in terms of percentages, for example, a significant loss might be 90%

TABLE 10.1
Risk Rectangle

Severity	Probability				
	Certain (5)	Likely (4)	Possible (3)	Unlikely (2)	Rare (1)
Extreme (5)					
Major (4)					
Moderate (3)					
Minor (2)					
Insignificant (1)					

of functionality and a severe loss might be 75%. Once these levels have been specified, the loss for a specific event can then be estimated and placed in the appropriate category. Perceptions from the *Generic* HTP show that this is the same process that is used to determine the values of the ranges for each of the categories in a categorized requirements in process (CRIP) chart (Section 11.5). For an example of how to assign numbers to the probabilities and using the risk rectangle to decide whether to purchase options in the stock market, see the *Systems Thinker's Toolbox* (Kasser 2018: Section 5.42).

10.2.1.1 The Flaw in the Risk Rectangle

Since the risk rectangle multiplies probability by severity to assign the level of risk, risks with low probability of occurrence but high severity of consequences tend to be ignored in favour of risks with higher products of the multiplication. This is a flaw in the use of the risk rectangle because risks with a catastrophic severity of impact must be mitigated, avoided or prevented irrespective of their probability of occurrence.

10.3 RISK MANAGEMENT

In the traditional management paradigm, risk management is treated as, and documented as, separate management and engineering processes. In the systems approach, it is built-in to project management and systems engineering. There are two types of risk management:

1. *Proactive:* when risks are identified and mitigated/prevented ahead of time. This type of risk management is characterized by normal working hours and projects meeting their budget and schedules.
2. *Reactive:* when the events occur, usually unforeseen, and need to be dealt with urgently. This type of risk management is characterized by lots of frenzied activity, overtime and projects exceeding their budgets and schedules.

The common elements in both types of risk management are problems that need remedying and decisions that need to be made.

10.3.1 Selected Myths of Risk Management

Perceived from the systems approach to risk management, the traditional approach to risk management contains a number of myths (Kasser 2018: Section 12.7) including:

- Risk management is a separate activity from design and project management. In the systems approach risk management is integrated into project management.
- Risk can be quantified as a single number; the product of the probability of occurrence of a mishap and the severity of the potential outcome. This use is widespread encouraged by the U.S. Department of Defense (DOD) in the form of the risk rectangle or traditional risk assessment matrix (Section 10.2.1). The single-number quantification arose because project managers (PM) and decision makers want simplicity when making high-risk decisions. The systems approach uses risk-attribute profiles (Kasser 2018: Section 9.1).
- Published project risk assessment models provide consistent and rationale measures of project risks. In actuality they do not.
- Projects with high cost contingencies succeed and do not have cost overruns. In actuality they do not.
- Maintain risk registers for all the risks. This can lead to a very large and unmanageable risk register. A better approach maintains the top 6–10 risks at each level, or ~7 ± 2 (Miller 1956).
- Risks are measured, in reality they are estimated with an accuracy that may not be accurate.

10.3.2 The Traditional Approach to Risk Management

The traditional approach to risk management is (DOD 2001):

1. *Risk identification:* identifying all the risks. Reviewing the work breakdown structure (WBS) elements down to the level being considered and identifying risk events.
2. *Risk analysis:* analysing each risk event to determine probability of occurrence and consequences/impacts, along with any interdependencies and risk event priorities.
3. *Risk mitigation and contingency planning:* planning mitigation actions and creating contingency plans. Translating risk information into decisions and actions (both present and future) and implementing those actions.
4. *Risk tracking:* tracking the risks. Monitoring the risk indicators and actions taken against risks.
5. *Risk controlling:* monitoring them and correcting deviations from planned risk actions.
6. *Risk communicating:* between the team and management. Providing visibility and feedback data internal and external to the project on current and emerging risk activities.

10.3.3 THE SYSTEMS APPROACH TO RISK MANAGEMENT

The systems approach to risk management is:

1. *Risk identification:* performed in the planning process as part of creating the work packages (WP) by examining each process in the WP and the product being produced by that process to identify the risks. In this context, a risk is anything that can negatively impact cost and schedule.
2. *Risk analysis:* consists of the following two parts:
    1. *Risk effect analysis:* performed from two perspectives: cause and effect.
        1. *Cause:* starting with a proposed failure, inferring the symptoms that could arise from that failure.
        2. *Effect:* starting with a symptom, deducing what could have caused it (root cause).
    2. *Risk impact analysis:* analysing each risk event to determine probability of occurrence and consequences/impacts, along with any interdependencies. The risk priorities will be based on the priority of the WPs and whether the WP is in the critical path.
3. *Risk mitigation and prevention planning:* planning mitigation actions and prevention. Translating risk information into decisions and actions (both present and future). This is done in the planning process as part of creating the WPs by inserting the prevention and mitigating activities into the appropriate WPs.
4. *Risk tracking:* monitoring the risk indicators and actions taken against risks.
5. *Risk controlling:* taking the planned corrective action in the appropriate WPs. Should an unforeseen event occur that will negatively impact the cost and schedule, the occurrence of the event identified the risk. The risk effect and impact analyses are performed and risk mitigation activities identified. A change request (Section 11.3.2) is then issued to modify the WPs and if accepted, is implemented.
6. *Risk communicating:* between the team and management. Providing visibility and feedback data internal and external to the project on current and emerging risk activities. This can be done using the enhanced traffic light (ETL) chart (Section 11.6.2) to summarize those activities.

10.4 RISKS BASED ON THE AVAILABILITY OF TECHNOLOGY

Technology is widespread in our civilization, so most projects will employ some kind of technology in whatever the project is creating. Technology has a lifecycle; it is developed through research, then moves into widespread use in many products, and then becomes obsolete and is replaced. Some projects use technology that already exists; some projects are dependent on technology that is still being developed. Shenhar and Bonen recognized that a single project management methodology would not work for technology in different states of development or existence.

They categorized projects into the following four types based on the state of the technology (Shenhar and Bonen 1997):

- *Type A: Low-Tech Projects:* projects that rely on existing and well-established technologies to which all industry players have equal access.
- *Type B: Medium-Tech Projects:* projects that rest mainly on existing technologies and incorporate a new technology or a new feature of limited scale.
- *Type C: High-Tech Projects:* projects in which most of the technologies employed are new, but existent — having been developed prior to the project's initiation.
- *Type D: Super-High-Tech Projects:* projects based primarily on new, not entirely existent, technologies.

The recommended management methodologies for the different types of projects are:

- *Type A: Low-Tech Projects:* The design cycle in these projects is usually characterized by a single pass through the system development process (SDP) because the final design of the system is usually frozen prior to the project initiation.

  The management style is a firm and formal style, in which no changes are allowed or required.
- *Type B: Medium-Tech Projects:* The design cycle in these projects usually consists of at least one iteration of the SDP. The management style of these projects can be described as moderately firm but more flexible than in Type A projects.

  The management style requires more communication with the customer (both formal and informal) since more trade-offs and changes are made.
- *Type C: High-Tech Projects:* The design cycle in these projects is also iterative, usually entails two or more cycles through the SDP, and the system design freeze takes place at a later stage than in Type B systems. It often occurs as late as the second quarter or the midpoint of the project.

  The management style of high-tech systems is moderately flexible, since many changes are expected and are a natural part of this type of system development. It involves intensive customer interaction and the use of multiple formal and informal communication channels.
- *Type D: Super-High-Tech Projects:* The design cycle in these projects requires extensive development both of the new technologies and the actual system being developed by the project. Their development frequently requires building an intermediate, small-scale prototype, on which new technologies are tested and approved before they are installed on the prototype. System requirements are hard to determine; they undergo enormous changes and involve extensive interaction with the customer. Obviously, the system functions are of similar nature–dynamic, complex, and often ambiguous. A super-high-tech system is never completed before at least two, but very often even four, iterations of the SDP are performed, and the final system design freeze is never made before the second or even the third quarter of the project.

  The management style of these projects is highly flexible to accommodate the long periods of uncertainty and frequent changes. Managers must live with continuous change for a long time; they must extensively increase interaction, be concerned with many risk mitigation activities and adapt a 'look for problems' mentality.

So, when architecting the project process, the technology risk must be assessed and the appropriate number of iterations of the SDP must be built into the PP.

10.4.1 THE TECHNOLOGY AVAILABILITY WINDOW OF OPPORTUNITY

National Aeronautics and Space Administration (NASA) and the U.S. DOD have many projects dependent on technology that is still being developed. NASA developed the technology readiness level (TRL) to minimize the risk of the technology not being available when needed (Mankins 1995). The DOD adopted the TRL with slight modifications. The TRL only covers the early stages of the technology lifecycle. PMs must consider the entire lifecycle because there is no point in incorporating obsolete or almost obsolete technology in a new product. The Technology Availability Window of Opportunity (TAWOO) (Kasser 2016, 2018: Section 8.12) shown in Table 10.2 is a six-state tool developed to provide the PM with such a tool. The TAWOO:

TABLE 10.2
The TAWOO States and Levels

TAWOO State	Level	Comments
6. Antique	12	Few if any spares available in used equipment market. Phase out products or operate until spares are no longer available.
5. Obsolete	11	Some spares available, maintenance is feasible.
4. Approaching obsolescence	10	Use in existing products but not in new products. Plan for replacement of products using the technology.
3. Operational	9	Available for use in new products (in general). System 'flight proven' through successful mission operations.
2. Development	8	Actual system completed and 'flight qualified' through test and demonstration.
	7	System prototype demonstration in an operational environment.
	6	System/subsystem model or prototype demonstration in a relevant operational environment.
1. Research	5	Component and/or breadboard validation in relevant operational environment.
	4	Component and/or breadboard validation in laboratory environment.
	3	Analytical and experimental critical function and/or characteristic proof-of concept.
	2	Technology concept and/or application formulated.
	1	Basic principles observed and reported.

Source: © 2016 IEEE. Reprinted, with permission, from Kasser, Joseph Eli. 2016. "Applying Holistic Thinking to the Problem of Determining the Future Availability of Technology." The IEEE Transactions on Systems, Man, and Cybernetics: Systems no. 46 (3):440–444.

- Allows PMs to determine if a technology is mature enough to integrate into the system under development *as well as* determining if the technology will be available for the operating life of the system once deployed.
- Extends the TRL to cover the whole product lifecycle including consideration of diminishing manufacturing sources and material shortages (DMSMS) at the end of the technology lifecycle.
- Is described in greater detail in the *Systems Thinker's Toolbox* (Kasser 2018: Section 8.12).

Consider the TAWOO from the appropriate progressive and remaining HTPs.

- *Temporal:* 'Although TRL is commonly used, it is not common for agencies and contractors to archive and make available data on the timeline to transition between TRLs' (Crépin, El-Khoury, and Kenley 2012). Perceptions from the *Temporal* HTP suggest that the data should be archived and used to estimate/predict maturity. If that data were available, one could infer from the *Scientific* HTP that one could consider the rate of change of TRL rather than a single static value at one particular time. For example, Figure 10.1 (Kasser 2016) shows that a technology was conceptualized in 1991 and the development was planned to advance one TRL each year starting in 1993 for production in 1999. However, the development did not go according to plan. The technology did not get to TRL 2 until 1995 advancing to TRL 3 two years later in 1997 and jumping to TRL 6 in 1998. So, can the technology be approved for a project due to go into service in 1999? It depends. If the project can use the technology at TRL 6, then yes. But, if the product using the technology is to go into mass production, the answer cannot be determined because there is insufficient information to predict when the technology will be at TRL 9. The PM will have to obtain more information about the factors affecting the rate of change in TRL to make a forecast as to the future.
- *Generic:* Perceptions from the *Generic* HTP indicate that projects use EVA (Section 11.4) and display budgeted/planned and actual cost information in graphs such as Figures 11.2 and 11.3 in which future costs are forecast.

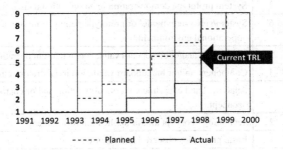

FIGURE 10.1 The TRL 1991–2001. (*Source:* © 2016 IEEE. Reprinted, with permission, from Kasser, Joseph Eli. 2016. "Applying Holistic Thinking to the Problem of Determining the Future Availability of Technology." The IEEE Transactions on Systems, Man, and Cybernetics: Systems no. 46 (3):440–444.)

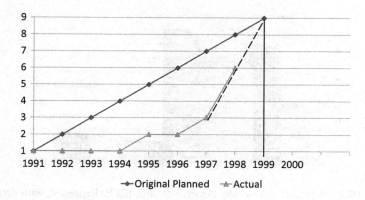

FIGURE 10.2 The dTRL. (*Source:* © 2016 IEEE. Reprinted, with permission, from Kasser, Joseph Eli. 2016. "Applying Holistic Thinking to the Problem of Determining the Future Availability of Technology." The IEEE Transactions on Systems, Man, and Cybernetics: Systems no. 46 (3):440–444).

- *Scientific:* Combine observations from the *Generic* and *Temporal* HTPs and display the rate of change of the TRL in the form of an EVA graph as shown in Figure 10.2 (Kasser 2016). When this is done, one additional significant item of information is obtained. Assuming nothing changes and progress continues at the same rate as in 1997–1998, the technology should reach TRL 9 by 1999. However, the reason for the rate of change between 1996–1997 and 1997–1998 is unknown. This provides the PM with some initial questions to ask the technology developers before making the decision to adopt the technology. The static single value TRL has become a dynamic TRL (dTRL) (Kasser and Sen 2013). The dTRL component would make adoption choices simpler. Prospective users of the technology could look at their need by date, the planned date for the technology to achieve TRL 9 and the past progress through the various TRLs. Then the prospective users could make an informed decision based on the graph in their version of Figure 10.2. If the rate of change in the dTRL predicts that the desired TRL will not be achieved when needed and they really needed the technology, they could investigate further and determine if they could help increase the rate of change of TRL.

Perceptions from the *Generic* HTP and insight from the *Scientific* HTP have conceptualized the use of a dTRL to help to predict when a technology will achieve a certain TRL. The need for a dTRL has been recognized in practice and there has been research into estimating the rate of change of technology maturity (El-Khoury 2012). The dTRL concept was used for quite a few years in the U.S. Aerospace and Defense industry beginning in the Strategic Defense Initiative era (early 1990s) and took the form of waterfall charts that tracked the TRL (Benjamin 2006).

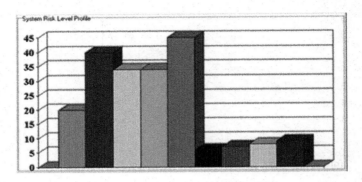

FIGURE 10.3 A project risk profile. (*Source:* © 2016 IEEE. Reprinted, with permission, from Kasser, Joseph Eli. 2016. "Applying Holistic Thinking to the Problem of Determining the Future Availability of Technology." The IEEE Transactions on Systems, Man, and Cybernetics: Systems no. 46 (3):440–444.)

10.5 RISK PROFILES

Risk profiles are:

- Attribute profiles (Kasser 2018: Section 9.1).
- An alternative to the risk rectangle (Section 10.2.1) that separates out the probability of occurrence and severity of impact into two separate charts:
  - *A probability of occurrence*: which plots the number of risks associated with each probability level.
  - *A level of severity of the impact*: which plots the number of risks associated with the level of severity of the impact.
- Can be considered as the 'A' column in a risk CRIP chart (Section 11.5) for a project plotted as a histogram.
- Plotted using the information in the WPs once the probability of occurrence and level of severity of the impact have been estimated for the WP.
- Shown in Figure 10.3.

Risk profiles may be used by the PM in several ways including:

- As selection criteria for two different alternatives. The selection criteria would be set up so that the alternative with the lowest risk profile is chosen.
- Selecting which risks must be mitigated and which risks must be prevented (the ones with the highest impact).
- Deciding if the project should be cancelled during the project planning state as being too risky.
- Comparing the risk profile of the project's system with a generic risk profile for the class of system to see if the project has a better than average chance of success.

10.6 RISK MITIGATION OR RISK PREVENTION

Some risks must be prevented while others may be mitigated or even ignored. For example:

- Truly catastrophic risks such as a nuclear power station meltdown must be prevented.
- Risks with high severity of impact and low probability of occurrence need to be mitigated or prevented.
- Risks with low probability of occurrence and low severity of impact are nuisances and may be ignored for a while, but should be mitigated at some future date.

10.7 CASCADING RISKS

Traditional risk management seems to consider single risks. However, in the systems approach, once the risk of failure is identified, the question, 'if this risk happens, will it stop the project?' must be posed. If the answer to the question is:

- *Yes:* ways of preventing or mitigating the risk need to be explored. For example, if an item of specialized equipment needs to be available for a certain WP, the consequences of not having that equipment when it is required need to be explored. The non-availability situation needs to be prevented or mitigated.
- *No:* the risk analysis has to determine the effect on the process to determine if the process can continue with acceptable performance. If it can't then ways of preventing or mitigating the risk need to be explored. If it can, then another round of risk analysis needs to be performed on this reduced performance process.

10.8 CONTINGENCIES AND CONTINGENCY PLANS

Contingencies are risks that cannot be prevented and must be mitigated. Contingency plans:

- Are the plans for mitigating the risk should it occur and turn it into an event.
- Are created before the risk occurs so that the way to resolve the problem posed by the events is documented.

When risks are identified in WPs during the project planning state (Section 5.3.1.2), a decision has to be made as to prevent or mitigate the risk. If the decision is:

- *To prevent the occurrence:* activities are added to WPs earlier in the schedule to prevent the occurrence of the risk.

- *To mitigate the risk:* a WP for developing the contingency plan is created. This WP should be performed in an earlier state of the project. The risk mitigation activities should either be in appropriate additional WPs or included in existing WPs.

10.9 THE ENGAPOREAN MCSSRP EXERCISE

The background and context for the progressive session exercises is in Appendix 1. The purpose of the exercise is:

1. To practice using the template for a student exercise presentation (Kasser 2018: Section 14.7).
2. To perform risk assessment.

10.9.1 THE REQUIREMENTS FOR THE EXERCISE

1. As PM, you are preparing for the systems requirements review (SRR).
2. Perform a risk assessment for each WP (Section 9.6) identifying probability of occurrence and severity of impact and mitigation strategy.
3. Update items 18 and 19 in each WP. Use that information to create:
   1. A risk rectangle (Section 10.2.1).
   2. Two risk profiles, one for probability and one for severity (Section 10.5).
4. Formulate the problem posed by the exercise using the problem formulation template (Section 3.7.1).
5. Make reasonable assumptions.
6. Format presentation according to the applicable parts of the sample four-part template for a management review presentation (Section 5.4.2.2.1).
7. Number the slides.
8. Prepare <5 min presentation that includes:
   1. The assumptions.
   2. The problem posed by the exercise formulated according to the problem formulation template (Section 3.7.1).
   3. Your technology assessment risk classification for the Multi-satellite Communications Switching System Replacement Project (MCSSRP) and justification (Section 10.4).
   4. Your assessment of the TAWOO state of the MCSSRP and justification.
   5. An example of the updated sections of one WP.
   6. The risk rectangle.
   7. The two risk profiles.
   8. Lessons learned.
   9. A compliance matrix (Kasser 2018: Section 9.5.2) for meeting the requirements for the exercise.
9. Save presentation as file '10-teamname'.pptx.

10.10 SUMMARY

This chapter focused on risk management is another important part of project management including risk prevention as well as mitigation.

REFERENCES

Benjamin, Daniel. 2006. Technology readiness level: An alternative risk mitigation technique. In *Project Management in Practice: The 2006 Project Risk and Cost Management Conference*, Boston, MA: Boston University Metropolitan College.

Crépin, Maxime, Bernard El-Khoury, and C. Robert Kenley. 2012. It's all rocket science: On the equivalence of development timelines for aerospace and nuclear technologies. In *The 22nd Annual International Symposium of the International Council on Systems Engineering*, Rome, Italy.

DOD. 2001. Program Manager's Guide for Managing Software, Draft 0.4, 2001.

El-Khoury, Bernard. 2012. *Analytic Framework for TRL-Based Cost and Schedule Models*. Cambridge, MA: Engineering Systems Division, Massachusetts Institute of Technology.

Hubbard, Douglas 2009. *The Failure of Risk Management: Why It's Broken and How to Fix It*. Hoboken, NJ: John Wiley & Sons.

Kasser, Joseph Eli. 2016. Applying holistic thinking to the problem of determining the future availability of technology. *The IEEE Transactions on Systems, Man, and Cybernetics: Systems* no. 46 (3):440–444. doi:10.1109/TSMC.2015.2438780.

Kasser, Joseph Eli. 2018. *Systems Thinker's Toolbox: Tools for Managing Complexity*. Boca Raton, FL: CRC Press.

Kasser, Joseph Eli, and Souvik Sen. 2013. The United States airborne laser test bed program: A case study. In *the 2013 Systems Engineering and Test and Evaluation Conference (SETE 2013)*, Canberra, Australia.

Mankins, John C. 1995. Technology Readiness Levels. Advanced Concepts Office, Office of Space Access and Technology, NASA.

Miller, George. 1956. The magical number seven, plus or minus two: Some limits on our capacity for processing information. *The Psychological Review* no. 63:81–97.

Shenhar, Aaron J., and Zev Bonen. 1997. The new taxonomy of systems: Toward an adaptive systems engineering framework. *IEEE Transactions on Systems, Man, and Cybernetics - Part A: Systems and Humans* no. 27 (2):137–145.

11 Successful Performance Monitoring and Controlling

This chapter explains how to monitor, control and communicate the performance of a project. The focus of performance monitoring is on the process and the product in the traditional project management paradigm. As the project proceeds along its timeline, the actual cost (AC) and schedule are compared with the estimated cost and schedule for that point in the time line. In many situations, especially during the long development times, the:

- *Customer:* makes periodic progress payments to the project. In this situation, since the acceptance tests are generally made at the end of the system development process (SDP), the suitability of the product for its mission is unknown for the time in which the bulk of the payments are made.
- *Project manager (PM):* provides the customer with minimal information to demonstrate the risk of non-compliance with the contract. The information is typically provided in the form of technical and management reports covering:
  - *Cost:* where the AC incurred by the project up to a point along the timeline is compared to the estimate of the amount of money that should have been spent on the project by that point on the timeline.
  - *Schedule:* where the actual time spent on the project is compared to the estimate of the time that should have been spent on the project up to that particular time.
  - *Intermediate products:* documents and lines of code produced, defects found, number of requirements satisfied, etc.
  - *Process:* degree of compliance to the appropriate required standards, capability maturity model, International Organization for Standardization (ISO) models, etc.

These measurements focus on the process and product and provide post facto information, namely they report on what has already happened with the result that:

- The remainder of the interdependent P's of a project (Chapter 2) tend to be ignored.
- Project management is reactive instead of being proactive.
- The PM is unable to tell the customer the exact percentage of completeness of the project at any point along the timeline until the project is almost completed (Kasser 1997).

The version of the systems approach discussed in this chapter incorporates the remaining five interdependent P's of a project allowing project management to be more proactive.

The chapter explains:

1. Ways to detect and prevent potential project overruns in Section 11.1.
2. Ways to detect impending project failure in Section 11.2.
3. Managing changes in project scope in Section 11.3.
4. Earned value analysis (EVA) in Section 11.4.
5. Categorized requirements in process (CRIP) charts in Section 11.5.
6. Enhanced traffic light (ETL) charts in Section 11.6.
7. Management by objectives (MBO) in Section 11.7.
8. Management by exception (MBE) in Section 11.8.

11.1 DETECTING AND PREVENTING POTENTIAL PROJECT OVERRUNS

The sooner a potential overrun condition can be identified, the greater the number of mitigating options. This technique for detecting and preventing potential overrun conditions is based on anticipatory testing which is designed to prevent and detect a potential overrun situation early in the development stage (Kasser 1995) and is performed as follows:

1. During the planning process by working backwards from the end of the project, identifying risks that could cause project overruns and preventing them from occurring in earlier work packages (WP)s.
2. During the project performance state of the project by monitoring the attributes of the project personnel (Section 11.2).
3. MBE during the project planning and performance states of the project (Section 11.8).
4. Reviewing periodic[1] WP summaries prepared by the task leads (TLs) containing a:
   1. Narrative description of the work to be accomplished in the WP.
   2. List of problems identified.
   3. List of problems solved and the solutions.
   4. Recommendations for improvements.
   5. List of open 'action items', who they are assigned to and completion dates.
   6. List of 'action' items closed out during the period.
5. Management by walking around (MBWA): meeting with the customer, TLs and task personnel to discuss the periodic summaries and any exceptions noted in MBE as part of or in addition to the periodic chunking status meetings (Section 5.6.1).
6. Holding effective meetings (Section 2.1.5) when necessary.

---

[1] Between the major milestones.

The PM evaluates:

1. The planned budget against the funds actually spent as a means of determining budget variance.
2. The planned schedule against the actual schedule as a means of determining schedule variance (SV).
3. Work throughput using CRIP charts (Section 11.5) which provide visibility into the progress of each WP and identify bottlenecks and other potential problems in time to take proactive steps to prevent problem escalation.
4. The state of problem mitigation and prevention using ETL charts (Section 11.6.2).

When a problem is detected, the PM is responsible for identifying the potential overrun conditions, the cause(s) and conceptual problem mitigation approaches. If the projected cost impact can be managed to less than 5% of the original nominal cost estimate, and the original WPs are essentially unchanged, the WPs should not be modified.[2] If the projected cost overrun exceeds 5% of the original cost estimate or the original WPs need to be substantially altered, the PM notifies the customer using the change request process (Section 11.3.2). Cooperative efforts between the PM and customer to evaluate various options and alternatives are more likely to succeed when all parties recommend the same corrective action. After some discussion, some kind of change will be approved. When that happens, the affected WPs should be revised to reflect the modified plan along with a revised estimated cost as part of the change management process (Section 11.3).

11.2 WAYS TO DETECT IMPENDING PROJECT FAILURE

There was evidence at least 20 years ago, which suggested that most projects did not fail due to the non-mitigation of technical risks (Kasser and Williams 1998, CHAOS 1995). However, project management literature does not seem to have incorporated ways of preventing those types of risks identified in the two references and other similar literature. This section discusses types of risks so they may be detected and mitigated. The risks were identified in case studies written by students in the Graduate School of Management and Technology at the University of Maryland University College (Kasser and Williams 1998).[3] These students wrote and presented term papers describing their experiences in projects that were in trouble. The papers adhered to the following instructions:

1. Document a case study. Students had to write a scenario for the paper based on personal experience.
2. Analyse the scenario.

---

[2] It is within the tolerance of the original estimate. This percentage may be changed by agreement between the customer and the PM.
[3] These students were employed in the workforce and were working towards their degree in the evening. Their employment positions ranged from programmers to PMs. Some also had up to 20 years of experience in their respective fields.

3. Document the reasons the project succeeded or ran into trouble.
4. List and comment on the lessons learned from the analysis.
5. Identify a better way with 20/20 hindsight.
6. List a number of situational indicators that can be used to identify a project in trouble or a successful project while the project is in progress.

The methodology:

- Summarized the student papers to identify common elements.
- Surveyed systems and software development personnel via the Internet to determine if they agreed or disagreed with the indicators.
- Summarized and analysed the results.

A total of 19 students produced papers that identified 34 different indicators. Each indicator identified was a risk or a symptom of a risk that can lead to project failure. Several indicators showed up in more than one student paper. A survey questionnaire was constructed based on the student provided risk-indicators[4] and sent to systems and software development personnel via the Internet. The survey asked respondents to state if they agreed or disagreed that the student-provided indicators were causes of project failure.[5] One hundred and forty-eight responses were received. The survey results were analysed in several different ways looking for the risk indicators with the most and the least agreement.

11.2.1 THE TOP TEN RISK-INDICATORS

The top ten risk-indicators survey respondents agreed with were:

1. *Poor requirements:* 97% of those surveyed agreed with this risk-indicator.
2. *Lack of or poor plans:* 95% of those surveyed agreed with this risk-indicator.
3. *Resources are not allocated well:* 92% of those surveyed agreed with this risk-indicator.
4. *Failure to validate original specification and requirements:* 91% of those surveyed agreed with this risk-indicator.
5. *Failure to communicate with the customer:* 88% of those surveyed agreed with this risk-indicator.
6. *Lack of management support:* 87% of those surveyed agreed with this risk-indicator.
7. *Political considerations outweigh technical factors:* 86% of those surveyed agreed with this risk-indicator.
8. *Unrealistic deadlines hence schedule slips:* 86% of those surveyed agreed with this risk-indicator.

---

[4] The students tended to state 'problems' using the semantics of 'solutions' or 'symptoms' rather than 'causes'.
[5] The authors of the survey recognized that there were other causes of (risks) and project failure and added an 'other' category to the survey questionnaire for 'write-in' risks.

Performance Monitoring and Controlling 285

9. *Failure to consider existing relationships when replacing systems:* 65% of those surveyed agreed with this risk-indicator.
10. *Lack of priorities:* 85% of those surveyed agreed with this risk-indicator.

11.2.2 THE SIX RISK-INDICATORS MOST OF THE RESPONDENTS DISAGREED

The six risk-indicators most of the respondents disagreed with were:

1. *Failure to reuse code:* A major advantage of object-oriented technology is the ability to lower costs by reusing code. Yet 73% of those surveyed did not agree with this risk-indicator.
2. *Hostility between developer and* independent verification and validation *(IV&V):* This risk indicator shows a team problem and results in less than optimal costs due to the lack of cooperation.
3. *There are too many people working on the project:* This risk-indicator is based on Brooks (1982) which describes the problems associated with assigning additional people to projects. However, only 24% of those surveyed agreed with this risk-indicator.
4. *Failure to use problem language:* The use of problem language was promoted as one of the major advantages by Ward and Mellor (1985). Yet, only 36% of the respondents agreed that it was a risk. Several did not know what the term meant.
5. *The quality assurance team is not responsible for the quality of the software:* As discussed in the study, this was the only indicator that should have shown disagreement.
6. *Client and development staff fail to attend scheduled meetings:* This is a symptom of poor communication between the client and the developer. In addition, while there are other communication techniques available, if meetings are scheduled, and not attended, negative messages are sent to the project personnel. Yet only 42% of the respondents agreed that this was a risk-indicator.
7. *Failure to collect performance and process metrics and report them to management:* If measurements are not made and acted upon, how can the process be improved? Yet only 48% of the respondents agreed that this was a risk-indicator.

The study was performed in 1997, so the Chaos study (CHAOS 1995) served as a reference. It had identified some major reasons for project failure. The five risk-indicators in this study that were chosen as the most important causes for project failure also appear on the Chaos list of major reasons for project failure. For this and other reasons listed in the study, this study supports the findings of the Chaos study.

Most of the top risk-indicators were not technical they were people-related and political. Yet the bulk of risk management for projects at the system and product layer (Hitchins 2006) focuses on the product and process and ignores the other four interdependent P's of a project (Chapter 2). Accordingly, in order to prevent the project from incurring cost and schedule overruns from the people, problem, prevention and political interdependent P's of a project, the person responsible for the quality of the project, not the PM, should perform an audit of the project before major milestones as a minimum to ensure that none of these risk indicators are present in the project.

For more information about the study and the use of the analysis tools to process the survey responses, see the *Systems Thinker's Toolbox* (Kasser 2018: Section 4.6.3).

11.3 MANAGING CHANGES IN PROJECT SCOPE

There will always be changes in a project for any number of reasons including a change in:

- Needs of the customer.
- Budget of the customer.
- Schedule precipitated by a delay.
- Technical approach to an activity which becomes evident as the work progresses and more detailed information becomes available.

These changes alter the scope of work which in turn impacts planned cost, schedule and technical aspects of the original WP. Often, changes to the staffing profile of assigned staff are needed to perform the modified task. It is important that changes be managed.

11.3.1 Effect of Change on Project

In traditional project management, the effect of a change on a project is said to impact the triple constraints of project management. This is because a change in any one causes changes in the other two. The systems approach considers quadruple constraints (Section 4.4.2).

11.3.2 Change Request Processing

The change control process is the same no matter the source or cause of the change if set up as shown in Figure 11.1 (Kasser 2002). A request for a change by the customer or a needed change due to some anticipated or unanticipated event occurring is processed through the same process. The process is as follows:

1. Log the change request and assign it an identification number.
2. Prioritize the change request.
3. Determine if a contradiction exists between this change request and the requirements and if so, resolve the contradiction.
4. Estimate impact on cost and schedule to implement due to the effect on the WPs.
5. Determine cost and schedule drivers for the impact.
6. Perform sensitivity analysis on the cost and schedule drivers to see if the impact can be reduced.

Performance Monitoring and Controlling

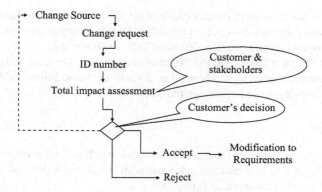

FIGURE 11.1 The systemic and systematic change control process.

7. Are cost drivers really necessary? Negotiate degree of change with the customer based on the sensitivity analysis. Sometimes customers ask for more than they really need, not realizing that they can get what they need in an affordable manner, but in order to get what they're actually asking for the cost and schedule will be severely impacted.
8. Make the decision as to accept or reject the change. Part of the decision to accept the change is to decide where and when it will be implemented. For example, in a software project where the product is being produced in sequential Builds, part of that decision is to decide in which Build the change will be implemented. One of the factors in this decision is the priority based on the urgency of the need for the change.
9. If the change is rejected, the requester is notified and the change request and impact analysis saved in case the decision needs to be revisited or understood in the future and the process ends here.
10. If the change is accepted, the appropriate project documentation in all three streams of work (Section 2.5.2) needs to be updated. For example:
    - *Development:* the requirements must be changed to reflect the new situation. This is done by adding, deleting or modifying (a combination of adding and deleting) requirements. The requirements change may affect the capability of components at various levels of the design.
    - *Management:* the project plan (PP) must be changed to show when and where the change will be implemented by changing the WPs for the individual tasks that are affected by the change.
11. If the additional work is:
    - Within 5% of the existing cost and schedule, modify the WP.
    - Within the scope of an existing WP, modify the WP.
    - Beyond the scope of an existing WP, initiate a new WP.
12. The change is implemented by adjusting the budget, schedule and any corresponding staffing levels. The PM will then have to manage to the new budget, staffing level and performance schedule.

13. Document decisions in project files as a reference in case a similar or identical change request shows up in the future. If the change request is rejected, the reasons for the rejection also need to be documented.
14. Adjust the plan to include the effect of the change. The cost and schedule will generally need to be changed as a result of the addition or deletion of work resulting from the change request.

11.4 EVA[6]

EVA is a project management tool for cost control. It is based on comparing the budgeted or planned cost and planned schedule with the actual values of money spent and time taken to do the work. It is used to:

1. Determine if the cost is under control.
2. Compare how the AC, scope and schedule of a project compares with the planned cost, scope and schedule.
3. Forecast if the project will overrun or underrun the planned budget at completion (BAC).
4. Forecast if the project will overrun or underrun the planned schedule at completion.
5. The traditional approach to EVA originated in 1966 when the U.S. Air Force mandated earned value management in conjunction with the other requirements on planning and controlling Air Force programmes. The requirement was entitled, the cost/schedule planning control specification (C/SPCS). The concept and its requirements have remained basically unchanged since then.

The systems approach integrates EVA into project management. The PM creates a PP in the project planning state of the project (Section 5.3.1.2). The contents of the plan include the estimated costs and schedules for each WP in the project as well as the estimated total cost and schedule for the project. The initial estimates provide the baseline costs and schedule. As time passes and each WP is completed, the AC, estimated and forecasted cost and schedule to completion are compared; variances noted and appropriate corrective action taken.

Activities earn value as they are completed; the earned value (EV) at a reporting milestone is the budgeted cost of the activity completed as of the time of the reporting milestone.

11.4.1 THE ELEMENTS OF EVA

In summary the elements of EVA are:

1. Scheduling.
2. Authorization of work.

---

[6] This section is a modified version of *The Systems Thinkers' Toolbox* (Kasser 2018: Section 8.3).

Performance Monitoring and Controlling

3. Variance analysis.
4. Estimate at completion (EAC).
5. Baseline maintenance and control.
6. Budgeting.
7. Organization.
8. Data accumulation and reporting.

11.4.2 The Terminology of EVA

The terminology generally used with EVA covers planning or estimating, project monitoring and controlling, and providing summaries of the state of the project as described in this section.

11.4.2.1 Planning or Estimating Terminology
- *BAC:* the total authorized budget for the project.

11.4.2.2 Project Monitoring and Controlling Terminology
- *AC:* also called *actual cost of work performed (ACWP):* the cost of all the work that was performed up to and including the reporting period.
- *EV:* also called *budgeted cost of work performed (BCWP):* the planned (not the actual) cost of the work that has been performed.
- *EAC:* the AC plus the *estimate to completion (ETC).*
- *ETC:* estimated or expected costs of completing the remainder of work on the project as of the reporting period.
- *Planned Value (PV)* also called *budgeted cost of work scheduled (BCWS):* the cost of all the work that was scheduled for the reporting period.
- *SV:* the difference between the actual and planned time taken to complete the project or an activity of a project.
- *Cost Variance (CV):* the difference between the cost actually incurred and the estimated cost.
- *Variance at Completion (VAC):* calculated as BAC minus EAC.

11.4.2.3 Indices and Summary Terminology
- *Cost Performance Index (CPI):* an estimate of the projected or AC of completing the project based on the performance to date.
- *Schedule Performance Index (SPI):* an estimate of the remaining time needed to complete the project.

11.4.2.4 EVA Calculations
The EVA elements and equations:

- Are summarized in Table 11.1.
- Can be adapted to give an ETL chart (Section 11.6), high-level summary or overview of the state of the project as defined in Table 11.2 (in the systems approach). The table also shows the suggested colours if the reporting presentation uses ETL charts to show the state of problems.

TABLE 11.1
EVA Elements and Equations

Term		Equation
Cost Performance Index	CPI	EV/AC
Cost Variance	CV	PV – AC
Cost Variance %	CV%	CV/EV
Estimate at Completion	EAC	AC + ETC
Earned Value	EV	BAC*AC/EAC
Schedule Performance Index	SPI	EV/PV
Schedule Variance	SV	EV – PV
Schedule Variance %	SV%	SV/PV
Variance at Completion	VAC	BAC – EAC

TABLE 11.2
EVA - Traffic Light Chart Performance Indicator Definitions

Index	Value	Condition	Traffic Light Chart
CPI	=1	AC = planned cost (no CV)	Green
	<1	Over budget	Yellow/red
	>1	Under budget	Blue
SPI	=1	On schedule (no SV)	Green
	<1	Behind schedule	Yellow/red
	>1	Ahead of schedule	Blue
CV	<0	Over budget	Yellow/red
	>0	Under budget	Blue
SV	<0	Behind schedule	Yellow/red
	>0	Ahead of schedule	Blue
VAC	<0	Projected underrun	Blue
	>0	Projected overrun	Yellow/red

A major difficulty with EVA is making reasonably accurate estimates of the amount of work completed. Perceptions from the *Quantitative* HTP point out that the estimate must be usable but does not necessarily have to be accurate to decimal points. The errors should tend to cancel out as the WP elements are rolled up. Ways of making estimates of the amount of work completed depend on the project and include:

1. Number of units completed as a percentage of the total number of units that have to be completed.
2. Number of WPs completed.

Performance Monitoring and Controlling

3. Amount charged to an activity. This is not a very good one, but if the budget for an activity is $50,000 and $25,000 has been charged to the activity, one hopes the activity can be estimated as being 50% complete.
4. Experience on similar projects.

The systems approach with its focus on products rather than process (activities) avoids many of these difficulties by focusing on tangible results rather than intangible activities.

11.4.3 Requirements for the Use of EVA in a Project

The requirements for the use of EVA in a project are as follows:

1. The project shall have a PP.
2. The work to be performed in the PP shall be split into activities.
3. The cost for each activity shall be estimated.
4. The schedule for each activity shall be estimated.
5. The project shall have an EVA system.
6. The project shall enter the following information into the EVA system in a timely manner.[7]
    1. Every activity.
    2. Estimated cost of every activity.
    3. ACs incurred during the performance of each activity.
    4. Estimated schedule for each activity.
    5. Actual time taken for the performance of each activity.
7. The EVA system shall allow the following comparisons between:
    1. AC and estimated costs.
    2. Actual time taken for each activity and estimated scheduled time.
8. The EVA system shall allow the following information to be projected forward to the date of completion:
    1. Cost.
    2. Schedule.
9. The EVA system shall provide the following information as a minimum:
    1. SV.
    2. AC.
    3. VAC.
    4. EV.
    5. BAC.
    6. CV.
    7. EAC.
    8. PV.

---

[7] The time shall be determined by the frequency of reporting as agreed to by the PM and the entity receiving the reports.

11.4.4 Advantages and Disadvantages of EVA

EVA has a number of advantages and disadvantages as described in this section.

11.4.4.1 EVA Advantages

The advantage of EVA is proactive project management. By comparing the difference between AC, estimated and projected cost and schedule to the completion date, it is possible to detect problems and compare the effect of different types of corrective action on cost and schedule.

11.4.4.2 EVA Disadvantages

The disadvantages of EVA include:

- EVA doesn't take quality into consideration; it only considers cost and schedule.
- EVA doesn't seem to take into effect changes to the project requirements and the consequent impact on cost and schedule other than by showing variances without changing the baseline. For example, if there was an approved change to the requirements which impacted the work during a reporting period than either the additional work was not planned but was performed or planned work was not performed because the change deleted the need to do the work. Accordingly, the variance is not real.
- On large projects a lot of data may need to be collected. However, these days most projects use some kind of software and so entering and retrieving information should not be too difficult.
- EVA needs to be integrated into the management planning, organizing, monitoring and control activities. If it is performed as an add-on for the purpose of meeting a contractual or other requirement, it will add work and accordingly inflate the cost of the project but will not provide any benefit at all.
- EVA only monitors two of the seven interdependent P's of a project (Chapter 2).

11.4.5 Examples of the Use of EVA

Consider the following two examples of the use of EVA.

1. The Master's Degree Project discussed in Section 11.4.5.1.
2. The Data Centre Upgrade Project discussed in Section 11.4.5.2.

11.4.5.1 The Master's Degree Project

Studying for a master's degree full-time is a project that will take a year. Extracting the cost information from the master's degree project WPs,[8] there are only five categories of estimated costs (simple numbers) which are:

---

[8] Not provided.

Performance Monitoring and Controlling

1. Tuition fees, $24,000.
2. Housing, $500 a month = $6,000.
3. Expenses (accommodation, books, food, clothes, phone, Internet, etc.), $1,000 a month = $12,000.
4. Travel, $2,000.
5. Contingency, $6,000.

The expenses are all lumped together. The number is an approximation based on other people's experience. There is little point in wasting time identifying the expenses for the individual items. There probably won't be any greater degree of accuracy. The contingency covers all other expenses, some of which may not be known at planning time.

The estimated payment schedule over 12 months is:

- Tuition paid in two equal instalments in months 1 and 7.
- Travel paid in month 1 (open return ticket).
- Other expenses paid monthly.

Total estimated expenses are $24,000 + $6,000 + $12,000 + $2,000 + $6,000 = $50,000.

The estimated payment schedule, excluding the contingency payments can be listed in a table as shown in Table 11.3. The monthly budget column shows how much is planned to be spent each month, and the cumulative budget column summarizes the cumulative planned expenses. This shows how much cash has to be in hand or in the bank at the beginning of each month without considering the contingency expenses. Table 11.3 provides no forecasting information. The same estimated planned budget can also be shown in a graphical format as shown in Figure 11.2 by plotting the planned cumulative monthly expenses (vertical axis) against the timeline (Kasser 2018: Section 8.15) (horizontal axis).

TABLE 11.3
Estimated Budget for Studying for a Master's Degree

Month	Monthly Budget	Cumulative Budget
1	$15,500	$15,500
2	$1,500	$17,000
3	$1,500	$18,500
4	$1,500	$20,000
5	$1,500	$21,500
6	$1,500	$23,000
7	$13,500	$36,500
8	$1,500	$38,000
9	$1,500	$39,500
10	$1,500	$41,000
11	$1,500	$42,500
12	$1,500	$44,000

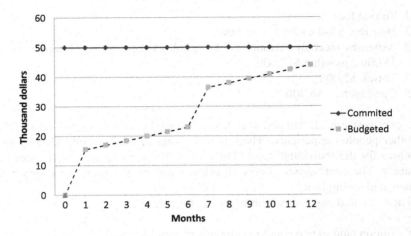

FIGURE 11.2 The Master's degree project projected cash flow.

Once the project starts, money is spent and so the amount of money spent each month is compared to the planned amount to see if the project spending is under control. The estimated amounts from Table 11.3 and the amounts actually spent are shown in Table 11.4. The monthly budget column shows how much is planned to be spent each month, and the cumulative budget column summarizes the cumulative planned expenses. The monthly spent column shows how much was spent each month and cumulative spent column summarizes the amount of spending up to that month. For example, in the first month the plan was to spend $15,500 but only $1,000 was actually spent. This is probably because a payment was delayed. Sure enough, in the second month, we can see that we planned to spend $1,500 for a cumulative spending of $17,000 and we actually spent $16,000 that month which brought the

TABLE 11.4
Actual Budget for Studying for a Master's Degree after 9 Months

Month	Monthly Budget	Cumulative Budget	Monthly Spent	Cumulative Spent
1	$15,500	$15,500	$1,000	$1,000
2	$1,500	$17,000	$16,000	$17,000
3	$1,500	$18,500	$1,500	$18,500
4	$1,500	$20,000	$4,000	$22,500
5	$1,500	$21,500	$500	$23,000
6	$1,500	$23,000	$200	$23,200
7	$13,500	$36,500	$13,000	$36,200
8	$1,500	$38,000	$2,000	$38,200
9	$1,500	$39,500	$1,300	$39,500
10	$1,500	$41,000		
11	$1,500	$42,500		
12	$1,500	$44,000		

Performance Monitoring and Controlling

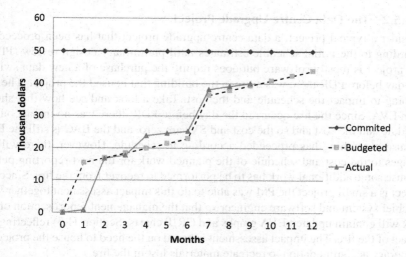

FIGURE 11.3 The Master's degree project financial status after 9 months.

cumulative amount spent up to $17,000 so this reflects the delayed tuition payment. The data in Table 11.4 shows that the actual amount of funds spent has varied slightly from the planned amount but the project is currently forecasted as finishing within budget.

The cost and expenditures for the first 5 months of the master's degree project provide an example of the use of EVA. The data shown in Figure 11.3 are summarized in Table 11.5. In accordance with the rules for EVA, the management reserve or contingency of $6,000 is not included in the data. That's why BAC has the value of $44,000. Since there are no additional expenditures, BAC remains at zero throughout the duration covered by the table.

TABLE 11.5
EVA for the First 5 Months of a Master's Degree

Month	1	2	3	4	5
BAC	$44,000.00	$44,000.00	$44,000.00	$44,000.00	$44,000.00
PV	$15,500.00	$17,000.00	$18,500.00	$20,000.00	$21,500.00
AC	$1,000.00	$17,000.00	$18,500.00	$22,500.00	$23,000.00
VAC	$0.00	$0.00	$0.00	$0.00	$0.00
EAC	$44,000.00	$44,000.00	$44,000.00	$44,000.00	$44,000.00
ETC	$43,000.00	$27,000.00	$25,500.00	$21,500.00	$21,000.00
EV	$1,000.00	$17,000.00	$18,500.00	$22,500.00	$23,000.00
CV	$14,500.00	$0.00	$0.00	−$2,500.00	−$1,500.00
CV%	1450.00%	0.00%	0.00%	−11.11%	−6.52%
SV	−$14,500.00	$0.00	$0.00	$2,500.00	$1,500.00
SV%	−93.55%	0.00%	0.00%	12.50%	6.98%

11.4.5.2 The Data Centre Upgrade Project

Consider a typical project, a data centre upgrade project that has been proceeding according to the initial schedule estimates until preliminary design review (PDR). The project is mostly software but does require the purchase of a new data switch. The day before PDR there was a fire in the building that housed the project. The fire is going to impact the schedule and the cost. Take a look and see how that shows up in EVA. Since the fire occurred the day before PDR, there was no impact on the schedule and the cost and so the cost and SVs are zero, and the BAC is still the ETC because everything has proceeded according to schedule. However, there will be changes to the cost and schedule of the planned work for the next reporting period because some additional work has to be performed to recover from the fire. Since the project is a small project the PM was able to do this impact assessment together with the chief system and software engineer so that the management report section of the PDR will contain updated EVA graphs and CRIP charts (Section 11.5) reflecting the impact of the fire. The impact assessment is based on the need to house the project in temporary accommodation to recreate materials lost in the fire.

The impact assessment estimated that it would take 2 months to restore the project to the state that it was at PDR. This would require a new WP to be inserted in the original schedule between the date of the PDR and the date of the critical design review (CDR). The estimated cost of the WP is $20,000. The project EVA status at PDR is shown in Figure 11.4. The ETC remains at $170,000, but the BAC has been increased by $20,000, so it's now equal to $190,000. The slope of the PV lines between PDR and CDR are different. This reflects the extra $20,000 estimated cost. However, the work following CDR to delivery readiness review (DRR) is expected to be the same as originally planned so the slope of the two lines remains the same. Accordingly, the project is currently scheduled to go over cost by $20,000 and incur a 2-month delay.

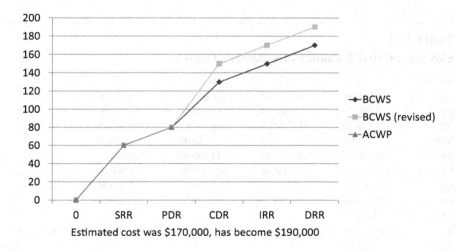

FIGURE 11.4 The data centre upgrade project EVA status at PDR.

Performance Monitoring and Controlling

TABLE 11.6
Notional ETL Chart Colours for Cost and Schedule Component of PDR

SRR	PDR		CDR (Planned)
	Last time	Current	
Green	Green	Yellow/red	Yellow/red

The ETL chart for cost and schedule component of PDR is shown in Table 11.6. The changes due to the fire can show up as green, yellow or red depending on the cost and SV thresholds set for the colour as shown in Table 11.26.[9]

Now, the deputy PM was on vacation at PDR. When the deputy PM returned from her vacation and found out what had happened, she produced a backup of all the work that she had kept on her personal portable disk drive. She believed in risk management, but had been unable to convince the PM to modify the configuration management (CM) system such as it was, to include an off-site backup. Her personal backup reduced the delay by 1 month and halved the cost of the delay. Accordingly, by the time the project got to CDR, it was only 1 month behind schedule and $10,000 over cost. The 1-month delay was due to the temporary use of furniture, verifying that the backup really was up-to-date, and associated activities with the temporary accommodation and then restoring the fire-damaged accommodation to workable condition. The revised project status at CDR is shown in Figure 11.5. The figure displays information with respect to the start of the project that is the original estimates. The figure does not show that the CDR cost and schedule were reduced by $10,000

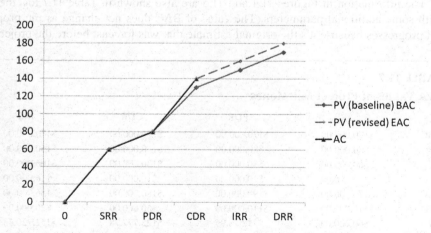

FIGURE 11.5 The revised data centre upgrade project EVA status at CDR.

---

[9] The table is located in the section of the chapter that discusses ETL charts, hence this forward reference.

and 1 month from the revised PDR estimates. This improvement could be shown in a number of ways, including:

- Two financial graphs:
  1. Showing the status with reference to the start of the project.
  2. Showing the change between PDR and CDR as in Figure 11.5.
- A compound line and bar chart such as Figure 8.7.
- Two ETL charts; the first with reference to the start of the project; the second showing the change between PDR and CDR.
- A table showing BAC and the changes in EAC and ETC.

One of the design elements approved at CDR was the necessary data switch. This was to be a commercial-off-the-shelf (COTS) device. The specific device was selected after a comparison of a number of COTS switches based on cost and delivery schedule. However, after CDR the proposed vendor went bankrupt. This meant that the selected switch was no longer available for use. As a result, the design team performed unplanned activities to locate and procure an alternative switch. These unplanned activities needed a new design WP to make any changes necessary in cable lengths, installation drawings, changes to test plans and procedures, etc. with a corresponding cost and schedule impact on the project.

The impact was a new WP for design had to be inserted into the schedule. This resulted in a delay of 1 month, and a cost escalation of $6,000, $1,000 for labour and an extra $5,000 for the next cheapest switch that could be delivered with the least amount of schedule delay. Consequently, the integration readiness review (IRR) was delayed by 1 month because the replacement switch had to be delivered and tested before IRR, and the EAC increased from $170,000 to $186,000 (9.4% increase) as shown in Table 11.7.

The information in Figures 11.4 and 11.5 are also shown in Table 11.7 together with some additional parameters. The value of BAC does not change as the project progresses because it's the original estimate that was forecast before the project

TABLE 11.7
EVA Values at Project Milestones

	SRR	PDR	CDR	IRR
BAC	$170,000.00	$170,000.00	$170,000.00	$170,000.00
PV	$60,000.00	$80,000.00	$140,000.00	$150,000.00
AC	$60,000.00	$80,000.00	$130,000.00	$166,000.00
VAC	$0.00	$20,000.00	$10,000.00	$16,000.00
EAC	$170,000.00	$190,000.00	$180,000.00	$186,000.00
ETC	$110,000.00	$110,000.00	$50,000.00	$20,000.00
EV	$60,000.00	$71,578.95	$122,777.78	$151,720.43
CV	$0.00	$0.00	$10,000	−$16,000
CV%	0.00%	0.00%	8.14%	−10.55%
SV	$0.00	−$8,421.05	−$17,222.22	$1,720.43
SV%	0.00%	−10.53%	−12.30%	1.15%

Performance Monitoring and Controlling

began. The project proceeded according to plan from the beginning to the PDR. However, the fire at the PDR resulted in additional work. That shows up in the change from $60,000 to $80,000 in the value of PV at PDR. The additional $20,000 shown in the VAC row means that EAC increased from $170,000 to $190,000 and the ETC is now calculated on the difference between the AC and the EAC. Similar changes occurred in PV and EV due to the events before CDR and IRR. CV shows up as negative in this table representing an overrun.

11.4.6 The Systems Approach Perspective on EVA

EVA seems to be designed to reduce project cost and schedule data to simple numbers. The terminology also seems to be confusing when first encountered. EVA, however, is mandated for use in U.S. government contracts so government contractors have to use it. However, in the commercial sector a simpler approach could be employed. Use the EVA numbers cost information but keep the schedule information in Gantt charts (Kasser 2018: Section 8.4). Only bring them together in an ETL chart during the management brief in the reporting meeting.

Financial information is only as good as the estimates and is only a part of the story. The systems approach to project management links all the work to requirements, manages change proactively, uses WPs and presents information in the simplest format and in the customer's language in accordance with the KISS principle (Kasser 2018: Section 3.2.3) thereby preventing the following top ten reasons as to why EVA does not work (Lukas 2008):

1. No documented requirements.
2. Incomplete requirements.
3. Work breakdown structure (WBS) not used or not accepted.
4. WBS incomplete.
5. Plan not integrated (WBS–Schedule–Budget).
6. Schedule and/or budget incorrect.
7. Change management not used or ineffective.
8. Cost collection system inadequate.
9. Incorrect progress.
10. Management influence and/or control.

Perceptions from the *Generic* HTP identify a similarity between these top ten reasons and the top risk-indicators for predicating project failure (Section 11.2.1).

11.5 CRIP CHARTS

The CRIP chart (Kasser 2018: Section 8.1) is a project management, systems and software engineering tool[10] that:

- Is used in monitoring and controlling projects.
- Provides a way to think about and measure technical progress.

---

[10] This section is an updated version of (Kasser 2015: Chapter 19) and slightly different to the version in (Kasser 2018: Section 8.1).

- Identifies potential problems in near real-time so as to be able to prevent or mitigate the problems before they occur (proactive risk management).

While simplistic approaches of tracking the realization activities of all the requirements or features such as feature-driven development (Palmer and Felsing 2002) can inform about the state of the realization activities, they cannot be used to estimate the degree of completion since each requirement or feature has a different level of complexity and takes a different amount of effort to realize. The need is for a measurement approach that can:

- Roll up the detailed information into a summary that can be displayed in one or two charts.
- Readily relate to the existing cost and schedule information.

The CRIP approach (Kasser 1999) meets that need by looking at the change in the state of a summary of the realization activities, which convert requirements into systems during the SDP from several HTPs. The summary information is presented in a table known as a CRIP chart which:

- Covers the entire SDP.
- Uses a technique similar to feature-driven development charts (Palmer and Felsing 2002).
- Provides simple summaries suitable for upper management uninterested in details.
- Indicates variances between plan and progress but not the causes of the variance. It is up to management to ask for explanations of the variances.
- Is based on the use of WPs in planning a project, hence have to be integrated into the SDP at the planning stage of a project.
- Is designed to be used in association with the quadruple constraints of scope, cost, people and schedule information shown in Figure 11.6.

This section first describes the CRIP chart and then gives examples of how it can be used.

11.5.1 THE FIVE-STEP CRIP APPROACH

The five-step CRIP approach is to:

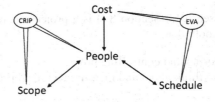

FIGURE 11.6 Measuring the quadruple constraints.

1. Identify categories for the requirements.
2. Quantify each category into ranges.
3. Categorize the requirements.
4. Place each requirement into a range in each category.
5. Monitor the differences in the state of each of the numbers of requirements in a range at the SDP formal and informal reporting milestones.

The first four steps take place before the systems requirements review (SRR). The last step, which takes place in all the states of the SDP following the SRR, is the key element in the CRIP approach because it is a dynamic measure of change rather than a static value. Consider each of the five steps.

11.5.1.1 Step 1: Identify Categories for the Requirements

Identify categories for the requirements. Typical categories are:

- *Complexity:* of the requirement.
- *Estimated cost:* to implement the requirement.
- *Firmness:* the likelihood that the requirement will change during the SDP.
- *Priority:* of the requirement.
- *Risk:* probability of occurrence, severity if it occurs, etc.

Each of the different categories may be considered as points on a perspectives perimeter of categories (Kasser 2018: Chapter 10).

11.5.1.2 Step 2: Quantify Each Category into Ranges

Quantify each category into no more than ten ranges. Thus, for:

- *Priority:* requirements may be allocated priorities between one and ten.
- *Complexity:* requirements may be allocated estimated complexities of implementation between 'A' and 'J'.
- *Estimated cost to implement:* requirements may be allocated estimated costs to implement values. For example, less than $100, between $100 and $500, between $500 and $1,000, etc.
- *Risk (schedule completion):* requirements may be awarded a value between one and five.

The ranges are relative, not absolute. Any of the several quantitative decision-making techniques for sorting items into relative ranges may be used (Kasser 2018: Section 4.6). The buyer/customer and supplier/contractor determine the range limits in each category.

A requirement may be moved into a different range as more is learned about its effect on the development or the relative importance of the need changes during the SDP. Thus, if the priority of a specific requirement or the estimated cost to implement changes between SDP reporting milestones changes, the requirement may be moved from one range to another in those categories. However, the rules for setting the range limits, and the range limits must not change during the SDP.

11.5.1.3 Step 3: Categorize the Requirements

Every requirement shall be placed in each category.

11.5.1.4 Step 4: Place Each Requirement into a Range in Each Category

Place each requirement into one range slot for each category. The information used to place the requirements into the ranges for the categories comes from the WPs in the PP. If all the requirements end up in the same range slot, such as all of them having the highest priority, the range limits should be re-examined to spread the requirements across the full set of range slots. If most of them end up in a single range slot, then that slot should be expanded into several slots. There is no need for the ranges to be linear. For example, ranges could be 1–10, 11–20, 20–30 etc., or 1–10, 11–15, 16–20, 21–30, etc. In the second case, it just means that the Range 11–20 has been split into two ranges.

Once the requirements have been categorized in ranges, an attribute profile (Kasser 2018: Section 9.1) for each category can be drawn in the form of a histogram (Kasser 2018: Section 2.9).

11.5.1.5 Step 5: States of Implementation

At any time during the SDP, each requirement shall be in one, and only one, of five CRIP states. The CRIP states of implementation of each requirement during the project are:

1. *Identified:* a requirement has been identified, documented and approved.
2. *In-process:* the supplier has begun the development activities to realize the requirement.
3. *Completed:* the supplier has completed development activities on the equipment that will perform the requirement.
4. *In test:* the supplier has started to test compliance to the requirement.
5. *Accepted:* the buyer has accepted delivery of the part of the system (a Build) containing the implementation of the requirement.

The summaries of the number of requirements in each state are reported at project milestones and reporting meetings.

11.5.2 Populating and Using the CRIP Chart

An unpopulated CRIP chart is shown in Table 11.8 where:

- The vertical axis of the chart is split into the ten ranges in the category (1–10).
- The horizontal axis of the chart is split into five columns representing the CRIP states of a project.

Each CRIP state contains three cells: planned 'P', expected 'E' and actual 'A', where:

1. *[P] Planned for next reporting period:* the number of requirements planned to be in the CRIP state before the following reporting milestone.

TABLE 11.8
An Unpopulated CRIP Chart

	Identified			In Process			Completed			In Test			Accepted		
Range	P	E	A	P	E	A	P	E	A	P	E	A	P	E	A
1															
2															
3															
4															
5															
6															
7															
8															
9															
10															
Totals															

2. *[E] Expected:* the number of requirements expected to be in the CRIP state, based on the number planned in the previous reporting milestone. This is a copy of the 'P' value in the CRIP chart for the previous milestone.
3. *[A] Actual:* the number of requirements actually in the CRIP state.

For the first milestone-reporting period, the values for:

- *[E]:* are derived from the PP for the time period.
- *[A]:* is the number actually measured at the end of the reporting period.
- *[P]:* is a number derived from the PP and the work done during the current reporting period.

As of the first milestone following the start of a project, the numbers in the 'P' column of a CRIP state of the chart at one milestone are always copied into the 'E' column of the same CRIP state in the chart for the next milestone. The 'A' and 'P' values reflect the reality. As work progresses the numbers flow across the CRIP states from 'Identified' to 'Accepted'.

At each reporting milestone, the changes in the CRIP state of each of the requirements between the milestones are monitored. The numbers of each of the requirements in each of the categories are presented in tabular format in a CRIP chart at reporting milestones (major reviews or monthly progress meetings). Colours can be used to draw attention to the state of a cell in the table. For example, the colours can be allocated[11] such that:

- *Violet:* shows realization activities for requirements in that range is well ahead of estimates.

---

[11] The quantitative numbers for the ranges would be agreed upon between the stakeholders, specified in the contract prior to the commencement of the project, and not changed during the SDP.

- *Blue:* shows realization activities for requirements in that range is ahead of estimates.
- *Green:* shows realization activities for requirements in that range is close to estimates.
- *Yellow:* shows realization activities for requirements in that range is slightly below estimates.
- *Red:* shows realization activities for requirements in that range is well under the estimates.

The range setting for the colours in the CRIP charts should be the same as the range settings using in the ETL charts (Section 11.6) for the project.

CRIP charts show that a problem might exist. Any time there is a deviation between 'E' and 'A' in a CRIP state, the situation needs to be investigated just as a deviation in an EVA chart (Section 11.4) has to be explained. A comparison of the summaries from different reporting milestones can identify progress and show that problems may exist. On its own, however, the CRIP chart cannot identify the actual problem; its purpose is to trigger questions to determine the nature of the actual problem.

The CRIP charts when viewed over several reporting periods can identify other types of 'situations'. While the CRIP chart can be used as a stand-alone chart, it should really be used together with EVA (Section 11.4) budget and schedule information. For example, if there is a change in the number of:

- *Identified requirements and there is no change in the budget or schedule:* there is going to be a problem. Thus:
  - If the number of requirements goes up and the budget does not, the risk of failure increases because more work will have to be done without a change in the allocation of funds.
  - If the number of requirements goes down, and the budget does not, there is a financial problem. However, if it is in the context of a fixed price contract it shows additional profit.
- *Requirements being worked on and there is no change in the number being tested:* there is a potential supplier management or technical problem if this situation is at a major milestone review.
- *Requirements being tested and there is no change in the number accepted:* there may be a problem with the supplier's process or a large number of defects have been found and are being reworked.
- *Identified requirements at each reporting milestone:* the project is suffering from requirements creep if the number is increasing. This situation may reflect controlled changes due to the change in the customer's need, or uncontrolled changes.

Since projects tend to delay formal milestones until the planned work is completed, the CRIP charts are more useful in the monthly or other periodic meetings between the formal major milestones.

11.5.3 Advantages of the CRIP Approach

The advantages of the CRIP approach include:

- Links all work done on a project to the customer's requirements.
- May be used at any level of system decomposition.
- Provides a simple way to show progress or the lack of it, at any reporting milestone. Just compare the 'E' and 'A' numbers and ask for an explanation of the variances.
- Provides a window into the project for upper management (customer and project) to monitor progress.
- Can indicate if lower-priority requirements are being realized before higher-priority requirements if priority is a category.
- Identifies the probability of some management and technical problems as they occur, allowing proactive risk containment techniques.
- May be built into requirements management, and other computerized project and development management tools.
- May be incorporated into the progress reporting requirements in system development contracts. Falsifying entries in the CRIP chart to show false progress then constitutes fraud.
- Requires a process. Some organizations don't have one, so they will have to develop one to use CRIP charts.
- Requires CM and change management (Section 11.3), which tends to be poorly implemented in many organizations. The use of CRIP charts will enforce good CM.

11.5.4 Disadvantages of the CRIP Approach

The CRIP chart approach has the following disadvantages, it:

- Is a different way of viewing project progress or lack thereof.
- Requires categorization of the requirements.
- Requires sorting of the requirements into ranges in each category.
- Requires prioritization of requirements if priority is used as a category, which it should be.
- Provides a window into the project for micromanagement by upper management (MBUM).

11.5.5 Examples of Using CRIP Charts in Different Types of Projects

The following sections of the chapter show how CRIP charts can indicate the technical progress of a project and identify potential problems using the following stereotype examples:

1. An ideal project discussed in Section 11.5.5.1.
2. A project with 'requirements creep' in Section 11.5.5.2.

3. A challenged project in Section 11.5.5.3.
4. A 'make up your mind' project in Section 11.5.5.4.

The projects are all identical until completion of SRR.

Since projects tend to delay formal milestones until the planned work is completed, the CRIP charts are more useful in the monthly or other periodic meetings between the formal major milestones. However, for the example of these stereotype projects, the CRIP charts are provided for the formal milestones since their names are widely known.

As an example, consider Federated Aerospace, a major government contractor with multiple simultaneous project and contracts. One of Federated Aerospace's projects provides examples of the use of CRIP charts as follows. Federated Aerospace organized a proposal team to bid on a Request for Proposal (RFP) issued by a Government agency. Upon receipt of the RFP, the project team identified 279 requirements in the document and created CRIP charts for several categories of requirements including cost. The proposal team estimated the costs to realize those requirements as part of the proposal effort. Once the costs were estimated, the proposal team defined ten ranges of costs and allocated each requirement into the appropriate range. The CRIP chart for the cost category at the completion of the proposal shown in Table 11.9 indicates that:

- The RFP contained 279 requirements.
- The requirements have been grouped into ten cost ranges. There are 86 requirements in Range 1, 73 requirements in Range 2, 23 requirements in Range 3 and so on.
- No further work is planned at this time as shown by the zero values assigned to the 'P' columns of the 'Identified' and the 'In process' states.[12]

TABLE 11.9
The CRIP Chart at the Completion of the Proposal

Range	Identified P	E	A	In Process P	E	A	Completed P	E	A	In Test P	E	A	Accepted P	E	A
1	0		86	0											
2	0		73	0											
3	0		23	0											
4	0		34	0											
5	0		26	0											
6	0		15	0											
7	0		8	0											
8	0		7	0											
9	0		5	0											
10	0		2	0											
Totals	0		279	0											

---

[12] This would change if Federated Aerospace wins the contract award.

Performance Monitoring and Controlling

Federated Aerospace's proposal was accepted, and the government awarded a contract to Federated Aerospace for the project. This example provides typical CRIP charts for the following six notional major milestones in the SDP (Section 5.4.2):

1. SRR.
2. PDR.
3. CDR.
4. Subsystem Test Readiness Review (TRR).
5. IRR.
6. DRR.

The Federated Aerospace development stream of activities in the project started by confirming that all the requirements in the RFP:

- Were understood by the Federated Aerospace project team.
- Had not changed since the RFP had been issued.
- Were complete; there were no additional or deleted requirements which would change the scope and cost of the contract.
- Were tagged with Acceptance Criteria.[13]

The CRIP chart at the start of the ideal project shown in Table 11.10, based on the information in the RFP CRIP chart in Table 11.9 indicated that there would be no planned change between the number of requirements identified at SRR and those identified in the RFP since:

- The numbers in each row of the 'E' column in Table 11.10[14] match those in the corresponding rows of the 'A' column in Table 11.9.
- The number in each row in the 'P' column in the 'Identified' state in Table 11.9 has been set to zero since there are no planned additional requirements.

Note, as of the first milestone following the start of a project, the numbers in the 'P' column of a state of the CRIP chart at one milestone are always copied into the 'E' column of the same state in the CRIP chart for the next milestone as shown in Table 11.11.

The CRIP chart at SRR shown in Table 11.11 indicates that the project has deviated from the baseline plan since:

- There are differences between the expected numbers and the actual numbers of identified requirements since there are differences between

---

[13] Not used in CRIP charts, but needed elsewhere.
[14] This is the only time in a project when 'A' column numbers are copied from one CRIP state in a CRIP chart at one reporting period to the CRIP chart of the following CRIP state.

TABLE 11.10
The CRIP Chart at the Start of the Ideal Project

	Identified			In Process			Completed			In Test			Accepted		
Range	P	E	A	P	E	A	P	E	A	P	E	A	P	E	A
1	0	86	0												
2	0	73	0												
3	0	23	0												
4	0	34	0												
5	0	26	0												
6	0	15	0												
7	0	8	0												
8	0	7	0												
9	0	5	0												
10	0	2	0												
Totals	0	279	0												

TABLE 11.11
The Ideal Project CRIP Chart at SRR

	Identified			In Process			Completed			In Test			Accepted		
Range	P	E	A	P	E	A	P	E	A	P	E	A	P	E	A
1	0	86	81	81											
2	0	73	78	78											
3	0	23	35	35											
4	0	34	30	30											
5	0	26	26	26											
6	0	15	20	20											
7	0	8	8	8											
8	0	7	7	7											
9	0	5	5	5											
10	0	2	2	2											
Totals	0	279	292	292											

the numbers in several rows of the 'A' column and the 'E' column of the 'Identified' state. For example, in:

- Range 1 an 'E' number of 86 became an 'A' of 81.
- Range 2 an 'E' number of 73 became an 'A' of 78.
- Ranges 4 and 6 also contain changes.[15]

---

[15] Upon investigation, it was found that the changes in the number of requirements are due to the clarifications that occurred during the requirements elicitation and elucidation process, a typical project happening.

Performance Monitoring and Controlling

- The total number of identified requirements has increased from 279 to 292.[16]
- The PM plans to work on all the requirements to put them into a development following the conclusion of the SRR since the numbers from the 'A' column in the 'Identified' state have been copied into the 'P' column of the 'In process' state.
- The PM does not plan to identify any new requirements between SRR and PDR since all the rows in the 'P' column of the 'Identified' state have been reset to zero.

The stereotype projects diverge after SRR. Consider how the CRIP charts provide early identification of the technical progress or lack of progress (an indication of a potential problem) in the milestone reviews of the stereotype projects using the ideal project as a reference.

11.5.5.1 The Ideal Project
The ideal project is the one in which everything happens according to the plan and there are no changes in the requirements during the SDP, such as in a short duration project or an educational example.

11.5.5.1.1 The Ideal Project CRIP Chart at PDR

The ideal project CRIP Chart at PDR shown in Table 11.12 indicates that the project is proceeding according to plan since:

- No additional requirements were levied on the system as indicated by the 0 in all the rows in the 'A' column in the 'Identified' state.

TABLE 11.12
The Ideal Project CRIP Chart at PDR

	Identified			In Process			Completed			In Test			Accepted		
Range	P	E	A	P	E	A	P	E	A	P	E	A	P	E	A
1	0	0	0	0	81	81	0								
2	0	0	0	0	78	78	0								
3	0	0	0	0	35	35	0								
4	0	0	0	0	30	30	0								
5	0	0	0	0	26	26	0								
6	0	0	0	0	20	20	0								
7	0	0	0	0	8	8	0								
8	0	0	0	0	7	7	0								
9	0	0	0	0	5	5	0								
10	0	0	0	0	2	2	0								
Totals	0	0	0	0	292	292	0								

---

[16] The cost and schedule was renegotiated as a result and is reflected in the updated cost and schedule summaries also presented in the SRR (not included herein).

- The system design state activities commenced as expected since the 'A' numbers in the 'In process' state match the 'E' numbers.
- The project does not plan to complete the development of any requirements by CDR since there is a 0 in all of the rows in the 'P' column in the 'Completed'. This is because the CDR will be held before the end of the 'In process' CRIP state.
- The project does not plan to identify any new requirements between PDR and CDR since all the rows in the 'P' column of the 'Identified' CRIP state remain at 0.

11.5.5.1.2 The Ideal Project CRIP Chart at CDR

The ideal project CRIP chart at CDR shown in Table 11.13 indicates that the project is still proceeding according to plan since:

- No additional requirements were levied on the system as indicated by the 0 in all the rows in the 'A' column of the 'Identified' state.
- The plan is for all the development activities to be completed by TRR since all the numbers in the 'A' column of the 'In-process' state have been copied into the 'P' column of the 'Completed' state.
- The project does not plan to identify any new requirements between CDR and TRR since all the rows in the 'P' column of the 'Identified' state remain at 0.

11.5.5.1.3 The Ideal Project CRIP Chart at TRR

The ideal project CRIP chart at TRR shown in Table 11.14 indicates that the project is still proceeding according to plan since:

TABLE 11.13
The Ideal Project CRIP Chart at CDR

	Identified			In Process			Completed			In Test			Accepted		
Range	P	E	A	P	E	A	P	E	A	P	E	A	P	E	A
1	0	0	0	0	81	81	81								
2	0	0	0	0	78	78	78								
3	0	0	0	0	35	35	35								
4	0	0	0	0	30	30	30								
5	0	0	0	0	26	26	26								
6	0	0	0	0	20	20	20								
7	0	0	0	0	8	8	8								
8	0	0	0	0	7	7	7								
9	0	0	0	0	5	5	5								
10	0	0	0	0	2	2	2								
Totals	0	0	0	0	292	292	292								

TABLE 11.14
The Ideal Project CRIP Chart at TRR

	Identified			In Process			Completed			In Test			Accepted		
Range	P	E	A	P	E	A	P	E	A	P	E	A	P	E	A
1	0	0	0	0	0	0	0	81	81	81					
2	0	0	0	0	0	0	0	78	78	78					
3	0	0	0	0	0	0	0	35	35	35					
4	0	0	0	0	0	0	0	30	30	30					
5	0	0	0	0	0	0	0	26	26	26					
6	0	0	0	0	0	0	0	20	20	20					
7	0	0	0	0	0	0	0	8	8	8					
8	0	0	0	0	0	0	0	7	7	7					
9	0	0	0	0	0	0	0	5	5	5					
10	0	0	0	0	0	0	0	2	2	2					
Totals	0	0	0	0	0	0	0	292	292	292					

- No additional requirements were levied on the system as indicated by the 0 in all the rows in the 'A' column of the 'Identified' state.
- The development activities for the system have been completed since the numbers in the 'A' column of the 'Completed' state match those in the 'E' column.
- Testing of all the requirements is expected to begin immediately following TRR since the numbers in the 'A' column of the 'Completed' state have been copied into the 'P' column of the 'In test' state.
- The project does not plan to identify any new requirements between TRR and the following milestone since all the rows in the 'P' column of the 'Identified' state remain at 0.

11.5.5.1.4 The Ideal Project CRIP Chart at IRR

The ideal project CRIP chart at IRR, shown in Table 11.15 indicates that the project is still proceeding according to plan since:

- No additional requirements were levied on the system as indicated by the 0 in all the rows in the 'A' column of the 'Identified' state.
- Testing has begun to verify that the system meets all the requirements since all the numbers in the 'A' column of the 'In test' state match those in the 'E' column.
- The PM is planning to integrate the system for successful acceptance by the customer before the DRR since the values in the 'A' column of the 'In test' state have been copied into the 'P' column of the 'Accepted' state.
- The PM does not plan to identify any new requirements between IRR and DRR since all the rows in the 'P' column of the 'Identified' state remain at 0.

TABLE 11.15
The Ideal Project CRIP Chart at IRR

	Identified			In Process			Completed			In Test			Accepted		
Range	P	E	A	P	E	A	P	E	A	P	E	A	P	E	A
1	0	0	0	0	0	0	0	0	0	0	81	81	81		
2	0	0	0	0	0	0	0	0	0	0	78	78	78		
3	0	0	0	0	0	0	0	0	0	0	35	35	35		
4	0	0	0	0	0	0	0	0	0	0	30	30	30		
5	0	0	0	0	0	0	0	0	0	0	26	26	26		
6	0	0	0	0	0	0	0	0	0	0	20	20	20		
7	0	0	0	0	0	0	0	0	0	0	8	8	8		
8	0	0	0	0	0	0	0	0	0	0	7	7	7		
9	0	0	0	0	0	0	0	0	0	0	5	5	5		
10	0	0	0	0	0	0	0	0	0	0	2	2	2		
Totals	0	0	0	0	0	0	0	0	0	0	292	292	292		

11.5.5.1.5 The Ideal Project CRIP Chart at DRR

The ideal project CRIP chart at DRR shown in Table 11.16 indicates that the project is still proceeding according to plan since:

- No additional requirements were levied on the system as indicated by the 0 in all the rows in the 'A' column of the 'Identified' state.
- The integrated system has been accepted by the customer as having met all its requirements as indicated by the match between each of the values of the rows in the 'A' column and the corresponding rows in the 'E' column in the 'Acceptance' state.

TABLE 11.16
The Ideal Project CRIP Chart at DRR

	Identified			In Process			Completed			In Test			Accepted		
Range	P	E	A	P	E	A	P	E	A	P	E	A	P	E	A
1	0	0	0	0	0	0	0	0	0	0	0	0	0	81	81
2	0	0	0	0	0	0	0	0	0	0	0	0	0	78	78
3	0	0	0	0	0	0	0	0	0	0	0	0	0	35	35
4	0	0	0	0	0	0	0	0	0	0	0	0	0	30	30
5	0	0	0	0	0	0	0	0	0	0	0	0	0	26	26
6	0	0	0	0	0	0	0	0	0	0	0	0	0	20	20
7	0	0	0	0	0	0	0	0	0	0	0	0	0	8	8
8	0	0	0	0	0	0	0	0	0	0	0	0	0	7	7
9	0	0	0	0	0	0	0	0	0	0	0	0	0	5	5
10	0	0	0	0	0	0	0	0	0	0	0	0	0	2	2
Totals	0	0	0	0	0	0	0	0	0	0	0	0	0	292	292

Performance Monitoring and Controlling

11.5.5.2 A Project with Requirements Creep

This section shows how the CRIP charts can indicate that a project has requirements creep. Assume that the project has completed the SRR shown in Table 11.11 and that the changes in the number of requirements identified occur during the system design state and the subsystem construction states between the SRR and TRR milestones.

The project has chosen to hold the milestone reviews as scheduled even though the work on the additional requirements may be out of phase with the original requirements. This is fine when the additional requirements can be realized without impacting the original planned work such as when the additional requirements are for additional functionality which can be provided independently and integrated into the system as a separate plug in.

11.5.5.2.1 The CRIP Chart for a Project with Requirements Creep at PDR

The CRIP chart for the project with requirements creep at PDR shown in Table 11.17 indicates:

- Twenty-six unexpected[17] additional requirements were identified between SRR and PDR resulting in the total number of requirements increasing from 292 to 318.
- Twenty of the unexpected requirements were identified in Range 1 shown by the 20 in the 'A' column of the 'Identified' state.
- Development activities have begun on these 20 requirements since the value of 81 in the 'A' column of the 'In process' state has become 101, namely the original 81 and the additional 20.

TABLE 11.17
The CRIP Chart for the Project with Requirements Creep at PDR

	Identified			In Process			Completed			In Test			Accepted		
Range	P	E	A	P	E	A	P	E	A	P	E	A	P	E	A
1	0	0	20	0	81	101	0								
2	0	0	0	0	78	78	0								
3	0	0	0	0	35	35	0								
4	0	0	0	0	30	30	0								
5	0	0	2	0	26	28	0								
6	0	0	0	0	20	20	0								
7	0	0	4	0	8	12	0								
8	3	0	0	0	7	7	0								
9	0	0	0	0	5	5	0								
10	0	0	0	0	2	2	0								
Totals	3	0	26	0	292	318	0								

---

[17] These were unexpected as far as the contractor was concerned.

- Two of the unexpected requirements were identified in Range 5 shown by the 2 in the 'A' column of the 'Identified' state.
- Development activities have begun on these two requirements since the value of 26 in the 'A' column of the 'In process' state has become 28, namely the original 26 and the additional two.
- Four of the unexpected requirements were identified in Range 7 shown by the 4 in the 'A' column of the 'Identified' state.
- Development activities have begun on these four requirements since the value of the 'A' column of the 'In process' state has become 12, namely the original 8 and the additional 4.
- Three of the additional requirements in Range 8 are expected to be identified after PDR as indicated by the 3 in row 8 of the 'P' column of the 'Identified' state.
- The PM does not plan to identify any new requirements in any of the other ranges between PDR and CDR since all the rows in the 'P' column of the 'Identified' state in those ranges remain at 0.

The impact of the additional work should also show on the schedules (Section 7.1) and EVA charts (Section 11.4).

11.5.5.2.2 The CRIP Chart for a Project with Requirements Creep at CDR

The CRIP chart for the project with requirements creep at CDR shown in Table 11.18 indicates:

- Two additional requirements were identified between PDR and CDR resulting in the total number of requirements increasing from 318 to 320.
- No additional requirements were levied on the system as indicated by the 0 in all the rows in the 'A' column of the 'Identified' state.

TABLE 11.18
The CRIP Chart for the Project with Requirements Creep at CDR

	Identified			In Process			Completed			In Test			Accepted		
Range	P	E	A	P	E	A	P	E	A	P	E	A	P	E	A
1	0	0	0	0	101	101	101								
2	0	0	0	0	78	78	78								
3	0	0	0	0	35	35	35								
4	0	0	0	0	30	30	30								
5	0	0	0	0	28	28	28								
6	0	0	0	0	20	20	20								
7	0	0	0	0	12	12	12								
8	0	3	2	2	7	7	9								
9	0	0	0	0	5	5	5								
10	0	0	0	0	2	2	2								
Totals	0	3	2	2	318	318	320								

Performance Monitoring and Controlling

- Only two of the three requirements in the 'E' column of Range 8 were actually identified as indicated by the 2 in row 8 of the 'A' column of the 'Identified' state.[18]
- The PM plans to start and complete the development activities on these additional two requirements in Range 8 as indicated by the '2' in the 'P' column of the 'In process' state and the 9 (7 + 2) in the 'P' column of the 'Completed' state in row 8.
- The development activity on the remaining requirements is progressing according to plan as shown by the entries in the 'In process' and 'Completed' states.
- The PM does not plan to identify any new requirements between PDR and CDR since all the rows in the 'P' column of the 'Identified' state remain at 0.

11.5.5.2.3 The CRIP Chart for a Project with Requirements Creep at TRR

The CRIP chart for the project with requirements creep at TRR shown in Table 11.19 indicates:

- Ten unexpected additional requirements were identified between CDR and TRR resulting in the total number of requirements increasing from 320 to 330.
- Four of the unexpected requirements were identified in Range 4 shown by the 4 in row 4 of the 'A' column of the 'Identified' state.

TABLE 11.19
The CRIP Chart for the Project with Requirements Creep at TRR

Range	Identified P	E	A	In Process P	E	A	Completed P	E	A	In Test P	E	A	Accepted P	E	A
1	0	0	0	0	0	0	0	101	101	101						
2	0	0	0	0	0	0	0	78	78	78						
3	0	0	0	0	0	0	0	35	35	35						
4	0	0	4	4	0	0	0	30	30	30						
5	0	0	6	0	0	6	6	28	28	34						
6	0	0	0	0	0	0	0	20	20	20						
7	0	0	0	0	0	0	0	12	12	12						
8	0	0	0	0	2	2	0	9	9	9						
9	0	0	0	0	0	0	0	5	5	5						
10	0	0	0	0	0	0	0	2	2	2						
Totals	0	0	10	4	2	8	6	320	320	326						

---

[18] It is possible that a change request was made for the third requirement and the request was rejected for some reason. The CRIP chart just indicates the change, but it does not provide reasons for the change.

- The PM plans to start the development activities on the additional four requirements in Range 4 as indicated by the '4' in row 4 of the 'P' column of the 'In process' state.
- Six of the unexpected requirements were identified in Range 5 shown by the 6 in row 5 of the 'A' column of the 'Identified' state.
- Development activities actually began on these six requirements as shown by the 6 in row 5 of the 'A' column in the 'In process' state.
- The PM plans to complete the development activities on these six requirements as shown by the 6 in row 5 of the 'P' column of the 'Completed' state.
- Development work began on two of the additional requirements in Range 8, as shown by the match between the numbers in row 8 of the 'E' and 'A' columns of the 'In process' state of Table 11.19.[19]
- The development work for the two additional requirements in Range 8 is not expected to be completed by the subsequent milestone since there is a 0 in row 8 of the 'P' column of the 'Completed' state.
- Development work on the remaining requirements is progressing according to plan and the project is expected to commence testing as shown by the matches between the entries in the 'A' columns of the 'Completed' state and the 'P' column of the 'In test' state.
- The PM does not plan to identify any new requirements between TRR and the following milestone since all the rows in the 'P' column of the 'Identified' state remain at 0.

11.5.5.2.4 The CRIP Chart for a Project with Requirements Creep at IRR

The CRIP chart for the project with requirements creep at IRR shown in Table 11.20 indicates:

- For a change, no unexpected additional requirements were identified between CDR and IRR resulting in no change in the total number of requirements.
- No additional requirements were identified since the values of all rows of the 'A' column of the 'Identified' state are 0.
- Development on the requirements in Range 4 proceeded according to plan by the match between numbers in each of the rows in the 'E' and 'A' columns of the 'In process' state of Table 11.20. The number 4 in the 'P' column of the 'Completed' indicates that the project development activities are planned to have been completed by the following milestone.
- Development on the requirements in Range 5 proceeded according to plan since the number 6 was copied from the 'P' column of the 'Completed' state in Table 11.19 to the 'E' and 'A' columns of the 'Completed' state of Table 11.20. The number 6 in the 'P' column of the 'In test' state indicates that the testing activities are planned to have begun by the following milestone.

---

[19] The reason for only starting work on two of the three could be that one was rejected for some reason, or that the system design state for meeting that requirement was deferred. The CRIP chart just indicates the variance without providing a reason.

Performance Monitoring and Controlling

TABLE 11.20
The CRIP Chart for the Project with Requirements Creep at IRR

	Identified			In Process			Completed			In Test			Accepted		
Range	P	E	A	P	E	A	P	E	A	P	E	A	P	E	A
1	0	0	0	0	0	0	0	0	0	0	101	101	101		
2	0	0	0	0	0	0	0	0	0	0	78	78	78		
3	0	0	0	0	0	0	0	0	0	0	35	35	35		
4	0	0	0	0	4	4	4	0	0	0	30	30	30		
5	0	0	0	0	0	0	0	6	6	6	34	28	34		
6	0	0	0	0	0	0	0	0	0	0	20	20	20		
7	0	0	0	0	0	0	0	0	0	0	12	12	12		
8	0	0	0	0	0	0	0	0	0	2	9	7	9		
9	0	0	0	0	0	0	0	0	0	0	5	5	5		
10	0	0	0	0	0	0	0	0	0	0	2	2	2		
Totals	0	0	0	0	4	4	4	6	6	6	326	318	326		

- Something has stopped the development of activities of the two requirements in row 8 as shown by the 0 in the 'E' and the 'A' columns of the 'In process' state and the 0 in the 'P' column of the 'Completed' state. However, this can only be seen when the two CRIP charts are compared directly. The CRIP chart does not provide a reason for the stoppage; it only provides the information that a stoppage has occurred.
- Nearly all the requirements that were planned to enter the 'In test' state have done so because most of the numbers in the 'P' column of the 'In test' state in Table 11.19 have been copied to the 'E' and 'A' columns in Table 11.20.
- There are some problems in starting to test the requirements in rows 5 and 8 since the numbers in the 'E' columns do not match those in the 'A' columns.
- The PM plans to catch up on these requirements as since there is a 6 in the 'P' column of row 5 and a 2 in the 'P' column of row 8 in the 'In test' state.
- The PM plans for the testing of all the requirements in the 'In test' state to be successfully completed and accepted by the next milestone as indicated by the match between the numbers in the 'A' column of the 'In test' state and the 'P' column of the 'Accepted' state.
- The PM plans to overcome the delays in testing the requirements in rows 5 and 8 by the next milestone as indicated by the match between the numbers in the 'E' column of the 'In test' state and the 'P' column of the 'Accepted' state.

11.5.5.2.5 The CRIP Chart for a Project with Requirements Creep at DRR

The CRIP chart for the project with requirements creep at DRR shown in Table 11.21 indicates:

- No additional requirements were identified.
- Development activities on the requirements in row 4 have been completed as planned, as indicated by the number 4 in the 'P' column of the

TABLE 11.21
The CRIP Chart for the Project with Requirements Creep at DRR

	Identified			In Process			Completed			In Test			Accepted		
Range	P	E	A	P	E	A	P	E	A	P	E	A	P	E	A
1	0	0	0	0	0	0	0	0	0	0	0	0	0	101	101
2	0	0	0	0	0	0	0	0	0	0	0	0	0	78	78
3	0	0	0	0	0	0	0	0	0	0	0	0	0	35	35
4	0	0	0	0	0	0	0	4	4	0	0	4	0	30	34
5	0	0	0	0	0	0	0	0	0	0	6	6	0	34	34
6	0	0	0	0	0	0	0	0	0	0	0	0	0	20	20
7	0	0	0	0	0	0	0	0	0	0	0	0	0	12	12
8	0	0	0	0	0	0	0	0	0	0	2	2	2	9	7
9	0	0	0	0	0	0	0	0	0	0	0	0	0	5	5
10	0	0	0	0	0	0	0	0	0	0	0	0	0	2	2
Totals	0	0	0	0	0	0	0	4	4	0	8	12	2	326	328

'Completed' state in Table 11.20 being copied into the 'E' and 'A' columns of Table 11.21. Moreover, the testing activities on those requirements have not only begun as indicated by the 4 in the 'A' column of the 'In test' state, they have been completed and accepted by the customer as indicated by the 34 in the 'A' column of the 'Accepted' state. This is 4 more than the expected 30 in the 'E' column of the state.

- The customer accepted all the requirements that had been tested except for two in row 8 as indicated by the number 2 in the 'E' and 'A' columns of the 'Accepted' State.
- The project expected that two requirements in row 8 would go into testing and they did, as indicated by the number 2 in the 'E' and 'A' columns of the 'In test' state.

11.5.5.3 The Challenged Project

Consider the challenged project, which is the same as the ideal project until the 'In test' state begins. Accordingly, the CRIP charts for the challenged project at SRR, PDR and CDR are the same as those for the ideal project shown in Tables 11.11–11.13, respectively. The project diverges from the ideal project after CDR so discrepancies can be seen when the TRR is held on the originally scheduled date.

11.5.5.3.1 The CRIP Chart for the Challenged Project at TRR

The CRIP chart for the challenged project at TRR shown in Table 11.22 indicates:

- No additional requirements were identified.
- Development activities in all requirement ranges except Range 6 have not been completed since the 'E' and 'A' values in row 6 of the 'Completed' state do not match.

Performance Monitoring and Controlling

TABLE 11.22
The CRIP Chart for the Challenged Project at TRR

	Identified			In Process			Completed			In Test			Accepted		
Range	P	E	A	P	E	A	P	E	A	P	E	A	P	E	A
1	0	0	0	0	0	0	40	81	41	81					
2	0	0	0	0	0	0	40	78	38	78					
3	0	0	0	0	0	0	5	35	30	35					
4	0	0	0	0	0	0	10	30	20	30					
5	0	0	0	0	0	0	10	26	16	26					
6	0	0	0	0	0	0	0	20	20	20					
7	0	0	0	0	0	0	6	8	2	8					
8	0	0	0	0	0	0	4	7	3	7					
9	0	0	0	0	0	0	1	5	4	5					
10	0	0	0	0	0	0	1	2	1	2					
Totals	0	0	0	0	0	0	117	292	175	292					

- The PM plans to catch up on the development activities as shown by the numbers in the 'P' column of the 'Completed' state.
- The PM is optimistic about commencing testing following the TRR as evidenced by the difference between numbers in the 'P' column of the 'In test' state and the numbers in the corresponding rows of the 'A' column of the 'Completed' state. The customer definitely needs to find out the reason for the optimism.

11.5.5.3.2 The CRIP Chart for the Challenged Project at IRR
The CRIP chart for the challenged project at IRR shown in Table 11.23 indicates:

- No additional requirements were identified.
- The project should not have transitioned into the subsystem testing state of the SDP because of the difference between the numbers in the 'E' and 'A' columns in the 'In test' state.
- The PM plans to catch up as shown by the numbers in the 'P' column of the 'In test' state.
- The project is still optimistic about completing the testing before DRR because the 'P' numbers in the 'Accepted' state match the 'E' numbers instead of the 'A' numbers in the 'In test' state. The customer definitely needs to determine the reasons for the optimism.

11.5.5.4 The 'Make Up Your Mind' Project

Consider the typical 'make up your mind' project, which is the same as the ideal project until SRR and diverges between SRR and PDR because the customer keeps changing their mind.

TABLE 11.23
The CRIP Chart for the Challenged Project at IRR

	Identified			In Process			Completed			In Test			Accepted		
Range	P	E	A	P	E	A	P	E	A	P	E	A	P	E	A
1	0	0	0	0	0	0	0	0	0	40	81	21	81		
2	0	0	0	0	0	0	0	0	0	30	78	48	78		
3	0	0	0	0	0	0	0	0	0	5	35	30	35		
4	0	0	0	0	0	0	0	0	0	8	30	22	30		
5	0	0	0	0	0	0	0	0	0	14	26	12	26		
6	0	0	0	0	0	0	0	0	0	0	20	20	20		
7	0	0	0	0	0	0	0	0	0	7	8	1	8		
8	0	0	0	0	0	0	0	0	0	6	7	1	7		
9	0	0	0	0	0	0	0	0	0	4	5	1	5		
10	0	0	0	0	0	0	0	0	0	0	2	2	2		
Totals	0	0	0	0	0	0	0	0	0	114	292	158	292		

11.5.5.4.1 The CRIP for the Typical 'Make Up Your Mind' Project at SRR

The CRIP chart for the 'makeup your mind' project at SRR is that same as that for the ideal project shown in Table 11.11.

11.5.5.4.2 The CRIP Chart for the Typical 'Make Up Your Mind' Project at PDR

The CRIP chart for the 'makeup your mind' project at PDR shown in Table 11.24 indicates:

- Fifty-six unexpected additional requirements were identified between SRR and PDR resulting in the total number of requirements increasing from 292 to 348.
- Twenty of the unexpected additional requirements were identified in Range 1 as indicated by the number 20 in the 'A' column of the 'Identified' state in row 1.
- Development activities have commenced on these requirements as shown by the difference in the numbers in the 'E' and 'A' columns in row 1 of the 'In process' state (20 + 80 = 101).
- Ten of the unexpected additional requirements were identified in Range 2.
- Development activities have commenced on these requirements as shown by the difference in the numbers in the 'E' and 'A' columns in row 2 of the 'In process' state.
- Fourteen of the unexpected additional requirements were identified in Range 3.
- The PM plans to identify five additional requirements in Range 4 following the PDR.[20]

---

[20] The change requests have been submitted.

TABLE 11.24
The CRIP Chart for the for the 'Makeup Your Mind' Project at PDR

	Identified			In Process			Completed			In Test			Accepted		
Range	P	E	A	P	E	A	P	E	A	P	E	A	P	E	A
1	0	0	20	0	81	101	0								
2	0	0	10	0	78	88	0								
3	0	0	14	0	35	49	0								
4	5	0	0	0	30	30	0								
5	9	0	2	0	26	28	0								
6	12	0	0	0	20	0	0								
7	0	0	4	0	8	12	0								
8	3	0	0	0	7	7	0								
9	0	0	6	0	5	11	0								
10	2	0	0	0	2	0	0								
Totals	31	0	56	0	292	326	0								

- Two unexpected requirements were identified in Range 5 and the PM plans to identify nine additional requirements following PDR.
- The PM plans to identify 12 additional requirements in Range 6 following PDR.
- Four of the unexpected additional requirements were identified in Range 7.
- Development activities have commenced on these requirements.
- The PM plans to identify three additional requirements in Range 8 following PDR.
- Six of the unexpected additional requirements were identified in Range 9.
- Development activities have commenced on these requirements.
- The PM plans to identify two additional requirements in Range 10 following PDR.
- Development activities are proceeding according to plan except for the requirements in Ranges 6 and 10.
- No development activities have started on the requirements in Range 6 as shown by the 20 in the 'E' column and the zero in the 'A' column of the 'In process' state in row 6.
- Development activities have proceeded as planned on the requirements in Range 7 since the values in the 'E' column and 'A' column of the 'In process' state in row 7 are the same.
- No development activities have started on the requirements in Range 10.
- The PM does not plan to complete the development activities in process before the CDR because every row in the 'P' column of the 'In process' and 'Completed' states have been set to 0. This could be because the project has been cancelled or for some other reason.

11.5.6 COMMENTS

Although written up for use with requirements, CRIP charts can also be used for Function points, features, use cases, scenarios, technical performance measures (TPM) and any other technical measurement that can be tracked across the SDP. When the A column in the CRIP chart is plotted as a histogram (Kasser 2018: Section 2.9), it shows an attribute profile (Kasser 2018: Section 9.1) for that attribute.

The rows containing 0s in the CRIP charts were gray to assist reading the tables, similarly, colours may be used to highlight rows where differences between expected and actual values appear. These colours should be synchronized with those in Section 11.6.1 and Tables 11.25 and 11.26.

In summary, CRIP charts help to identify poor management, negative politics (Section 2.2.2) and other factors (Section 11.1) that cause cost and schedule overruns in time to impose corrective action.

11.6 TRAFFIC LIGHT AND ETL CHARTS

Traffic light charts are tools that summarize progress and project status in a simple easily understood manner. The systems approach to project management enhances the traditional single snapshot traffic light chart by adding the time dimension to enhance the information provided by the chart. Perceptions from the *Generic* HTP indicate that this is similar to CRIP charts (Section 11.5) and the dynamic technology readiness level (dTRL) (Section 10.4.1).

11.6.1 TRAFFIC LIGHT CHARTS

Traffic light charts are tools that show summaries in the colours of a stoplight or traffic light. For example:

- *Green:* represents the data being summarized is good.
- *Red:* indicates that something is seriously wrong or broken.
- *Yellow:* indicates that something needs to be watched.
- *Blue:* is used to indicate that the data being summarized is much better than good.

TABLE 11.25
Information Represented by Colours in Different Types of Traffic Light Charts

Situation	Red	Yellow	Green	Blue
Project management	Behind schedule	On schedule but needs watching	On schedule	Ahead of schedule
Risk management	Very risky	Medium risk	Low risk	No risk
Number of defects being produced	Much more than expected, corrective action needed yesterday (exceeding tolerance limit)	More than expected, corrective action needed soon (approaching tolerance limit)	As expected (within tolerance)	Fewer than expected

Performance Monitoring and Controlling

TABLE 11.26
Values for Traffic Light Colours

Colour	Schedule	Budget	Problems
Blue	Ahead of schedule by $X\%$	Well within budget by $A\%$	None
Green	On schedule ($\pm S\%$)	Within budget ($\pm M\%$)	None
Yellow	Slightly behind schedule by $Y\%$	Slightly over budget by $B\%$	At least 1 minor problem
Red	Well behind schedule by $Z\%$	Well over budget by $C\%$	At least 1 major problem

Table 11.25 provides examples of the types of information traffic light charts that can be used to represent information in different situations. The colours can be added to existing tables, inserted into summary tables or used to tag other types of graphics. For example, in project management the colours:

- Provide a quick overview of the status of the project.
- Are based on combination of the status of project schedule, budget and problems or risks at the reporting milestone where the contribution of each status is defined at the start of project to match cost/schedule charts. In addition, the definition of the A, B, C, S, M, X, Y and Z values and the definition of a major and a minor problem shown in Table 11.26 need to be clarified at the start of the project. These values shall not change during the lifetime of the project.
- *Blue and Green:* signify that combination of (an 'and' function) all three items, project schedule, budget and problems or risks are above the value defined in Table 11.26.
- *Yellow and Red:* signify that least one of the three items, project schedule, budget and problems or risks (an 'or' function) is below the value defined in Table 11.26. Note the $-P$, $-B$ and $-S$ suffixes identify which of the items contributed to the colour.

11.6.2 ETL Charts

ETL charts are a tool that:

- Can provide information that is not presently available in most projects.
- Can potentially reduce the time spent in meetings.
- Is based on adding perceptions of the project from the *Temporal* HTP to the basic traffic light charts (Section 11.6.1).

As an example, consider the monthly project status meeting at the Systems Engineering and Evaluation Centre (SEEC) in UniSA scheduled for February 2005 wherein several PMs were to report on the status of their project. One PM prepared a project status presentation covering his 11 projects containing 126 PowerPoint slides of project status information, continuity and other relevant information. The agenda list of projects is shown in Figure 11.7 as a typical bulleted list. Each project was to be introduced with an overview chart containing a traffic light coloured block

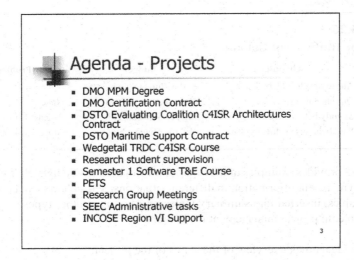

FIGURE 11.7 A typical project status meeting agenda.

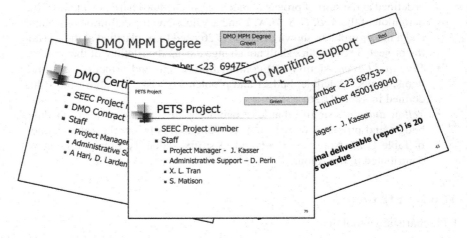

FIGURE 11.8 Some of the projects with traffic lights.

as shown in deck of cards style Figure 11.8. The approach provides the traffic light summary when the project is up for discussion.

Now consider an alternative approach using the systems approach. Present the list of projects as an agenda slide in the form of the table shown in Figure 11.9. That is not much of an improvement until you add the traffic light blocks in a third column as shown in Figure 11.10 and include hot links to each of the projects.

The traffic light chart allows the discussion to immediately focus on the projects in trouble (in red) hence is ideal for MBE and MBUM. The chart in Figure 11.10 can be enhanced by adding perceptions from the *Temporal* HTP as shown in the ETL chart in Figure 11.11 in the form of two additional columns reminding the viewers of

Performance Monitoring and Controlling

#	Project
1	DMO MPM Degree
2	DMO Certification Contract
3	DSTO Evaluating Coalition C4ISR Architectures Contract
4	DSTO Maritime Support Contract
5	Wedgetail TRDC C4ISR Course
6	Research student supervision
7	Semester 1 Software T&E Course
8	PETS
9	Research Group Meetings
10	SEEC Administrative tasks
11	INCOSE Region VI Support

FIGURE 11.9 The agenda in tabular form.

#	Project	Status
1	DMO MPM Degree	Green
2	DMO Certification Contract	Yellow -P
3	DSTO Evaluating Coalition C4ISR Architectures Contract	Yellow -P
4	DSTO Maritime Support Contract	Red -BS
5	Wedgetail TRDC C4ISR Course	Blue
6	Research student supervision	Green
7	Semester 1 Software T&E Course	Yellow -P
8	PETS	Green
9	Research Group Meetings	Green
10	SEEC Administrative tasks	Green
11	INCOSE Region VI Support	Yellow -PB

FIGURE 11.10 The agenda in tabular form with traffic light blocks.

the state of the project at the prior reporting period and showing expected state of the project at the following future reporting period in a similar manner to the CRIP chart (Section 11.5) (*Generic* HTP). New information that was not previously available can now be seen at the high-level summary. For example:

- *Projects 1, 6, 8 and 9:* are conforming to the planned schedule and budget and are not facing any problems, namely they are proceeding according to the nominal status.
- *Projects 3 and 7:* have been facing minor problems are expected to face a major problem during the next reporting period.

#	Projects	Last time	Current	Next
1	DMO MPM Degree	Green	Green	Green
2	DMO Certification Contract	Yellow -P	Yellow -P	Red-P
3	DSTO Evaluating Coalition C4ISR Architectures Contract	Yellow -P	Yellow -P	Green
4	DSTO Maritime Support Contract	Red -BS	Red -BS	Red -BS
5	Wedgetail TRDC C4ISR Course	Green	Blue	Green
6	Research student supervision	Green	Green	Green
7	Semester 1 Software T&E Course	Yellow -P	Yellow -P	Green
8	PETS	Green	Green	Green
9	Research Group Meetings	Green	Green	Green
10	SEEC Administrative tasks	Green	Green	Yellow -P
11	INCOSE Region VI Support	Yellow -PB	Yellow -PB	Yellow -PB

FIGURE 11.11 The ETL chart.

- *Project 4:* has been facing minor problems is expected to overcome them and return to a nominal state during the next reporting period.
- *Project 5:* exceeded expectations during this reporting period and is expected to return to nominal status during the next reporting period.
- *Project 11:* has been facing problems and they are ongoing.

These perceptions from the *Generic* HTP can be generalized as shown in Figure 11.12 where:

- *Project 1:* is proceeding according to nominal status.
- *Project 2:* has identified that a major problem or risk is expected to arise during the next reporting period. The project may need additional resources to remedy the problem.

#	Projects	Last time	Current	Next
1	Project Ho-hum	Green	Green	Green
2	Project Oh oh	Yellow -PB	Yellow -P	Red
3	Project Catching up	Yellow -P	Yellow -P	Green
4	Project Replace manager	Red -BS	Red -BS	Red -BS
5	Project Very happy customer	Green	Blue	Blue
6	Project Completed	Green	Green	N/A
7	Project Promote manager	Red -P	Yellow -P	Green
8	Project Watch this person	Yellow –BS	Green	Blue
9	Project No risk management	Green	Red -P	Red -P
10	Project Took course in risk management	Green	Green	Yellow -P
11	Project Manager doing risk management	Yellow -P	Yellow -P	Yellow -P

FIGURE 11.12 Project trends using the ETL chart.

- *Project 3:* seems to be overcoming its problems and returning to nominal status.
- *Project 4:* has been facing one or more major problems and no change is expected. This could be due to (1) poor management, (2) being underfunded so there is no funding to complete the remaining work, or some other reason to be determined.
- *Project 5:* has a very happy customer since it is ahead of the nominal status and is expected to remain so in the next reporting period. This situation may be because the project was overfunded and given more than enough time to complete, or has a smart PM.
- *Project 6:* has been nominal and is now complete on schedule and within budget.
- *Project 7:* had major problems during the previous reporting period which are being tackled successfully and the project is expected to return to nominal status by the next reporting period. This PM may be a candidate for promotion or a candidate to take over other projects in trouble and get them back on track.
- *Project 8:* has gone from experiencing budget and schedule problems in the previous reporting period to nominal status and the PM is anticipating being well within the budget and schedule in the next reporting period. This PM may be a candidate for promotion
- *Project 9:* was nominal in the last reporting period but is currently experiencing one or more major problems, which are expected to continue into the next reporting period. If the major problem was not anticipated in the previous status report then the situation needs to be investigated.
- *Project 10:* which has been nominal since the last reporting period (or earlier) is expecting to experience minor problems in the next reporting period. The PM may have taken a continuing education course and started to practice risk management.
- *Project 11:* which has been experiencing minor problems since the last reporting period (or earlier) is expecting to continue to experience such problems. If there is no corresponding adverse budget and schedule impact (shown in Gantt charts (Kasser 2018: Section 8.4)) and EVA charts (Section 11.4) accompanying the status report, the PM may be practicing risk management (Section 10.3) and bringing it to the attention of senior management.

In summary the ETL chart:

- Provides information that was not currently available.
- Allows positive MBUM by delving into the reasons for the deviation from the planned status.
- Is an ideal tool for MBE (Section 11.8).
- Helps to identify poor management, negative politics (Section 2.2.2) and other factors (Section 11.1) that cause cost and schedule overruns in time to impose corrective action.

11.6.2.1 Creating an ETL Chart for Use in a Presentation

The process for creating an ETL chart is as follows[21]:

1. Gain consensus from the stakeholders as to the values of the parameters for the placement of information into the colours shown in Table 11.26 at the start of the project.
2. Create a table listing the projects in an appropriate order (e.g. by cost, by closing date, in alphabetical order) in the format shown in Figure 11.11.
3. For each project, determine the traffic light colour according to the specification for the colour in Table 11.26.
4. Copy the information from the 'current' column in the ETL chart presented at the previous status report into the 'last' column[22].
5. For each project, estimate the status of the project at the next reporting period and determine the traffic light colour according to the specification for the colour in Table 11.26.
6. Save the table.
7. Copy the table into the status report PowerPoint presentation.
8. Add hot links to the first slide of each of the projects as shown in Figure 11.11.

The ETL chart should look like Figure 11.11 but with different colours in the 'last', current' and 'next' columns.

11.6.2.2 Adding Even More Information

The ETL chart shown in Figure 11.12 can be improved using an idea from the CRIP charts (Section 11.5) (*Generic* HTP) to show how the current situation compares with the predicted situation as of the last report presentation as shown in Figure 11.13. The current reporting period has been separated in two parts:

1. *Expected from last time:* copied from the 'next' column in the last report presentation.
2. *Actual:* as achieved by the work done during this reporting period.

For example,

- *Project 1:* was expected to be green and is green.
- *Project 2:* was expected to be in the green but is still yellow (problems).
- Differences can be seen in some of the other projects, which require explanation at the reporting presentation.

---

[21] Steps 1 and 2 are done once at the start of the project.
[22] Not applicable for first reporting period.

Performance Monitoring and Controlling

#	Projects	Last time	Current Expected	Current Actual	Next
1	Project Ho-hum	Green	Green	Green	Green
2	Project Ohoh	Yellow -PB	Green	Yellow -P	Red
3	Project Catching up	Yellow -P	Yellow -P	Yellow -P	Green
4	Project Replacemanager	Red -BS	Red -BS	Red -BS	Red -BS
5	Project Very happy customer	Green	Blue	Blue	Blue
6	Project Completed	Green	Green	Green	N/A
7	Project Promote manager	Red -P	Yellow -P	Green	Green
8	Project Watchthis person	Yellow –BS	Green	Green	Blue
9	Project Norisk management	Green	Red -P	Red-P	Red -P
10	Project Tookcourse in risk management	Green	Green	Green	Yellow -P
11	Project Managerdoing risk management	Yellow -P	Yellow -P	Yellow-P	Yellow -P

FIGURE 11.13 Final version of ETL chart.

11.7 MBO

MBO (Mali 1972) is a compound tool:

- For planning a project.
- Monitoring the progress towards completion.
- Was popular as a management approach in the 1960s.

MBO as introduced was unnecessarily complex (subjective and objective). However, the core concepts in MBO are simple and manageable, namely:

- Set realistic objectives or goals for every activity.
- Monitor progress towards those objectives as the project proceeds.
- Take corrective action if the objectives are not going to be met.

As perceived from the *Generic* HTP, these core concepts are actually embodied in standard project management and EVA (Section 11.4) and built into the SDP waterfall chart template (Kasser 2018: Section 14.9). The SDP is divided into a number of states, each of which has to accomplish predefined objectives by the milestone at the end of the state. For example, the objectives of the system requirements state are to produce the system requirements document (SRD) at the SRR, which takes place at the end of the state.

MBO was originally introduced by Peter Drucker as a systems approach to managing an organization (Drucker 1954) where the objectives of one layer in the hierarchy are set to accomplish and support the objectives of the higher level of the organization, a similar (*Generic* HTP) concept to the 'Do Statement' (Chacko 1989, Kasser 2018: Section 3.2.1).[23] In the language of systems, when setting objectives, the objectives of the subsystem shall support the objectives of the system.

---

[23] The 'Do Statement' only provides a tool for aligning objectives.

11.7.1 MBO in the Planning State of a Project

MBO relates to activities and objectives, it does not involve schedules and timelines. The process for using MBO in the planning state of a project is as follows:

1. Process architect a set of states and milestones appropriate to the project for the project (Section 5.4) and the number of iterations of the SDP (Section 10.4). There are no uniformly agreed-to names for milestones. Milestones can be major and minor, formal and informal.
2. If the project is complex, use the waterfall chart (Royce 1970, Kasser 2018: Section 14.9) with its built-in states and milestones as a template. Since the scope of projects varies, the mixture of major and minor, formal and informal, milestones also vary. Each project will have its own mixture, similar to those of similar projects. Moreover, some states may have waterfalls of their own.
3. Determine the objectives to be completed by the last major milestone (the last state) in the project. These may include events to be achieved and products to be produced.
4. Working back from the last major milestone, determine the objectives for each state of the SDP.
5. Determine the sub-objectives for each objective. A useful tool for doing this is the 'Do Statement' (Chacko 1989, Kasser 2018: Section 3.2.1).
6. Create the WPs (Section 8.19) and project network of activities that will produce each objective and sub-objective in a sequential orderly manner. Useful tools for doing this planning process are Gantt charts (Kasser 2018: Section 2.12), PAM network charts (Kasser 2018: Section 2.12.2) and PERT charts (Kasser 2018: Section 8.10) and project management software tools.

11.8 MBE

MBE is a tool that:

- Allows the span of control to be widened by allowing the manager to ignore projects that are performing to the PVs (the norm).
- Sets tolerance limits on the parameters of an object (process or product being developed) and ignores the object if the limits are not exceeded.
- Allows delegating decisions to the lowest level, giving employees responsibility, as long as the consequences of the decision do not cause the limits to be exceeded.

11.8.1 The Key Ingredients in MBE

The key ingredients in MBE are (Bittel 1964):

- *Selection:* pinpoints the criteria that will be measured. In the systems approach, these are the objectives determined by MBO (Section 11.7).
- *Measurement:* assigns value to past and present performance.

Performance Monitoring and Controlling

- *Projection:* analyses the measurements and projects future performance.
- *Observation:* informs management of the state of performance.
- *Comparison:* compares performance with limits or boundaries and reports the variances to management.
- *Decision-Making:* prescribes the action to be taken to:
  1. Bring performance back within the limits.
  2. Adjust the limits or expectations.
  3. Exploit opportunity.

11.8.2 Advantages and Disadvantages of MBE

The advantages include:

- Reduces the amount of monitoring and controlling performed by upper management.
- Upper management need not spend time on projects that are progressing normally.
- Undesirable situations are indicated by the limits being exceeded so that upper management can intervene.

The disadvantages include:

- Open to MBUM once a parameter exceeds the limit threshold.
- Mistakes in setting limits may cause problematic situations to be unnoticed until it is too late to take corrective action, or may flag too many alarms.
- Can be confused with 'management without caring' resulting in low morale in organizational units that do not raise exceptions as in the C3I Group Case (Kasser 2018: Section 13.1.2).
- It is based on the assumption that only upper management can deal with a problematic situation. However, this disadvantage may be overcome if the PM can produce a plan for dealing with the problem that satisfies upper management.

11.8.3 Using MBE

There are two states in MBE, namely:

1. *The planning state:* in which the limits are set. The objectives of each WP are to produce the product within the estimated cost and schedule. The upper- and lower-level threshold levels for each objective are determined. The upper and lower threshold levels for cost management may be the best- and worst-case cost and schedule estimates or may be assumed by using a number of standard deviations. However, as with most other things there is a trade-off; the smaller the limit, the more exceptions will need to be investigated. The larger the limit, the fewer investigations; but there will be a greater chance that a serious situation will be missed.

2. *The monitoring and controlling states:* in which the limits should not be exceeded. Tools used in MBE in this state include process control trend charts (Kasser 2018: Section 2.15), CRIP charts (Section 11.5) and ETL charts (Section 11.6). Should the limits be exceeded, corrective action may need to be taken.

11.9   THE ENGAPOREAN MCSSRP EXERCISE

The background and context for the progressive session exercises is in Appendix 1. The purpose of the exercise is:

1. To practice using the template for a student exercise presentation (Kasser 2018: Section 14.7).
2. To update the project information from Exercise 10.9 after the occurrence of an unanticipated event.

11.9.1   THE REQUIREMENTS FOR THE EXERCISE

Sometime between the SRR and the PDR an event occurred which may change the cost, schedule and staffing. Accordingly,

1. Select the event at random as discussed in Appendix 2.
2. Assume that the CRIP chart for the MCSSRP is the same as the CRIP chart for the ideal project shown in Table 11.11.
3. Prepare a presentation according to the instructions in Appendix 3.
4. Save presentation as file '11-teamname'.pptx.

11.10   SUMMARY

This chapter explained how to monitor, control and communicate the performance of a project incorporating the remaining interdependent P's of a project allowing project management to be more proactive. The chapter explained ways to detect and prevent potential project overruns, ways to detect impending project failure, managing changes in project scope, EVA, CRIP charts, ETL charts, MBO and MBE.

REFERENCES

Bittel, Lester R. 1964. *Management by Exception; Systematizing and Simplifying the Managerial Job.* New York: McGraw-Hill.
Brooks, Fred. 1982. *The Mythical Man-Month Essays on Software Engineering, Reprinted with corrections.* Boston, MA: Addison-Wesley Publishing Company.
Chacko, George K. 1989. *The Systems Approach to Problem Solving.* New York: Prager.
CHAOS. *The Chaos study.* The Standish Group 1995 [cited March 19, 1998]. Available from www.standishgroup.com/chaos.html.
Drucker, Peter Ferdinand. 1954. *The Practice of Management.* New York: Harper.
Hitchins, Derek K. 2006. *World class systems engineering - The five layer Model* [Web site] 2000 [cited 3 November 2006]. Available from www.hitchins.net/5layer.html.

Kasser, Joseph Eli. 1995. *Applying Total Quality Management to Systems Engineering.* Boston, MA: Artech House.
Kasser, Joseph Eli. 1997. What do you mean, you can't tell me how much of my project has been completed? In *INCOSE 7th International Symposium*, Los Angeles, CA.
Kasser, Joseph Eli. 1999. Using organizational engineering to build defect free systems, on schedule and within budget. In *PICMET*, Portland, OR.
Kasser, Joseph Eli. 2002. The cataract methodology for systems and software acquisition. In *SETE 2002*, Sydney, Australia.
Kasser, Joseph Eli. 2015. *Perceptions of Systems Engineering* (Vol. 2). Charleston, SC: Solution Engineering: Createspace.
Kasser, Joseph Eli. 2018. *Systems Thinker's Toolbox: Tools for Managing Complexity.* Boca Raton, FL: CRC Press.
Kasser, Joseph Eli, and Victoria Regina Williams. 1998. What do you mean you can't tell me if my project is in trouble? In *First European Conference on Software Metrics (FESMA 98)*, Antwerp, Belgium.
Lukas, J. Anthony. 2008. Earned value analysis - Why it doesn't work. In *2008 AACE International Annual Meeting*, Toronto, ON.
Mali, Paul. 1972. *Managing by Objectives.* New York: John Wiley & Sons Inc.
Palmer, Stephen R., and John M. Felsing. 2002. *A Practical Guide to Feature - Driven Development.* Upper Saddle River, NJ: Prentice Hall.
Royce, Winston W. 1970. Managing the development of large software systems. In *IEEE WESCON*, Los Angeles, CA.
Ward, Paul T., and Stephen J. Mellor. 1985. *Structured Development for Real-Time Systems.* Upper Saddle River, NJ: Yourdon Press.

12 The Human Element

This chapter focuses on the human element; the most important attribute of project management because people perform the project, cause, prevent or mitigate problems and, take part in the processes that produce the products. Once the project has begun, project managers (PMs) spend most of their time dealing with people. The outcomes of these interactions with people affect the cost and the schedule and are reflected in the Gantt and EVA charts presented at the milestone reviews. This chapter explains the following aspects of the human element:

1. Time management in Section 12.1.
2. People management in Section 12.2.
3. Conflict in Section 12.3.
4. Leadership in Section 12.4.
5. Negotiation in Section 12.5.
6. Managing stakeholders in Section 12.6.

12.1 TIME MANAGEMENT

The PM's most important resource is time; accordingly managing time proactively and productively is a habit that needs to be developed. The PM is often multitasking monitoring and controlling the activities in different work packages (WPs) and dealing with interruptions. Some of the activities being multitasked may take less than a day, others may take more than a day, for example, if a WP is scheduled to take 3 weeks, the PM will spend a little time every now and again during those 3 weeks on that WP. Time management of these activities is basically a five-step daily process, namely:

1. *Reviewing the previous day's activities:* identifying which of them were completed, and estimating what needs to be done to complete the remaining activities.
2. *Prioritization:* list the activities that need completion. For each activity, ask yourself three questions:
   1. When is the activity supposed to be completed?
   2. How long will it take to complete this activity?
   3. What will happen if I don't complete the activity on time?
3. *Procrastination*: for each activity, if there is:
   1. More than enough time to complete it, it is not urgent so put it on your procrastination list.[1]
   2. Enough time or a less than enough time to complete it, put it on your to-do list for the day.

---
[1] The list of task to do when you get around to them.

4. *Planning:* create a daily work plan by scheduling the activities on the to-do list in order of urgency in blocks of 15-30 min. Note activities may take more time than 15 or 30 min so they can run sequentially. Schedule in the 80% factor (Malotaux 2010) so you don't use planned lollygagging time to make up for the under estimate of the time it will take to do the activity.
5. *Performing and lollygagging:* take up the remainder of the day. Take breaks every 45 min or so to lollygag, power nap, consume some refreshment or network (in person, by telephone or email). Let those breaks allow your subconscious to think about the activity (Section 9.1.4). Work-related emails should not be dealt with during this time; they should be scheduled as an activity. This is your down time.

The first four steps should not take more than the first 15-30 min of the day. Reviewing the previous day's activities at the start of the day instead of reviewing the activities at the end of the previous day allows your subconscious some time to think about what needs to be done to complete the remaining activities.

Table 5.9 is also a useful tool for individual time management as well as project time management.

12.2 PEOPLE MANAGEMENT

While PMs may not have the power to grant financial rewards, they do have the ability to recommend them and accordingly should understand the organization's reward and recognition system, how their team members benefit and how they benefit from the system.

12.2.1 REWARDS AND RECOGNITION

Rewards and recognition are a fundamental part of the politics of the organization. PMs can and should:

- Provide some types of non-financial rewards and recognition (Section 2.1.4.5).
- Use the 'Thank You' (Kasser 2018: Section 8.13).
- Perform performance evaluations.

12.2.1.1 Performance Evaluations

'The ways in which people are paid, the measures by which their performance is evaluated, and so forth – are the primary shapers of employees' values and beliefs' (Hammer and Champy 1993: p. 75). Winning (world-class) organizations need to focus on individual excellence and reward individuals for their achievements and the risks that they are willing to take (Harrington 2000). According to Henry Ford, 'If an employer urges men to do their best, and the men learn after a while that their best does not bring any reward, then they naturally drop back into "getting by"' (Ford and

Crowther 1922: p. 117).[2] Yet performance evaluations were discouraged (Deming 1986) for many reasons including:

- *Measurements are subjective:* subjective measurements demoralize people so don't bother to make such measurements. If indeed measurements are subjective, then the search should be started for objective measurements.
- *Measurements are made based on arbitrary goals:* since the goals are arbitrary, they may not be achievable or desired by the employee. The goals should be set in a participative manner with the employee contributing, understanding the need for and taking ownership of the goals. They will then cease to be arbitrary.
- *The system is at fault and people's performance cannot improve within the boundaries of the system:* Deming's 'Red Bead Experiment' is often quoted to reinforce this interpretation (Deming 1986). However, Deming's comments about changing the system have been conveniently forgotten.
- *Half the people will always be performing below average:* is a shield for poor performance. The fallacy in the argument is the definition of average. Since the average in one organization can be better or worse than the average in another organization, there is no reason why the average of an organization cannot be raised.

So, the problem is not the performance evaluation, it is in the way it is implemented. A systematic reward and recognition system (SRRS) (Kasser 1996, 2013) is bidirectional and performed by people on people, using processes and evaluation criteria created by people. The aim of the SRRS is to:

1. Reinforce behaviour that is in accordance with the values of the organization (Harrington 1995: p. 469).
2. Gently encourage employees to move toward the Theory Y end of the continuum of motivation (Section 2.1.4.2). The keys to developing an effective performance process are (Harrington 2000):
   - Measure the right things or selecting the right evaluation criteria.
   - The employee and manager agreeing to the performance standards ahead of time, usually as goals to be achieved by the next reporting period.
   - Ongoing measurement and feedback.
   - Conducting formal evaluations following the achievement of a milestone, not according to a periodic calendar date.

The evaluation criteria chosen to achieve this purpose are critical. Consider using the following (Kasser 1995):

- *The seven habits of highly effective people* (Covey 1989): based on the demonstration of the habits.

---

[2] This observation is valid for the factory environment and the type of worker.

- *Individual contributions to their project:* based on the contribution of each member of the team to the development of the product and the improvement of the process. Both attributes may be assessed by managers and peers.
- *Team spirit:* based on how each member of the team works together with the team and contributes to the success of the team. These attributes are assessed by peers.
- *Contribution to company growth and reputation:* based on volunteer work on proposals and professional societies, adopted suggestions for process improvement in areas within and outside the person's work area; how they grow and improve other people in the project; letters of commendation and awards from customers and other sources external to the organization.
- *Personal growth:* based on demonstrated willingness to learn and develop (Hammer and Champy 1993: p. 189), evidenced by courses taken, conferences attended, technical journal articles published and conference papers presented.
- *People skills:* negotiation skills (Section 12.5) and other such skills for working with people (Lewicki and Litterer 1985).

Reward and recognition is an ongoing process. Evaluation of personnel takes place at appropriate times. The evaluation criteria must be posted and known to all employees. An employee should never be evaluated against the 'broader picture' or undocumented criteria. Once a manager has evaluated an employee against unspecified criteria, even if the employee doesn't file a complaint, that employee will no longer trust the manager, because even if mutually agreed goals are subsequently set, there is no guarantee that the manager will not repeat the process at the next performance evaluation and evaluate the employee against another set of unspecified criteria.

When an evaluation is made on each criterion, the reason for the evaluation is documented (also important for legal and regulatory compliance reasons). The grading of the employee with respect to the evaluation criteria must be objective and fair. Evaluation may be made by several different people and in that case, the results for any specific criterion are a weighted sum of all the evaluations. The evaluations for each criterion (if performed by different people) and the normalized results may be plotted for each employee as a bar chart as shown in Figure 12.1. Each criterion also has upper and lower limits just like a statistical process control chart (*Generic HTP*). The upper and lower limits are set so that normal behaviour is within the limits.

If the SRRS is working correctly, most evaluations should fall within the upper and lower limits showing that the process is in control. Any situation in which an employee receives evaluations outside the limits is to be investigated by an independent organizational element to ensure the evaluation was fair and determine the reason for the exceeding of the limit. Exceeding the upper limit may show excellence or; falling below the lower limit may show poor performance, managerial vindictiveness or something entirely different. Each instance has to be investigated. Accordingly, since Figure 12.1 shows the employee has exceeded the upper limit for Evaluation Criterion E7, the reason needs to be determined and noted for possible reward and recognition.

The Human Element

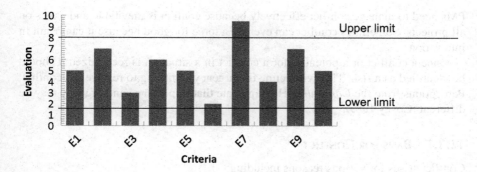

FIGURE 12.1 A performance evaluation chart.

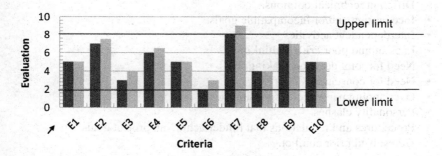

FIGURE 12.2 Performance evaluation chart for two time periods.

The evaluations also have to be checked over time to learn if there is an abnormal pattern. For example, there may be a supervisor who never gives a certain employee a good evaluation. The mediocre evaluation may be out of phase with other elements of the evaluation at that time or with the employee's performance history. In today's litigious society, these checks are becoming important to ensure that the evaluations reflect the real performance. Today's technology can perform 'pattern checking' on evaluations to weed out this situation. A typical performance evaluation chart for two time periods is shown in Figure 12.2. Note while performance is within the normal range, the second evaluation shows an improvement in some of the evaluation criteria and no deterioration in others.

12.3 CONFLICT

Conflict is an active disagreement between people:

1. With opposing opinions.
2. And the rules, culture or other factors that govern behaviour in various situations.

PMs need to manage conflict effectively because conflict is inevitable and exists on all projects. Sometimes conflict can even be a force for good because it can result in innovation.

Once a conflict or a potential for a conflict in a situation is recognized it should be identified as a risk. This will define the urgency of failing to resolve the conflict. Perceptions from the *Continuum* HTP indicate that importance and urgency may be different (Covey 1989).

12.3.1 Basis for Conflict

Conflict arises for various reasons including:

- Ambiguous personnel roles.
- Communications barriers.
- Different technical opinions.
- Inconsistent and/or incompatible goals.
- Interdependent activities.
- Lack of and poor critical thinking.
- Need for joint decision-making.
- Need for consensus.
- Overlapping personnel responsibilities.
- Personality clashes.
- Procedures and regulations that hinder achieving project goals.
- Unresolved prior conflict.

12.3.2 Response to Conflict

Conflict on a project is generally an undesirable situation. When stated in these terms, the response to conflict is a variation of the generic problem-solving process (Section 2.4) where the undesirable situation is the conflict and the feasible conceptual future desirable solution (FCFDS) is the lack of conflict.

12.3.2.1 Gaining an Understanding of the Situation

The first step in resolving conflict is gaining an understanding of the situation, namely understand the:

- Nature of the conflict: e.g. logical (technical) and emotional (personal).
- People involved in the conflict.
- A useful systems thinking tool for gaining an understanding of the situation when everyone is panicking is STALL (Section 2.1.8.3).

12.3.3 Perceptions of Conflict

Perceiving conflict resolution from various HTPs:

- *Big Picture:* conflict takes place in a situation.

- *Continuum:* ways of resolving conflicts are situational. This means that there isn't a single way of resolving conflicts that is the optimal way of resolving every conflict. The PM needs to recognize the type of situation in which the conflict exists and determine and use the appropriate way of resolving the conflict which include:
  - *Accommodating:* allowing parties to give up their needs and wishes to accommodate the other person.
  - *Collaborating:* working together with the personnel involved to find a mutually beneficial solution, often a win-win solution.
  - *Competing:* a win-lose approach in which one person puts forward their resolution to the conflict at the expense of other resolutions.
  - *Compromising:* working together with the personnel involved to find a middle ground in which each party is partially satisfied. Leads less than optimal solutions and lose-lose solutions.
  - *Delaying:* postponing conflict by introducing a cooling off period to give the parties a chance to work it out for themselves.
  - *Denying:* denying that the conflict even exists and ignoring or avoiding it.
  - *Dictating:* asserting one viewpoint at the potential expense of another, such as majority rule or a person with authority dictates the resolution and the other personnel accept it.
  - *Smoothing:* suppressing the conflict, a short-term approach to buy time for a more complete resolution.
  - *Win–wining:* resolving the conflict so that everyone gets something. Since different people want different things, what is a desired outcome of the conflict for one person may be a Don't Care outcome for another person (Chacko 1989, Kasser 2018: Section 3.2.2).
- *Functional:* negotiation (Section 12.5) is a useful tool in resolving conflicts.
- *Quantitative:* the Thomas-Kilmann Conflict Mode Instrument (TKI®) assesses an individual's behaviour in conflict situations in two dimensions: assertiveness and cooperativeness in five conflict-handling modes (accommodating, avoiding, collaborating, competing and compromising) (Thomas and Kilmann 1974).

12.4 LEADERSHIP

The role of a leader is to define the journey; the role of manager is to make sure that the journey takes place in the most expeditious manner. There is no reason why a single person cannot take up both roles on a project. A good manager may even delegate[3] leadership in particular situations to the member of the team who has the expertise to lead the project through a particular series of activities. The attributes that make people follow leaders or accept their decisions include:

---

[3] Formally or informally.

- *Power:* the authority to make and enforce decisions derived from their position in the organization.
- *Influence:* derived from respect for knowledge and skills, and perceived expertise and perceived ability to provide opportunities for growth and development.
- *Trust:* based on the leader's history of keeping their word, promises and other types of ethical behaviour as well as the willingness to accept bad news without 'killing the messenger'.

12.4.1 Types of Power and Influence

Perceptions from the *Continuum* HTP identified different types of power and influence including:

- *Coercive:* based on motivating the members of the team by punishment and fear of losing status, positions or bonuses. A Theory X type of leader (Section 2.1.4.1).
- *Connection:* based on formal and informal connections with influential or important people.
- *Expert:* earned if the team respects the leader's skills and knowledge in resolving both technical and organizational problems.
- *Referent:* based on charisma, personality, charm, interpersonal communications skills, listening skills and the perception that the leader 'cares' about the members of the team.
- *Reward:* the ability to provide positive reinforcement of behaviour (financial as well as non-financial).

Perceptions from the *Scientific* HTP infer that the effect of the different types of power and influence and motivation might be as shown in Table 12.1. The 'Don't Care' (Chacko 1989, Kasser 2018: Section 3.2.2) in the table means that it won't have much effect on the Theory X team and the Theory Y won't care about it.

TABLE 12.1
Where to Apply Power and Influence

Type of Power and Influence	Theory X	Theory Y
Coercive	Use	Do not use
Connection	Use	Don't Care
Expert	Don't Care	Use
Referent	Don't Care	Use
Reward	Use	Don't Care

12.4.2 Types of Authority

Authority can be:

- Verbal or written: where the perception is that written authority is more powerful than verbal authority.
- Delegated.

12.4.3 Leadership Styles

Perceptions from the *Continuum* HTP indicate there is a range of different leadership styles. No single style is correct for all situations so the manager needs to be able to use the style that fits the specific situation. There are a number of widely recognized styles with some variations, including:

1. Persuading.
2. Telling.
3. Selling.
4. Participating.
5. Delegating.
6. Situational.

12.4.3.1 Persuading
- Is used in situations where the members of the team are not motivated to do the necessary activities.
- Is often coercive; leaders make decisions unilaterally, without much (or any) input from the members of the team who must do as instructed.
- Is often results-based; managers are not concerned with how things get done, as long as they get done well and in the quickest way possible.

12.4.3.2 Telling
- Is used in situations where the members of the team are motivated but unable to take responsibility for the necessary activities perhaps due to lack of experience and limited knowledge, e.g. Type I personnel (Section 2.1.2.1).
- Can employ coaching; managers have a hands-on approach to coaching and encouraging new skill sets in members of the team while getting the project completed.
- Can employ mentoring; managers allow subordinates to make the majority of decisions, with management providing guidance when needed.
- Provides guidance, direction and input into the decision-making process regarding:
  - What activities will be done.
  - How activities will be done.
  - Where the activities will be done.
  - When the activities will be done.

12.4.3.3 Selling
- Is used in situations where the members of the team can take responsibility for the necessary activities but are not motivated to do so.
- Can be used to sell concepts and proposals to:
  - Upper management.
  - Customers.
- Develops 'buy-in' and acceptance.
- Explains:
  - Why decisions are made.
  - How followers will benefit.
  - Where suggestions made by followers fit in.
- Is used to develop followers.

12.4.3.4 Participating
- Is used in situations where the members of the team can take responsibility for the necessary activities and are motivated to do so.
- Can be:
  - *Collaborative:* managers offer the members of the team an opportunity to engage in decision-making so that all decisions are agreed upon by the majority.
  - *Consultative:* managers consult the members of the team but make the final decision.
  - *Mentoring:* managers allow subordinates to make the majority of decisions, with management providing guidance when needed.
- Uses team's ideas and expertise to formulate:
  - Strategies.
  - Develop decisions.
- Turns the role of the leader into a facilitator.
- Requires regular feedback regarding the performance of the team.

12.4.3.5 Delegating
- Is used when the members of the team are motivated and have the ability to perform activities with minimal supervision.
- Depends on:
  - The clarity of the delegation instructions (minimal confusion).
  - A minimal amount of micromanagement. Once an activity is delegated, the leader should monitor but not get involved.
- Turns the role of the leader into:
  - Mentoring.
  - Monitoring.
  - Observing.
  - Providing resources when scheduled.
  - Protecting followers from distractions and organization make work.

The Human Element

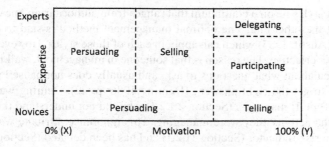

FIGURE 12.3 Application of different styles of management.

12.4.3.6 Situational

Perceptions from the *Scientific* HTP infer that the situational use of the different leadership styles:

- Can be represented in a matrix where the vertical axis is expertise and the horizontal axis is motivation; 0% (Theory X) (Section 2.1.4.1) at one end and 100% (Theory Y) at the other as shown Figure 12.3.[4]
- Will also depend on the leader's power and influence (Section 12.4.1). For example:
  - Leaders who use the persuading style generally employ coercive power and influence.
  - Leaders without the power of reward will only succeed when leading a Theory Y team.
  - Leaders who employ coercive power should not manage Theory Y teams.
- Managers who have a tendency to micromanage should not use the delegating style.

Each style fits a specific situation and has its own strengths and weaknesses. Accordingly, a good manager needs to be able to assume the style that fits the situation, recognize when the situation has changed and adopt the appropriate style for the new situation. For example, when dealing with:

- *Theory X subordinates:* (Section 2.1.4.2) or when the schedule does not permit time for discussion, the manager should use an authoritarian style.
- *Theory Y subordinates:* (Section 2.1.4.2) the manager should use a more participative style.
- *Subordinates who lack knowledge and experience:* if the schedule permits, the manager should use a coaching style as well as the authoritarian style.

---

[4] Note that the sizes of each leadership style area are governed by the size of the wording in the area rather than the boundary where the leadership style can be best applied.

The first six styles lie on a continuum that ranges from authoritarian one end to very laid-back at the other end. The optimal management method is said to be MBWA (Peters and Austin 1985) which can apply in each of those styles. However, 'MBWA is hardly ever effective. The reason is that someone in management, walking around, has little idea about what questions to ask, and usually does not pause long enough at any spot to get the right answer'. (Deming 1986: p. 22). Deming was probably discussing Type II managers (Section 2.1.2.1) who did not understand the technical aspects of the activity they were managing. This ignorance attribute was also recognized as a risk-indicator (Section 11.2.1) and has been debated (Section 2.1.1.4).

12.5 NEGOTIATION

PMs negotiate with customers, suppliers, team members and other stakeholders. Negotiation:

- Takes place in the context of a conflict generally in the form of a dialogue.
- Can only be successful if both parties in a conflict are willing to negotiate.

12.5.1 NEGOTIATION POSITIONS

When entering into a negotiation you should always have two positions for each item being negotiated:

1. The starting position which you share with your opponent which states:
   1. What you want.
   2. What you're willing to offer in return.
2. A desired position which you do not share with your opponent which is:
   1. What you really want which should be less than your starting position.
   2. What you are really willing to offer which should be more than your starting position.

These two criteria should enable you to move away from your starting position and still come away from the negotiation with what you really want.

12.5.2 NEGOTIATION OUTCOMES

There are three possible outcomes from a negotiation:

1. *Win-win:* both parties come out of the negotiation with what they really want.
2. *Win-lose:* one party comes out of the negotiation with what it really wants the other does not.
3. *Lose-lose:* neither party comes out of negotiation with all of what they really want.

If the negotiation covers a number of different items or issues, it is generally possible to achieve a win-win outcome by recognizing that the different parties want different outcomes.

12.5.3 Negotiating Styles

Different people have different negotiating styles. Perceptions from the *Generic* HTP show that these are similar to the styles adopted in conflict resolution, which should not be a surprise since negotiation is one of the ways of resolving conflicts. One set of five negotiating styles is (Shell 2006):

1. *Accommodating:* enjoying solving the other party's problems and preserving personal relationships.
2. *Avoiding:* disliking negotiating and not doing it unless warranted.
3. *Collaborating:* enjoying negotiations that involve solving tough problems in creative ways.
4. *Competing:* enjoying negotiations because they present an opportunity to win something.
5. *Compromising:* eager to close the deal by doing what is fair and equal for all parties involved.

12.5.4 Negotiating Tips

Here are some tips for negotiating.

- Understand what is being negotiated and why.
- Determine both your negotiating positions for all issues and items to be negotiated.
- Understand what your opponent is likely to really want to determine if the negotiation has a chance of being successful.
- Understand what items or issues your opponent does not care about.
- Understand the cultural differences between the parties, and how they might affect the negotiation.
- Find out the negotiating style of your opponent and adopt the same style or delegate the negotiation to somebody who can adopt the same style.
- Be prepared to walk away[5].
- Make sure person who has the final say on the agreement is present at the negotiation.[6] If that person is not present, the negotiations cannot be concluded.
- When negotiating multiple items or issues, don't agree on one item at a time. It leaves no scope for give and take on the last few items, and that is where negotiations can break down.
- Speak your opponent's language or delegate the negotiation to somebody who can.
- Read up on the topic if you are going into serious negotiations.

---

[5] This is a great tactic when shopping in a location where bargaining is expected. I've noticed asking prices dropping markedly the closer I get to the door.
[6] Or reachable by telephone from the negotiation site in real time. If you do not want to reach an agreement at the meeting, make sure that person is absent.

Understanding both wants and Don't Cares can be critical to achieving a successful negotiation. A simple example might be a husband-and-wife negotiating the type of new car to purchase. The husband wants a sports car and doesn't care about anything else; the wife wants a red car and doesn't care about anything else. She has correlated the family needs for that car with the different types of cars and knows that any type of car will meet the family needs. Accordingly, a successful win-win outcome of the negotiation is an agreement to purchase a red sports car.

12.6 MANAGING STAKEHOLDERS

> Stakeholder management is the process of managing the expectation of anyone that has an interest in a project or will be affected by its deliverables or outputs.
>
> **Project Smart (2013)**

Managing stakeholder concerns can be considered as a process containing the following activities:

1. Identifying the stakeholders.
2. Identifying the influence and authority of each stakeholder.
3. Identifying the areas of concern of each stakeholder.
4. Addressing the areas of concern of each stakeholder.
5. Converting stakeholder concerns to requirements.
6. Informing the stakeholders how their areas of concern were considered.
7. Gaining stakeholder consensus on the outcome.
8. Maintaining stakeholder interest and consensus.

There are four major difficulties in managing stakeholders:

1. *Identifying the stakeholders:* the literature generally advises that the way to identify stakeholders is by looking at the formal and informal relationships envisioning the stakeholder environment as a set of concentric circles. The inner circles stand for the most important stakeholders who have the highest influence (Recklies 2001), while the set of concentric circles identifies categories of stakeholders:
    - It is not that helpful in determining which of them has a stake in a specific project.
    - It does not provide any information about the nature of the relationships, nor how to manage them.

   In general, the literature is helpful but incomplete. Once stakeholders have been identified, things to note include:
    - *Name:* to be remembered and mentioned in communications.
    - *Position and title:* indicates their level of power and interest.
    - *Concern.* How much they care about the project, ranging from uninterested to their career depends on the project being successful; this is an indication of how much help they might be able to provide if the project runs into trouble.

- *Importance to the project:* namely the effect that loss of their support would have on the project.
- *How they can impact the project:* e.g. affect funding and resources.
- *What they get out of the project:* namely the benefits to them.
- *Level of engagement:* how often they need to be contacted.

2. *Making sure the important stakeholders are identified:* one way of doing this is to use the Nine System Model (Kasser 2019: Chapter 6) to identify the stakeholders.
3. *Satisfying stakeholders:* Making sure the stakeholders are satisfied irrespective of their importance.
4. *Managing stakeholder expectations:* stakeholders often don't have any idea of the cost of implementing their goals. Sometimes stakeholders who have expensive goals don't provide any funding to the project. One way of managing this situation is to make use of the configuration control board (CCB) or change control board and run all stakeholder requirements and change requests (formal and specially informal) through the board to estimate the cost and schedule impact of the proposed desire and let the customer decide if they want to pay for them (Section 11.3.2).

Perceptions from the *Generic* HTP indicate that:

- Most customers are generally facing problems similar to ones that other people are facing or have faced and have already created solutions.
- The concept of inheritance from those solutions to similar problems may be used to assist customers in identifying what they need and what they want.

Perceptions from the *Continuum* HTP indicate that there may be a difference between what customers need and what customers want. In the ideal world, what customers want is what they need. In the real world, the PM may be faced with one or more of the following:

1. Customers who *know* what they need.
2. Customers who *don't know* what they need.
3. Customers who *want* what they need.
4. Customers who *don't want* what they need.
5. Customers who *want* what they *don't* need.
6. Customers who *don't want* what they *don't* need.[7]

Inferences from the *scientific* HTP indicate a relationship between what customers want, what customers need and the structure of the problem they pose as shown in Figure 12.4. When customers know what they want and also know what they need, the PM is faced with a well-structured problem (Section 3.4.5.1). On the other hand, when the customers don't know what they want and also don't know what they

---

[7] And so the PM doesn't have to deal with this category.

Customers		Know what they **need**	
		Yes	No
Know what they want	Yes	Well-structured problem	Ill-structured problem
	No	Ill-structured problem	Well-structured problem

FIGURE 12.4 The relationship between what customers need, want and the structure of the problem.

need, the PM is also faced with a well-structured problem albeit a different two-part well-structured problem. This two-part well-structured problem is to:

1. Determine what the customers need.
2. Convince the customers of what they want, which should be what they need.

Perceptions from the *Operational* HTP indicate that the ultimate goal in managing stakeholders is to satisfy all stakeholders' expectations. However, in practice, generally, all stakeholders' expectations cannot be completely fulfilled. Thus, the goal in managing stakeholders often ends in a form of negotiated agreement with the stakeholders. That is to say, the difficulty in managing stakeholders is not about how to meet all the stakeholders' requests, but help all the stakeholders gain maximal satisfaction at the same time. Achieving stakeholder satisfaction is a continual activity for over the entire project lifecycle.

Achieving one stakeholder's satisfaction doesn't always mean that another stakeholder has to sacrifice. In general, stakeholders have different concerns and a final win–win agreement can often be achieved after several rounds of discussion or negotiations.

12.7 THE MULTI-SATELLITE COMMUNICATIONS SWITCHING SYSTEM REPLACEMENT PROJECT EXERCISE

The background and context for the progressive session exercises is in Appendix 1. The purpose of the exercise is:

1. To practice using the template for a student exercise presentation (Kasser 2018: Section 14.7).
2. To update the project information from Exercise 11.9 after the occurrence of an unanticipated event.

12.7.1 THE REQUIREMENTS FOR THE EXERCISE

Sometime between the preliminary design review (PDR) and the critical design review (CDR) an event occurred which may change the cost, schedule and staffing. Accordingly,

1. Select the event at random as discussed in Appendix 2.
2. Prepare a presentation according to the instructions in Appendix 3.
3. Save presentation as file '12-teamname'.pptx.

12.8 SUMMARY

This chapter discussed the human element; the most important attribute of project management because people perform the project, cause, prevent or mitigate problems and take part in the processes that produce the products. Once the project has begun, PMs spend most of their time dealing with people. The outcomes of these interactions with people affect the cost and the schedule and are reflected in the Gantt and EVA charts presented at the milestone reviews. This chapter explained time management, people management, conflict, leadership, negotiation and managing stakeholders.

REFERENCES

Chacko, George K. 1989. *The Systems Approach to Problem Solving*. New York: Prager.
Covey, Steven R. 1989. *The Seven Habits of Highly Effective People*. New York: Simon & Schuster.
Deming, W. Edwards. 1986. *Out of the Crisis*. Cambridge, MA: MIT Center for Advanced Engineering Study.
Ford, Henry, and Samuel Crowther. 1922. *My Life and Work*. Reprint Edition 1987 Ayer Company, Publishers, Inc. ed. New York: Doubleday Page & Company.
Hammer, Michael, and James Champy. 1993. *Reengineering the Corporation*. New York: HarperCollins.
Harrington, H. James. 1995. *Total Improvement Management the Next Generation in Performance Improvement*. New York: McGraw-Hill.
Harrington, H. James. 2000. Was W. Edwards Deming wrong? *Measuring Business Excellence* no. 4 (3):35–41.
Kasser, Joseph Eli. 1995. *Applying Total Quality Management to Systems Engineering*. Boston, MA: Artech House.
Kasser, Joseph Eli. 1996. There's no place for managers in a quality organization. In *9th Annual National Conference on Federal Quality*, Washington, DC.
Kasser, Joseph Eli. 2013. *A Framework for Understanding Systems Engineering*. 2nd edition. Charleston, SC: Solution Engineering: CreateSpace Ltd.
Kasser, Joseph Eli. 2018. *Systems Thinker's Toolbox: Tools for Managing Complexity*. Boca Raton, FL: CRC Press.
Kasser, Joseph Eli. 2019. *Systemic and Systematic Systems Engineering*. Boca Raton, FL: CRC Press.
Lewicki, Roy J., and Joseph A. Litterer. 1985. *Negotiation*. New York: IRWIN.
Malotaux, Neils. 2010. Predictable Projects delivering the right result at the right time. EuSEC 2010 Stockholm: N R Malotaux Consultancy.
Peters, Tom, and Nancy Austin. 1985. *A Passion for Excellence*. New York: Warner Books.
Project Smart. 2013. *Stakeholder management managing expectations* 2013 [cited 31 October 2013 2013]. Available from www.projectsmart.co.uk/stakeholder-management.html.
Recklies, Dagmar. 2013. *Stakeholder management* 2001 [cited 31 October 2013]. Available from www.themanager.org/Resources/Stakeholder%20Management.htm.
Shell, Richard G. 2006. *Bargaining for Advantage*. New York: Penguin Books.
Thomas, Kenneth W., and Ralph H. Kilmann. 1974. *Thomas–Kilmann Conflict Mode Instrument*. Tuxedo: XICOM.

13 Ethics

This chapter discusses the ethics of project management[1] because as a project manager (PM) you may find yourself in a situation in which you think something is wrong. Perceptions from the *Operational* and *Functional* HTPs note what PMs do when everything is above board. Perceptions from the *Continuum* HTP note issues that arise when you discover an activity, or are asked to participate in an activity that you feel is wrong or illegal. This chapter:

- Explains personal, organizational and professional ethics.
- Explains the ethical dilemma and the consequences of speaking out according to your conscience.
- Concludes with some personal lessons learned by the author.

13.1 PERSONAL ETHICS

Personal ethics deals with the way you relate to the organization and to the people you work with. There are two golden rules within an organization (Kasser 2018: Section 8.5) as:

1. *The Golden Rule Governing Behaviour:*
   - Is based on a part of Hillel's response to the pagan who wanted an instant summary of Judaism. As reported in the Babylonian Talmud Tractate Sabbath, Hillel summed up Judaism as, 'What is hateful to thee, do not unto thy fellow; this is the whole law' (Rodkinson 1903). Hillel was summarizing the Torah's social behaviour laws by personalizing the commandment, 'Do not stand idly by while your neighbour is hurt' (Leviticus 19:16).
   - Is a systems thinking tool because people are part of the system. It is the fundamental principle of Judaism, and should also be a code of behaviour for any manager or leader. The rule is stated in a negative way. Sometimes the rule has been stated in a positive way. There been many commentaries over the years on the difference between stating the rule in a positive and negative manner. In summary, the negative manner is realistic, the positive manner (treat others as you'd like them to treat you) is idealistic. One of the most important and effective uses of this principle is the 'Thank You' (Kasser 2018: Section 8.13).

   Effective leaders often embody this principle by not asking their followers to do something they would not do themselves.

---

[1] This chapter is an updated version of Kasser (1995).

2. *The Golden Rule pertaining to funding* is often stated as, 'he who has the gold makes the rules' (Augustine 1986). This rule:
   - *Reflects the real world:* projects exist because they're funded, that means the person who is funding the project[2] makes the rules and holds the final authority with responsibility for changes.
   - *Deals with satisfying the customer:* your internal principles (Covey 1989) must not be violated by any process you have to endure to satisfy the customer during the course of your career in an organization. The customer may be external or internal (Kasser 1995).

13.2 ORGANIZATIONAL ETHICS

The goal of a commercial organization is to make a profit within the operational framework of its environment, the law, suppliers, customers and its employees. However, there are often:

- Tendencies to take shortcuts to maximize profit.
- Personal biases.
- Plain ignorance.

which may lead to law breaking. The organization tries to minimize these tendencies and comply with the law by setting up an 'Ethics Program'. The purpose of this ethics programme seems to be to protect the organization from getting caught breaking the law. The best way to avoid getting caught breaking the law is not to break it in the first place. The organizational ethics programme is designed to:

1. Comply with the law.
2. Protect the organization from harm.

13.3 PROFESSIONAL ETHICS

Deciding what to do in a situation in which you think something is wrong is not an easy choice. In situations in which death or serious bodily harm has occurred or will occur, the decisions tend to be clear. Situations that are less clear pose much more difficult problems, and are probably numerous.

13.4 PERSONAL INTEGRITY

Personal integrity deals with:

- *Trust and Reputation:* being believed in and depended upon.
- *Internal principles:* and the compromising thereof.

---

[2] The customer in this book.

For example, in early 1985, Roger Boisjoly, a senior scientist at Morton Thiokol, predicted a space shuttle launch explosion based on the effect of the ambient temperature at the launch site on seals within the launch vehicle. He worked within his organization to effect a change, even unsuccessfully trying to prevent the launch of Challenger the night before the mission was to begin. History records that the launch vehicle exploded a minute or so after lift-off on 28 January 1986. Once Boisjoly testified before the Rogers Commission and told the whole story as he saw it, he became a whistle-blower and suffered retaliation (Boisjoly and Curtis 1990).

13.5 THE ETHICAL DILEMMA

When you discover an activity, or are asked to participate in, an activity that you feel is wrong, or illegal, remember the golden rules. You have several options. Before making the decision, as a minimum, you may want to ask yourself the following questions (Martin 1993: p. 208) as far as they are applicable:

- If I know what's going on, can I live with it?
- If it affects the bottom line, will I still do it?
- If it's not expressly forbidden, will I permit it?
- Is it right to make a profit when someone else's survival is at stake?
- How will I feel about myself at the end of the day if I do something that my conscience tells me is unethical?
- Do I condone or ignore unethical behaviour in my company as long as I don't do it personally?
- Does the importance of the individual in my organization determine how I will deal with an unethical act he or she commits?

13.5.1 THE ISSUES

Before you take any action, consider the following four and any other pertinent issues:

1. The law as discussed in Section 13.5.1.1
2. Your motives as discussed in Section 13.5.1.2
3. The company's ethics policy as discussed in Section 13.5.1.3.
4. The consequences of your actions as discussed in Section 13.5.1.4.

13.5.1.1 The Law

As an employee, you are regarded as an agent of the corporation that employs you. You have a duty to:

- Obey the directions of your employers.
- Act in your employer's interest in all matters relating to your employment.
- Refrain from disclosing confidential information, which, if revealed, might harm your employer.

- However, in the corporate environment, the law:
  - Does not require you to carry out illegal or immoral commands.
  - Does not authorize you to reveal such commands to the public.
  - Will not protect you if you do so (although that may be changing).
  - Will not protect you when you refuse to break it.
  - Will penalize you if you are caught breaking it.

If you are prepared to stick by your principles, and you are working on a U.S. government contract, you may have redress under The False Claims Act, 31 U.S.C. § 3729 et seq. The False Claims Act, in general covers fraud involving any federally funded contract or programme with the exception of tax fraud. The law:

- Grew out of the frauds against the government in the 1860s.
- Assesses damages of up to three times the dollar amount of the fraud (treble damages), and fines of up to $10,000 for each false claim submitted to the government.
- Empowers citizens to act on behalf of the government and sue the wrongdoer. This unique mechanism is known as *Qui Tam*. As a result of amendments in 1986, if a False Claims suit is initiated by a private party, that party is then eligible to receive between 15% and 30% of the total recovery.
- Seems to have been effective. *Qui Tam* actions returned nearly $1 billion to the Federal Government between 1986 and 1994 (TAF 1995). Defence contractors have aggressively lobbied the U.S. Congress to weaken the law.

13.5.1.2 Your Motives

Make sure you know why you are doing what you are doing, because your motives may be questioned. You may be doing it:

- *For the money:* If you are hoping to make your fortune by way of the *Qui Tam* provision in the False Claims Act, watch out! If you delay reporting the matter so that the amount of the fraud increases, you may be at risk for knowing and not speaking out in a timely manner.
- *For revenge:* Others will be able to see through your attempts at revenge.
- *Because you feel it is the right thing to do:* You may have to prove it.

13.5.1.3 The Company's Ethics Policy

In effective companies, management has established and communicated a clear set of business ethics for employees to follow and has created an environment that encourages employees to uphold those ethics (Martin 1993: p 201). The true test of the policy, however, is what happens when someone tries to invoke it. In the real world, based on prior case studies (Jos, Tompkins, and Hays 1989), *once you report a problem*, irrespective of what the ethics policy states, *there is a high probability that your career in that organization is over.*

If ethics are to be taken seriously in an organization, then the organization must:

- Define the actions the organization deems as unacceptable.
- Develop a method for reporting violations.
- Develop a consistent method of enforcement.
- Punish the violators.
- Develop a process to show that justice has been carried out.
- Protect the reporters of violations.

13.5.1.4 The Consequences of Your Actions

Currently, the typical initial response to the issue is to ignore the charges (Ettorre 1994) and focus instead on the person making the charges (you). Anticipate and prepare for retaliation. Retaliation may take several forms including those listed below (Devine and Aplin 1988):

- *Make the dissenters the issue instead of their message:* The organization will attack your motives, professional competence, credibility or anything else that will serve to cloud the issue. You will also be labelled as a troublemaker.
- *Isolate the dissenters:* You will be transferred away from the situation, and isolated from any sympathetic colleagues.
- *Make you the scapegoat:* They will put you in charge of solving the problem and deny you the resources you need to do the job. You will of course fail, and they will then fire you for cause, namely, failure to perform your assignment in a satisfactory manner.
- *Eliminate your job:* You will receive a layoff notice even if the company is hiring at the time. This will send a message to your colleagues.

13.5.2 THE APPROACH TO SOLVING THE ETHICAL PROBLEM

Before you do anything irreversible, make use of your problem-solving skills, as discussed below[3] to:

1. Analyse the situation as discussed in Section 13.5.2.1.
2. Identify appropriate lessons learned as discussed in Section 13.5.2.2.
3. Develop alternative decisions as discussed in Section 13.5.2.3.
4. Determine the probable outcome of each alternative decision as discussed in Section 13.5.2.4.
5. Evaluate the alternatives as discussed in Section 13.5.2.5.
6. Decide what to do and how to go about it as discussed in Section 13.5.2.6.

---

[3] Note this is an instance of the generic problem-solving process.

13.5.2.1 Analyse the Situation

The decision you make will depend on the particular situation. For example, the activity may be:

- *Illegal:* For example, bribery; fraud, such as lying, charging to the wrong contract cost account number on a government contract or falsification of test results on government contracts; discrimination based on race, religion or sex. You have to know which laws are broken in order to file a complaint with the government. The complaint must be filed with the agency authorized to enforce the law.
- *Wrong:* For example, poor management is not illegal, yet results in waste which reduces shareholder dividends. Often laws are introduced after someone has reported that something is wrong because the wrongness is considered bad enough to be illegal. For example, grass roots organizations have accomplished changes in the environmental laws.

At this time, try to determine what you want to achieve. For example, some goals might be:

- Right the wrong.
- Stop the illegal acts.
- Obtain retribution.
- Some or all of the above.

13.5.2.2 Identify Appropriate Lessons Learned

Apply the lessons learned concept. To anticipate what might happen to you, try to find out what happened to people who followed the company's ethics policy in the past and reported violations.

13.5.2.3 Develop Alternatives

Assuming you want to do something about the situation, the alternatives seem to be:

1. Keeping your mouth shut or doing nothing.
2. Taking part as requested.
3. Finding another job and walking away.
4. Reporting the situation in an anonymous manner.
5. Reporting the situation within the department.
6. Reporting the situation within the organization.
7. Reporting the situation outside the organization.

13.5.2.4 Determine the Probable Outcome of Each Alternative Decision

This is risk management. Examine the consequences of each alternative on your own self-esteem. You will have to live with yourself afterwards. The factors you have to consider are real and important, and include:

- *Anonymity:* In many instances, you can file a report anonymously. If you decide to file anonymously, then make the report as thorough as you can because you will not get the chance to talk with the investigating authority. Note that anonymity may be difficult to preserve for various reasons, so develop a contingency plan for the day you are identified.
- *Loyalty:* There are several loyalties within the organization. Your loyalty in order of priority is probably to the:
  - Team you are working with.
  - Department.
  - Department manager.
  - Division.
  - Organization.

When you decide to report something, either within or outside of the organization, your loyalty will be questioned and you will be treated as a traitor.

- *Family:* If you decide to report the situation, retaliation will affect your family, directly or indirectly. You need to think about how they will react. It is very important that you discuss the matter with them. A supportive family will make up for a lot of external retaliation.
- *Friends:* If you decide to report the situation, *you are going to find out who your friends really are.* This factor may be the only reward you get out of the whole episode apart from retaining your self-respect.
- *Finances:* If you decide to report the situation, you will probably have to find another job. Even if you then take the company to court and achieve a settlement, you will still have to live in the meantime. The company will try to delay reaching a settlement as long as they can in the hope that either you will give up or the statute of limitations will kick in and there will then be nothing you can do about it in a legal manner. You must consider how you will pay your mortgage, food bills, children's tuition and all other bills. Since the situation is going to be stressful, you may end up with medical bills as well. And, if you are unemployed, you may not have any medical insurance.

13.5.2.5 Evaluate the Alternatives

Think long and hard. One of the major risks you will face is family breakup as a result of the pressures that will be brought to bear.

- *Blowing the whistle:* After reviewing the issues discussed above, you make the decision that in order to live with yourself, you have to report the situation. Gather the evidence and do it while the activities are in progress, don't wait until afterwards.
- *Consult with a specialist:* When faced with an issue regarding reliability, you consult with a reliability engineer. When faced with a production problem, you consult a production engineer, so when faced with a legal

issue, consult an attorney or an alternate source of help, with the utmost alacrity. There are several organizations, which may be able to help you, if you are working on a government contract. They can be found on the Internet.

If you want to locate an attorney, ask around outside the organization and get a reference to an attorney who specializes in these matters. If you are in the United States, your local public library should have a copy of the multivolume Martindale-Hubble legal directory, which lists:

- Most if not all law firms in the United States.
- A brief biography of their partners and their specialties.
- If you can't locate a lawyer through your personal network, contact the Bar Association referral service in your area. An attorney can:
  - Determine if the activity is against the law.
  - Alert you to any laws you might break in documenting the activity and in collecting evidence.
  - Help protect you from corporate retaliation, or expiry of the statute of limitations, by anticipating the organization's probable responses and advising you accordingly.

Lawyers, however, tend to view problems within the legal framework. Deciding to blow the whistle is a question of your own integrity, so while the attorney will advise you, in the end, the decision is yours, and yours alone.

13.5.2.6 Taking Action

If you decide to file a report then this is what you need to do.

1. Open a diary.
2. Gather the evidence.
3. Follow the organization's procedure for reporting the activity.
4. Consider external options.

Consider each of them.

13.5.2.6.1 Open a Diary

Open a diary and write everything in it. Use a bound book rather than loose pages and a binder or an electronic file. The diary will serve the same purpose as the engineering notebook you use to record action items and notes during technical and staff meetings. Record meetings using a micro-recorder or your phone to remind yourself of who said what to whom and when. Document:

- Who said what to whom.
- Who was present at the time.
- References to documents or other evidence you have found or have heard about.

Ethics

If you think you will need to prove you wrote something on a certain date, mail a copy to yourself, but don't open the letter when it arrives. The postmark on the sealed envelope will be proof. If you are really concerned, register the letter.

13.5.2.6.2 Gather the Evidence

You will need as much evidence about the situation as you can obtain. This may take the form of documentation or prototypes. If you can't obtain all the evidence, then try to find out where it can be obtained. However, *do not do anything illegal to obtain the data you need*. If the matter does end up in court, evidence obtained in an illegal manner tends to be ruled as inadmissible. Gather what you need before you file the report, you may be cut off from your source material once the report is received.

13.5.2.6.3 Follow Procedure

Follow the organization's procedure for reporting the activity, i.e. work within the system. If there is an ethics policy, follow it. If not, discuss the matter with a supervisor or manager you feel you can talk to. Note in your diary what you said to whom and when and what they replied. If you don't feel that you can talk with anyone in your division, locate someone at corporate headquarters. Report the activity in writing as well as verbally and stick to the facts. Write your report in the same way as you'd write any technical report.

Enclose copies of technical data and supporting information as appendices in your report. Keep a copy of every item included. Make sure there are plenty of copies of your report with fellow employees or friends you trust. Don't seek publicity, but at the same time, don't acquiesce to the organizational request to keep the matter confidential. Confidentiality works in their favour, not yours. If someone within the organization asks you about the matter, be truthful. Don't sabotage yourself.

Keep emotions out of the picture and never, ever, threaten anyone. Politely request answers by specific dates. If you do not get a timely answer, work your way up the chain of command, but only advance if you seem to be encountering delays. Remember the law has statute of limitations.

Use your diary to document what happens to you after you make your initial complaint, your situation may end up in court. However, going to court is usually a lose–lose situation. If you can't resolve the matter within the organization, you will have to go outside the organization. Where to go depends on the activity and your employment contract.

13.5.2.6.4 Consider External Options

If you are working on a government contract, signs must be posted to tell you where to report violations of the law. Let your conscience be your guide after careful and deliberate thought. If you do file a complaint with a government agency, there will be an initial dialogue. In the case of the Equal Employment Opportunity Commission, the intake officer will tell you what to expect, describe the process and give you an optimistic estimate of the time you can expect the process to take. After the initial dialogue, 6 months or more may pass before you hear anything. If you telephone an agency hotline, you will be contacted by someone from the agency Inspector General's office and once again, the procedure will be lengthy.

When the dust clears and you feel like filing a lawsuit, check your employment contract to determine if the issue has to be settled by arbitration or mediation.

13.6 LESSONS LEARNED

The remembered lessons learned as I wrote this chapter include[4]:

1. My experience matched the literature.
2. I realized that I had experienced variations of this process five times in three different countries (United States, Australia and Singapore) and had experienced the same consequences (Section 13.5.1.4) more or less. In each instance, the organization protected its own and was not interested in righting the wrong.
3. The brain tends to forget the experience very quickly. Which is why, I did not recognize each situation for what it was[5] and suffered through the process four additional times.
4. It is difficult to get an external perspective when caught up in the situation. Some of the situations I was in lasted for years, yet I only saw the pattern when I looked at my notes while preparing this chapter.
5. In thinking about why I went through this process so many times, I articulated my forgotten thoughts after the first experience. I'm Jewish, and one of the laws I studied in my bar mitzvah portion was, 'Do not stand idly by while your neighbour is hurt' (Leviticus 19:16). So, when I saw that something was wrong, I spoke out and suffered accordingly.
6. I can't share these experiences of organizational politics, bullying, harassment and retaliation to encourage others to speak out in this book because the organizations involved would never give the permission for the stories to be told.
7. It is probable that organizations don't want to admit any liability to the whistle-blower for fear of a lawsuit. That is understandable, but the situation can be taken care of by making an agreement that it will be corrected, and the miscreants punished in return for the whistle-blower agreeing not to file a lawsuit.
8. Confidentiality agreements protect the organization and the miscreants. If you are going to speak out, do not sign a confidentiality agreement. First of all, the organization can argue that filing a lawsuit violates that agreement, and secondly, if you did publish your story, you are in the wrong for violating the confidentiality agreement.
9. Remember you will be attacked to sidetrack you from the issues involved.
10. The overall cost to the organization would be lower if the miscreants were removed from their positions of authority.

---

[4] Perhaps my experiences of unethical behaviour and negative politics (Section 2.2.2) are the reason why I focused my research on developing tools which make it difficult for the miscreants and their buddies in middle management to hide what is going on from upper management until it is too late; tools such as the CRIP charts (Section 11.5), the ETL charts (Section 11.6) and the use of the *Temporal* HTP to compare the performance evaluations over multiple time periods and investigate differences (Section 12.2.1.1).

[5] Except in the last instance.

Ethics

If you decide that something is wrong, and having read this chapter still want to do something about it, you will go through pain and suffering but you will come out of it a better person.[6] Courage is doing what needs to be done even or especially when you don't want to do it.

13.7 THE MULTI-SATELLITE COMMUNICATIONS SWITCHING SYSTEM REPLACEMENT PROJECT EXERCISE

The background and context for the progressive session exercises is in Appendix 1. The purpose of the exercise is:

1. To practice using the template for a student exercise presentation (Kasser 2018: Section 14.7).
2. To address ethical issues related to project closeout.

13.7.1 THE REQUIREMENTS FOR THE EXERCISE

After 5 years in development, the Multi-satellite Communications Switching System Replacement Project (MCSSRP) is nearing completion. The PM has put together the categorized requirements in process (CRIP) chart shown in Table 13.1 while preparing for the delivery readiness review (DRR). The project has no funding to continue past DRR and does not expect the customer to provide any more funds. The PM proposes to ask his golfing buddy, the customer's contracting officer's technical representative (COTR) to agree to:

TABLE 13.1
The CRIP Chart for the MCSS Project at DRR

Range	Identified P	E	A	In process P	E	A	Completed P	E	A	In test P	E	A	Accepted P	E	A
1	0	0	0	0	0	0	0	0	0	0	0	21	0	101	81
2	0	0	0	0	0	0	0	0	0	0	0	20	0	78	58
3	0	0	0	0	0	0	0	0	0	0	0	10	0	35	25
4	0	0	0	0	0	0	0	0	0	0	0	0	0	30	34
5	0	0	0	0	0	0	0	0	0	0	0	0	0	34	34
6	0	0	0	0	0	0	0	0	0	0	0	0	0	20	20
7	0	0	0	0	0	0	0	0	0	0	0	0	0	12	12
8	0	0	0	0	0	0	0	0	0	0	0	0	0	9	7
9	0	0	0	0	0	0	0	0	0	0	0	0	0	5	5
10	0	0	0	0	0	0	0	0	0	0	0	0	0	2	2
Totals	0	0	0	0	0	0	0	0	0	0	0	51	0	326	278

[6] I think I did.

1. Waive completing the incomplete requirements.
2. Not present a CRIP chart at the DRR.
3. Sign off on the project as being satisfactorily completed.

The team takes on the role of a task leader (TL) to whom the PM has confided his proposal.

1. The requirements for the exercise are to:
2. Formulate the problem posed by the exercise using the problem formulation template (Section 3.7.1).
3. Examine the CRIP chart shown in Table 13.1.
4. Make reasonable assumptions.
5. Discuss:
    1. The situation from an ethical perspective.
    2. How the PM could bury the fact that the system would not meet its requirements in the DRR earned value analysis (EVA) reports.
    3. How the CRIP chart makes it difficult to bury the fact that the system would not meet its requirements at DRR.
6. Number the slides.
7. Prepare <5 min presentation that includes:
    1. The problem posed by the exercise formulated according to the problem formulation template (Section 3.7.1).
    2. A summary of the ethical and other issues in the situation.
    3. The decision to be made including listing the alternatives.
    4. The specific selection criteria for the decision.
    5. The decision of what to do about the situation.
    6. The benefits of the CRIP chart.
    7. Lessons learned.
    8. A compliance matrix (Kasser 2018: Section 9.5.2) for meeting the requirements for the exercise.
8. Save presentation as file '13-teamname'.pptx.

13.8 SUMMARY

This chapter discussed the ethics of project management because you may find yourself in a situation in which you think something is wrong. This chapter explained personal, organizational and professional ethics, the ethical dilemma and the consequences of speaking out according to your conscience and concluded with some personal lessons learned by the author.

REFERENCES

Augustine, Norman R. 1986. *Augustine's Laws*. New York: Viking Penguin Inc.
Boisjoly, Russell P., and Ellen Foster Curtis. 1990. Roger Boisjoly and the challenger disaster: A case study in management practice, corporate loyalty and business ethics. In *Business Ethics*, edited by W. Hoffman Michael and Jennifer Mills Moore. New York: McGraw-Hill.

Covey, Steven R. 1989. *The Seven Habits of Highly Effective People*. New York: Simon & Schuster.
Devine, Thomas M., and Donald G. Aplin. 1988. Whistleblower protection-The gap between the law and reality. *Howard Law Journal* no. 31:223.
Ettorre, Barbara. 1994. Whistleblowers: Who's the real bad guy. *Management Review* no. 83:18–23.
Jos, Philip H., Mark E. Tompkins, and Steven W. Hays. 1989. In praise of difficult people: A portrait of the committed whistleblower. *Public Administration Review* no. 49 (6):552.
Kasser, Joseph Eli. 1995. Ethics in systems engineering. In *NCOSE 5th International Symposium*, St. Louis, MO.
Kasser, Joseph Eli. 2018. *Systems Thinker's Toolbox: Tools for Managing Complexity*. Boca Raton, FL: CRC Press.
Leviticus 19:16. *The Five Book of Moses*. Mt Sinai.
Martin, Don. 1993. *Teamthink: Using the Sports Connection to Develop, Motivate and Manage a Winning Business Team*. New York: Dutton.
Rodkinson, Michael L., ed. 1903. *Tractate Sabbath in The Babilonian Talmud*. Boston, MA: New Talmud Publishing Company.
TAF. 1995. *The False Claims Act: Information Sheet*. edited by Legal Center. Washington, DC: Taxpayers Against Fraud.

Appendix 1
The Engaporian Multi-Satellite Operations Control Centre Communications Switching System Replacement Project

1.1 CONTEXT

The class exercise is set in 2010 in the fictitious Federated Aerospace that won the contract to implement the Engaporian[1] Multi-satellite Communications Switching System Replacement Project (MCSSRP). The complexity of the MCSSRP is represented by the categorized requirements in process (CRIP) chart at the start of the project shown in Table 11.10.

The exercise in each session is to:

1. Develop parts of a project plan (PP) for the MCSSRP in an incremental manner.
2. Direct and control the execution of the MCSSRP.

Each team will perform project management on the same project to allow students to compare their project management with that of other teams. To make the simulation interesting, there will be instructor-provided differences between the teams, and a number of different unforeseen events will occur in the second half of the class.

1.2 SCOPE OF EFFORT

The scope of effort is determined by the number of people in the team and the number of hours students are expected to invest in the class. Within this constraint, presentations are expected to contain a representative sample of information showing that the topic knowledge acquired during a session has indeed been applied to the MCSSRP.

---

[1] A fictitious country.

1.3 PART I ACTIVITIES

Student teams will prepare and present sections of a PP for the MCSSRP. The teams will add elements during each session as follows:

- *Session 5:* creates partial work packages (WPs) of at least 8 products using Table 5.1 as a template in Section 5.12.
- *Session 6:* adds staffing information to the WPs in Section 6.4.
- *Session 7:* creates the schedule for the project based on the staffing information and adds the schedule information to the WPs in Section 7.8.
- *Session 8:* costs the project and adds the cost information to the WPs in Section 8.9.

1.4 PART II ACTIVITIES

A primary activity performed by project managers is coping with change. These exercises allow students to experience events that affect cost and schedule and experience the replanning that needs to be done.

The purpose of the second part of each change management exercise is to demonstrate that students have learnt what to do to cope with changes during project execution. Consequently, reasonable assumptions as to the scope and extent of the impact of these events can be made as long as the events affect cost and schedule.

- *Session 9:* adjusts the project due to a cost reduction in Section 9.6.
- *Session 10:* adjusts the project due to an unanticipated event in Section 10.9.
- *Session 11:* adjusts the project due to an unanticipated event in Section 11.9.
- *Session 12:* adjusts the project due to an unanticipated event in Section 12.7.

Appendix 2
Change Management Events

Each change management presentation covers the period of activity between the previous milestone review and the one being made. During that period of activity one of the following events occurred (as determined by the drawing of a numbered slip during class and provided to team leaders).

1. Expected resources were not available; project was delayed by two time periods.
2. Critical component delivery was late by one time period.
3. Critical component delivery was early by one time period.
4. Chief systems engineer resigned.
5. Company won a major contract for new and exciting project, 50% of managerial staff applied for transfer to new project.
6. Company won a major contract for new and exciting project, 50% of all technical staff applied for transfer to new project.
7. Company won a major contract for new and exciting project, 50% of all technical and managerial staff applied for transfer to new project.
8. No major glitches, project proceeds according to plan. Note: any team that gets this number in a session is disqualified from receiving this number in a subsequent session.
9. Engineering was implemented smartly reducing the critical path by 20%.
10. Last milestone review presentation generated 20 review item discrepancies (RID).
11. Customer informed you that the remaining project schedule is to be speeded up (reduced in time) by 25%.
12. Expected resources were available; project was early by one time period.
13. Customer's budget has been reduced by 25% for rest of project.
14. Customer's budget has been reduced by 35% from previous milestone to this milestone.
15. Another company project ended; you had your choice of up to three additional junior personnel.
16. Customer changed requirements to increase number of inputs to the switch by 50%.
17. Two junior personnel quit.
18. Project manager was severely injured in an automobile accident and was on medical leave for ten time periods.
19. Poor engineering resulted in delay of five time periods in task requiring most time.
20. Poor engineering resulted in delay of five time periods in most costly task.

21. Innovative engineering reduced project costs by 10%.
22. Customer cancelled the project (only applicable to Session 12).
23. Vendor/manufacturer of the most critical component went bankrupt and cannot deliver (only applicable *after* Session 10).

Notes.

1. Not all numbers may be drawn each session.
2. Each team shall draw a different number.
3. Numbers shall be retired after use.

Appendix 3
Presentation Guidelines and Requirements

The presentation guidelines and requirements are:

1. Make reasonable assumptions and appropriate supporting rationale.
2. Formulate the problem posed by the exercise using the problem formulation template (Section 3.7.1).
3. Format presentation according to the applicable parts of the sample four-part template for a management review presentation (Section 5.4.2.2.1).
4. Number the slides.
5. Do not repeat the detailed content of the presentation from last time.
6. Assumptions must be realistic and consistent with information presented in prior sessions.
7. Some event mitigation techniques may not be reused by another team when the same event shows up in a later session; instructor will notify students at time of presentation.
8. Do not repeat statements made in previous presentations (your teams' or other teams'), refer to the statements instead.
9. You may have to backfill something (provide information that wasn't there before) into your project plan to cope with events.
10. A time period will depend on the team project's way of measuring time (e.g. weeks, months, etc.)
11. The team can choose (and state in the presentation) when the event took place during the reporting period, i.e. start, sometime in the middle, end, except as instructed.
12. Consider at least two alternative approaches to dealing with the event.
13. Select one approach.
14. Update the work packages affected by the event.
15. Update the project summary information.
16. Prepare <5 min presentation that includes:
    1. Event number, event and assumed impact of event in the introduction to the presentation.
    2. Previous milestone as a baseline and summary of effect of the event.
    3. Outline the two approaches and state why the selected one was chosen.
    4. Present appropriate cost, schedule, ETL and CRIP charts.
    5. The process the team went through to perform the exercise.

6. Lessons learned in exercise. Lessons learned must come from experience, not as quoted topics from book. For example, 'a plan is important' is a book topic while 'a plan helped us manage our time because …' can be a lesson learned.
7. A compliance matrix (Kasser 2018: Section 9.5.2) for meeting the requirements for the exercise.

REFERENCE

Kasser, Joseph Eli. 2018. *Systems Thinker's Toolbox: Tools for Managing Complexity*. Boca Raton, FL: CRC Press.

Appendix 4
Staffing Information

1. The scope of the resumes is limited to providing information pertaining to the performance of personnel within Federated Aerospace.
2. Information about their past employment has not been provided.
3. The staffing shortage is acute and unless the right mix of staff is hired and brought up to speed, the project will fall behind schedule.
4. Some staff assist other projects and corporate activities when needed (e.g. writing proposals) and providing specialty knowledge that does not warrant a full-time person on a project (e.g. reliability).
5. Federated Aerospace has grown over the years via mergers and acquisitions. As such, while there has been some degree of consolidation of personnel classifications, there is still a mixture of job titles in the various projects, a legacy from the former organizational structures.
6. Current salary ranges in Federated Aerospace are as shown in Table A4.1.
7. Resumes are in alphabetical order to facilitate finding and comparing information.
8. Resumes are based on edited open position descriptions advertised on various job sites on the Internet in November 2007, not on any specific individuals.

TABLE A4.1
Systems Engineering Salary Ranges

Labour Category	Years of Professional Experience	Salary Range[a]
Senior	Eight or more	$40 – 75,000
Mid-Level	Three to seven	$25 – 45,000
Junior	Less than three	$20 – 35,000

[a] Demonstrated competence and higher degrees tend to place an employee in the upper part of the range.

Appendix 5
Resumes for the Session 6 Exercise

5.1 SANDRA ANSCOMB

Position: Systems Analyst
   Duties:

- System-level requirements development, management and verification.
- Development of products such as system-level performance and interface specifications, systems engineering management plans, requirements traceability matrices (RTM), and concept of operations documents.
- Manage the requirements identification and development process by planning, organizing and monitoring information that affects product design (e.g. customer needs, regulatory requirements, performance, cost constraints, project capability and capacity).
- Validate that customer needs, requirements, expectations and constraints are satisfied in the various product definition states (e.g. requirements, architectures, detail designs) and in the end product through the use of analysis, simulation, incremental reviews, demonstrations and operational tests.
- Verify that each state of the product design is compliant with previous states by conducting and documenting tests, demonstrations, simulations, analyses or inspections.

Skills and competencies

- Ability to understand the big picture and the interrelationships of all positions and activities in the organization, including the impact of changes in one area on another area.
- Ability to see and understand the interrelationships between components of systems and plans and anticipate future events.
- Eight years of experience with modelling, simulation and performance analysis.

Education

- PhD in Operations Research 1995.
- Federated Aerospace short course on Advanced Systems Engineering, 2005.
- Prof Kasser's Requirements Workshop, 2005.

Federated Aerospace comments

- Joined in 2005.

5.2 SARA ARNOLD

Position: Systems Engineer
Duties:

- Ensure the logical and systematic conversion of customer or product requirements into total systems solutions that acknowledge technical, schedule and cost constraints.
- Perform functional analysis, timeline analysis, detail trade studies, requirements allocation and interface definition studies to translate customer requirements into hardware and software specifications.
- Ensure analysis is performed at all levels of total system product, including concept, design, fabrication, test, installation, operation, maintenance and disposal.

Skills and competencies

- Has an understanding of systems engineering principles.
- Knowledge of DOORS and CORE.
- Familiar with standard MS Office applications.
- Excellent communication skills.
- Demonstrated ability to prepare and deliver technical presentations to a high standard.
- Demonstrated exemplary report writing skills.
- Good interpersonal skills.
- Able to work as part of a multidisciplinary, cross-organizational, project team.

Education

- Master of Science in Systems Engineering 2005.
- Federated Aerospace short course on Advanced Systems Engineering, 2005.
- Prof Kasser's Requirements Workshop, 2005

Federated Aerospace comments

- Joined in 2002.

5.3 JACK APPLEBEE

Position: Junior Systems Engineer
Duties:

- Supports senior staff as needed.
- Research standards for compliance.

- Write testing software routines as needed.
- Support configuration management.

Skills and competencies

- Excellent communication skills.
- Likes to work with people.
- Good organization and documentation abilities.
- Computer literate – word, spreadsheet.
- Programs in Basic and C++.

Education

- Higher National Diploma in Electrical Engineering – Willesden College of Technology 2006.

Federated Aerospace comments

- Joined 2007.

5.4 EILEEN BROWN

Position: System Engineer
  Duties

- Perform a full range of technical analyses and investigations, including evaluation and selection of technical options.
- Prepare reports with suggested recommendations to support the resolution of problems and the improvement of engineering capability.
- Prepare a full range of specifications, develop designs in line with specifications and all quality and technical standards and carry out design appraisals in assigned areas to ensure all standards are maintained.
- Perform investigations into a full range of problems, issues or developments, and develop and prepare solutions, individually or as a member of a project team.
- Define testing approaches, specify tests for subsystems and evaluate and make recommendations based on results, including making suggestions to improve test/diagnosis processes.
- Lead engineering support activities in the field (and carry out the most complex activities) for specific subsystems to ensure the operability of company products.
- Plan, control and deliver assigned projects or WPs.
- Provide assistance to cross-functional teams, ensuring requirements are delivered to plan.
- Perform team planning, risk management and quality activities in assigned areas in line with all requirements and processes.

- Provide effective support to customers in assigned areas, delivering solutions to meet the customer's needs and requirements as well as to the programme plan.
- Provide help and direction to less experienced team members to support both their effective working and their development.
- Manage and report against the team's work programmes, including analysis of risk.

Skills and competencies

- Demonstrated analytical and problem-solving skills.
- An understanding of the system design process (including system and subsystem specification, system test, integration and certification).
- Good verbal and written communications.

Education

- Higher National Diploma, endorsements in Testing and Evaluation, Craft College of Technology, 1999.
- Federated Aerospace short course on Advanced Systems Engineering, 2005.
- Prof Kasser's Requirements Workshop, 2005.

Federated Aerospace comments

- Joined 1999.

5.5 RICK BLAINE

Position: Project Systems Engineer
Duties:

- Review bid and tender documentation.
- Attend meetings with clients.
- Present system developments and technical progress.
- Ensure technical scope of supply is defined and agreed.
- Establish and coordinate overall system design, interfaces, configuration management and document control.
- Technical follow up of system suppliers and resolution of system and interface issues.
- Prepare for, and chair the internal Project System Design Review followed by client attended System Design Review.
- Initiation and verification of all integration testing procedures.
- Initiate actions between interfacing contractors to ensure timely interface resolution.
- Follow up the technical and commercial aspects resulting from interface resolution.

Appendix 5

- Work closely with other members of the project management team in order to expedite and verify such resolutions.
- Ensure that appropriate qualification testing is performed where necessary.

Skills and competencies

- Able to work with and within a team of engineers having multiple disciplines.
- Self-motivated, and able to self-manage tasks whilst undertaking a significant workload.
- Able to communicate effectively.
- Has an aptitude for learning new technology, techniques, methods and principles.
- Able to produce novel ideas and concepts.
- Computer literate.
- Capable of multitasking and working to agreed schedules.
- Able to understand how systems solutions are defined in the presence of equipment performance and environmental constraints.
- Demonstrated willingness to share knowledge.

Education

- Higher National Certificate in Engineering, 1995.
- Federated Aerospace short course on Advanced Systems Engineering, 2005.
- Prof Kasser's Requirements Workshop, 2005.

Federated Aerospace comments

- Joined 2000.
- Received recognition for outstanding direction and performance of systems engineering and architecting.
- Developed computer system procurement plan that saved customer $1.5 million.
- Was recognized as an outstanding facilitator in field testing NMA's 'Leading Process Improvement' course (Feb/Mar 2003).

5.6 TED CHEN

Position: Systems Engineering Manager
Duties:

- Identify and provide skilled engineering resources according to project needs.
- Plan and prioritize use of engineering resources with respect to project requirements.
- Manage performance and development of staff in the project.

- Identify skill gaps and implement the necessary plans to fill these gaps by recruitment or other means.
- Provide technical leadership for the engineering teams.
- Maintain high standards of technical discipline.
- Keep up-to-date on technology and trends within the marketplace and share this within Federated Aerospace.
- Provide the necessary procedural framework to enable project needs to be effectively controlled: e.g. design reviews, sign-off criteria, requirements documentation, FMEA, engineering change control and fault reporting systems.
- Plan and monitor the project budget for headcount, materials and purchased services.
- Ensure that overspends are escalated to senior engineering management level.
- Develop engineering cost estimates for systems engineering activities to support bid win activities.
- Travel to customer facilities.

Skills and competencies

- Demonstrated leadership and people management skills in a design and development environment.
- Self-motivated with the ability to motivate others.
- Capability to resolve conflicts and set engineering priorities to meet overall company objectives.
- Willingness and ability to drive problem-solving activities.
- Good communication skills.
- Fluent in Mandarin, English and French.

Education

- Bachelor of Science in Mechanical Engineering, Singapore Ford University, 1990.
- Master of Science in Transportation Studies, Natural University of Singapore, 1995.
- Fellow of the IET.
- Member of the National Management Association.

Federated Aerospace comments

- Joined 2000.
- Received recognition for outstanding direction and performance of systems engineering and architecting.
- Keeps a project to cost while satisfying customers.

5.7 DAVID DRAKE

Position: Senior Systems Engineer
  Duties:

- Review and approve system requirements and customer manuals to ensure they meet the expectations and uses of the customer and are in accordance with quality system, regulatory and business requirements.
- Define characterization, verification and validation tests to confirm the behaviour of the system meets specification, user needs and intended use.
- Interact with hardware, software, firmware and systems groups across Federated Aerospace working in the design and development of products.
- Provide technical support for technicians performing testing.
- Assure root cause analysis and resolution of all issues raised during testing.
- Review/prepare documentation for regulatory submissions and product release.
- Participate in root cause issue analysis and ensure that resolutions are carried forward into projects.
- Support business quality management systems in the implementation and training of related quality systems practices and policies.
- Participate in and/or lead continuous improvement activities to drive quality improvements in systems requirements analysis, design verification and design validation.

Skills and competencies

- Demonstrated technical writing skills.
- Well-developed communication skills.
- Adept at using a computer.
- Strong influence management skills.
- Experience working with cross-functional teams.
- Experience in problem-solving methodologies (8-D, Root Cause Analysis, etc.).
- Background in quality systems and corrective and preventive actions implementation.
- Ten years' experience in systems engineering discipline and use of systems engineering methodologies.

Education

- Bachelor of Science in Electrical Engineering - Hypothetical University, 2000.
- Master of Science in Systems Engineering - Loyola Technical University, 2008.
- Member Society for Quality Engineering.

Federated Aerospace comments

- Joined 2009.

5.8 HECTOR HUBERT

Position: Junior Engineer
Duties:

- Configuration control of design control documentation.
- Technical writing (i.e. user manuals, standard operating procedures, product updates in marketing materials).
- General engineering tasks including support with customer trials, and trade show technical support.

Skills and competencies

- Ability to multitask, pay attention to and create detailed paperwork.
- Excellent troubleshooting skills.

Education

- Bachelor of Engineering – West Sheffield University 2006.

Federated Aerospace comments

- Joined 2006.

5.9 ANTHONY JAMES

Position: Systems engineer
Duties:

- Ensure the logical and systematic conversion of customer needs into total systems solutions that acknowledge technical, schedule and cost constraints.
- Perform functional analysis, timeline analysis, detail trade studies, requirements allocation and interface definition studies to translate customer requirements into hardware and software specifications.
- Ensure analysis is performed at all levels of total system product including concept, design, fabrication, test, installation, operation, maintenance and disposal.

Skills and competencies

- Have a thorough understanding of systems engineering principles.

- Working knowledge of industry-wide accepted tools and techniques.
- Familiar with standard MS Office applications.
- Demonstrated excellent communication skills.
- Able to prepare and deliver technical presentations to a high standard.
- Demonstrated exemplary report writing skills.
- Good interpersonal skills.
- Experienced in participating in multidisciplinary and cross-organizational project teams.
- Experience in a defence systems engineering environment.

Education

- Bachelor of Technology – Oxbridge Cotton University 2000.
- Master of Science in Systems Engineering, Magdelen University, 2004

Federated Aerospace comments

- Joined 2004.

5.10 ELMER KOWALSKI

Position: System Engineer
Duties:

- Provide technical expertise in support of systems engineering for the specific project area.
- Provide technical guidance to subordinate and inexperienced engineers, and participate in on-the-job training for engineering staff.
- Maintain a good working relationship with customers/subcontractors in particular liaison on engineering aspects while not giving too much information to actual/potential competitors.
- Interpret system/subsystem architecture design.
- Define and design system/subsystems.
- Analyse and confirm system, subsystem or equipment test results.
- Define and specify equipment.
- Manage engineering WPs including work placed on contractors or suppliers and ensure timely completion of tasks within cost and schedule.
- Advanced state of the art systems studies with particular emphasis on innovative ideas and concepts.

Education

- Higher National Diploma, Larden College of Technology, 2008.

Federated Aerospace comments

- Joined 2008.

5.11 ROGER LAPIN

Position: Reliability Engineer
Duties:

- Develops test procedures to verify the reliability of a new design or an update to an existing product.
- Tests products including environmental stress tests, product acceptance tests, life tests, etc.
- Communicates test results to other groups as required, both written and verbal.
- Assists design engineers with professional opinions, reference materials, testing methods, etc. as and when needed.

Skills and competencies

- Able to perform computer analyses such as MTBF prediction programs.
- Familiar with databases and statistical analysis programs.
- Able to design and carry out tests and collect and report on results in a concise and accurate manner.
- Good interpersonal skills allowing him interface on a professional level with a wide variety of technical disciplines that exist within the company.
- Able to work with minimal supervision.
- Computer literate with various software programs such as Windows, Excel and Word.
- Able to program test equipment in Basic.

Education

- BSEE – Romulus University, 2002.

Federated Aerospace comments

- Joined 2004.

5.12 VICTOR MASLOW

Position: Senior Systems Engineer
Duties:

- Lead systems engineering aspects of development on tenders and projects.

- Ensure that systems engineering processes are employed throughout the project lifecycle.
- Support overall technical leadership of multidisciplined engineering team(s).
- Requirements management of the technical aspects of the project.
- Manage technical aspects of subcontractor relationships.
- Support bid preparation.

Skills and competencies

- Proven experience in a systems engineering role working on complex defence-related projects.
- Conversant in the specification of communications equipment for telemetry, tracking and control links.
- Demonstrated interpersonal skills which have enabled effective leadership of technical staff, internal support departments and customers.
- Practical knowledge of requirements management.
- Experienced in working with stakeholders (internal and external) to elicit, analyse and negotiate requirements.
- Ability to support design reviews internally, with subcontractors and with customers.
- Experience in application of validation and verification techniques.
- Working knowledge of project management, requirements and design tools (Microsoft Project and DOORS).

Education

- BSEE, University of Engaporia, 2000.
- Master of Science in Special Studies, University of the North, 2005.

Federated Aerospace comments

- Joined in 2005.

5.13 PITA MOON

Position: Junior Engineer
Duties:

- Support senior staff as needed.
- Support configuration management.

Skills and competencies

- Likes to work with people.
- Computer literate – Word and spreadsheet.
- Programs in Basic and C++.

Education

- Higher National Diploma in Electrical Engineering – Willesden College of Technology 2006.

Federated Aerospace comments

- Joined 2007.
- Two poor performance evaluations.
    - Behaves like a Type I person.
    - Works at the pace of a startled snail.
    - Cannot be motivated to put in more than the minimum required level of effort.
    - Poor memory; has to be told everything over and over again.

5.14 CLARK MORSE

Position: Senior Systems Engineer
Duties:

- Provide technical leadership for project.
- Serve as technical liaison to customers, consolidate and negotiate customer requirements and vehicle-specific interfaces.
- Approve specification release and change, supports configuration management, prototype vehicle build support, system tuning and troubleshooting.
- Design and optimize simultaneous system performance goals through requirements flow down.
- Effectively integrate all aspects of a design using multidisciplinary engineering.
- Tune vehicle-steering characteristics, diagnose and correct deficiencies in designs and prototypes.
- Understand and write system and subsystem specifications.
- Assist in system testing and verification activities.
- Assist in maintaining cost controls.
- Stay abreast with new technology.

Skills and competencies

- Excellent problem-solving techniques.
- Enjoys driving and evaluating vehicles.
- Knowledge of system synthesis, analysis and verification techniques.
- Six years' experience in systems or automotive engineering.
- Excellent communication skills.
- Likes to work with people.
- Good organization and documentation abilities.
- General competency in mechanical, electronics, controls, computer hardware and software engineering.

- Background in automotive engineering and chassis systems (steering, braking, suspension).
- Strong computer skills, spreadsheets (Microsoft Excel) and Matlab.

Education

- BSEE – Hypothetical University, 2002.
- Federated Aerospace course on Interpersonal Skills Development 2008.
- Chartered Engineer.
- Member of the IET.

Federated Aerospace comments

- Joined 2006.

5.15 SUSAN RICHARDSON

Position: Senior Systems Engineer
Duties:

- Support design teams engaged in the development of transportation vehicles.
- Perform functional and use case analyses, allocation and documentation of requirements, assessment of functional and survivability performance.
- Perform trade studies.

Skills and competencies

- Experience in the application of a formal systems engineering process to the product development lifecycle.
- Understands MOD conventions for requirements development and documentation.
- Experience in the application of a CMMI-compliant process.

Education

- Master of Science in Automobile Engineering, University of West Croydon, 2008.

Federated Aerospace comments

- Joined in 2008.

5.16 JOHN SEVERN

Position: Principal Systems Engineer
Duties:

- Responsible from conception to completion, for the systems engineering aspects of projects.
- Work closely with the project management team to achieve successful business acquisition (normally by competitive tender).
- Lead systems engineering aspects of development on tenders/projects typically worth about £1–15m.
- Ensure systems engineering processes are employed throughout the project lifecycle and provide technical input to project estimating/controlling processes.
- Provide overall technical leadership of multidisciplined engineering team(s).
- Requirements management of the technical aspects of the project and managing technical aspects of subcontractor relationships.

Skills and competencies

- Proven experience in a systems engineering role working on complex defence-related systems/platforms.
- Conversant in the specification of receivers, transmitters, signal processing and high-level software as applied to equipment and systems in the DC-3 GHz frequency range.
- Good interpersonal skills which allow effective leadership of technical staff (leading design reviews), subcontractors and customers.
- Understanding of requirements management including experience of working with stakeholders (internal and external) to elicit, analyse and negotiate requirements.
- Experience in validation and verification techniques.
- Working knowledge of project management, requirements and design tools, i.e. Microsoft Project, DOORS, etc.
- Experience of working on safety-related applications.
- Experience of defence contracting and the bid process.

Education

- BSEE – Hypothetical University, 2003.
- Federated Aerospace course on Interpersonal Skills Development 2008.
- Federated Aerospace course on Project Management 2009.
- Prof Kasser's Requirements Workshop 2006.
- Chartered Engineer.
- Member of the IET.

Federated Aerospace comments

- Joined 2000.

5.17 SYLVIA SIDEWINDER

Position: Systems Engineer
Duties:

- Perform system and subsystem architecture and design for systems.
- Define system interface and software requirements for software specifications.
- Synthesize requirements into system test plans and procedures.
- Coordinate integrated testing activities.

Skills and competencies

- Has an understanding of systems engineering principles.
- Knowledge of DOORS and CORE.
- Familiar with standard MS Office applications.
- Excellent communication skills.
- Demonstrated ability to prepare and deliver technical presentations to a high standard.
- Demonstrated exemplary report writing skills.
- Good interpersonal skills.
- Able to work as part of a multidisciplinary, cross-organizational, project team.

Education

- BSEE, Acme Loyola University, 2002.
- Federated Aerospace short course on Advanced Systems Engineering, 2005.
- Prof Kasser's Requirements Workshop, 2005.

Federated Aerospace comments

- Joined in 2002.

5.18 PHILIP SMITH

Position: Senior Systems Engineer
Duties:

- Provides direction, coordination and control on assigned systems projects, including management of specific WPs for scope, cost and schedule in accordance with established systems engineering process.
- Leads system architecting, technical requirements development, system interconnect/interface design coordination with design engineering and the customer, integration testing, and customer integration/certification support.

Skills and competencies

- Knowledge of avionics and cockpit operations.
- Excellent written and oral communications.
- Experience in using the following tools MS Office, Access, Project Scheduler and DOORs.

Education

- Bachelor of Science in Electrical Engineering - Hypothetical University, 1985.
- Master of Science in Aeronautics Engineering - Acme University, 2009.
- Senior Member Society for Aeronautical Engineering.

Federated Aerospace comments

- Joined 2003.

5.19 MARTIN SOAMES

Position: Mechanical Systems Engineer
  Duties:

- Independently apply engineering skills to solve complex problems and obtaining innovative and cost-effective solutions requiring the development or sustaining of new or improved techniques, procedures or products.
- Plan, conduct and evaluate approaches to meeting the project objectives in a timely fashion.
- Lead the design, analysis and mathematical modelling to develop innovative and cost-effective client/product/research solutions.
- Identify, organize and document requirements.
- Coordinate the assembly of prototypes, products and systems suitable for testing.
- Conduct tests, document test results and develop client presentations.
- Lead the design of proper testing procedures.
- Evaluate progress in design, development and testing.
- Modify methods and designs where necessary.
- Ensure timely corrective actions on all assigned defects and issues.
- Lead and/or participate in technical audits and in reviews of requirements, specifications, designs, code and other artefacts as needed.
- Ensure assigned project commitments are agreed, reviewed and met.
- Produce and iterate the development plan that best meet the project objectives and deliverables in a timely fashion.
- Present solutions effectively to clients or other technical and management audiences.

- Leads the contribution to design standards and support design reuse (best practices, etc.).
- Maintain familiarity with company technology, organization and business.

Skills and competencies

- Knowledge of machining.
- Knowledge of electrical circuits.

Education

- BS in Mechanical Engineering Argyle and Sutherland University 2001.

Federated Aerospace comments

- Joined 2008.

5.20 MARK TIME

Position: Principal Systems Engineer
  Duties:

- Direct the writing of technical procedures to cover staging of equipment through final test for implementation at the customer, including basic software and configuration development, installation, connection to the customer's infrastructure, final software update and account activation.
- Act as Federated Aerospace's engineering/staging lead and technical problem-solver during staging and implementation of the system at the customer site.

Skills and competencies

- Experienced in the use of systems engineering process/tools across the full systems engineering lifecycle.
- Ability to communicate effectively with Federated Aerospace's systems engineering and test and release staff.
- Previous experience of developing IT or software-based systems in a deployed-on-land environment.

Education

- BSEE – Hypothetical University, 2003.
- Federated Aerospace short course on Advanced Systems Engineering, 2005.
- Master of Science in Engineering Management, 2007.
- Federated Aerospace course on Interpersonal Skills Development, 2008.
- Federated Aerospace course on Project Management, 2009.

- Prof Kasser's Requirements Workshop 2006.
- Chartered Engineer.
- Member of the IET.
- Member of INCOSE.

Federated Aerospace comments

- Joined 2008.

5.21 ANTHONY ZIMBALIST

Position: Test engineer
Duties:

- Design of test strategies.
- Test planning and execution.
- Design and construction of test configurations.
- Requirements testing.
- Regression testing.
- Integration testing.
- Verification of defect fixes.

Skills and competencies

- Good communication and team skills.
- Able to work with developers.
- System testing experience.
- Experience of testing complex products.

Education

- Bachelor of Technology, Redbrick University, 2003.
- Member of ITEA.

Federated Aerospace comments

- Joined 2004.

Author Index

A

Ackoff, R. L., 12, 13, 50, 56, 59, 66, 94
Adams, S., 100, 113
Addison, H. J., 13, 66, 94
Allison, J. T., 79, 80, 94
Andrew, C., 13, 50
Aplin, D. G., 357, 365
Arnold, E., 100, 113
Arthur, L. J., 139, 193
Atkinson, R., 10, 22, 50
Augustine, N. R., 27, 28, 50, 107, 113, 213, 227, 354, 364
Austin, N., 10, 52, 19, 346, 351
Avison, D., 89, 94, 149, 193

B

Barry, K., 69, 94
Barry, M., 90, 94
Beckman, S. L., 90, 94
Benjamin, D., 275, 278
Bittel, L. R., 330, 332
Boisjoly, R. P., 355, 364
Bonen, Z., 272, 279
Boyd, J., 136, 193
Brooks, F., 214, 227, 253, 266, 285, 332
Brown, T., 90, 91, 94
Burns, R., 257, 258, 266

C

Carroll, L., 36, 50
Chacko, G. K., 153, 193, 329, 330, 332, 341, 342, 351
Chakrabarti, A. K., 112, 113
Champy, J., 336, 338, 351
CHAOS, 8, 283, 285, 332
Checkland, P., 89, 94
Churchman, C. W., 13, 50, 153, 193
Clark, W., 1, 8, 118, 193
Cleland, D. I., 152, 193
Collins, H., 91, 94
Colwell, B., 80, 94
Covey, S. R., 21, 50, 255, 266, 337, 340, 351, 354, 365
Crépin, M., 274, 278
Crosby, P. B., 43, 50, 239, 248
Cross, N., 90, 94
Crowther, S., 336, 351
Curtis, E. F., 355, 364

D

DeMarco, T., 200, 210
Deming, W. E., 1, 8, 49, 50, 101, 106, 113, 248, 337, 346, 351
Denzler, D. W. R., 238, 248
DERA, 139, 141, 193
Descartes, R., 14, 50
Devine, T. M., 357, 365
DOD, 270, 278
Domb, E., 69, 94
Douty, H. M., 121, 194
Drucker, P. F., 1, 8, 47, 50, 329, 332

E

Edward, K. A., 102, 113
Eichhorn, R., 11, 50
Elder, L., 11, 52
El-Khoury, B., 274, 275, 278
ElMaraghy, H., 79, 80, 94
ElMaraghy, W., 79, 80, 94
ETA, 21, 22, 50
Ettorre, B., 357, 365
Exodus, 101, 113

F

Facione, P., 11, 50, 51
Farnham, D. T., 1, 8
Farson, R. E., 41, 52
Fayol, H., 17, 51, 98, 100, 113
Fazar, W., 121, 194
Felsing, J. M., 300, 333
Fischer, A., 74, 79, 94
Fisher, G. L., 157, 193
Fitzgerald, G., 89, 94, 149, 193
Flood, R. L., 50
Ford, H., 336, 351
Ford, W., 61, 94
Foreman, R., 10, 19, 52
Frank, M., 21, 51
Funke, J., 74, 79, 94

G

Gallagher, B., 23, 51
Gharajedaghi, J., 12, 51
Goldratt, E. M., 219, 227
Goleman, D., 21, 51
Gomez, E., 79, 95

Gordon, G., 157, 193
Gray, C. F., 104, 113, 132, 151
Gray, C. J., 18, 22, 53
Greiff, S., 74, 79, 94

H

Hague, R., 43, 51
Hammer, M., 336, 338, 351
Hari, A., 233, 248
Harrington, H. J., 1, 8, 106, 113, 336, 337, 351
Harrop, M., 43, 51
Hays, S. W., 356, 365
Herzberg, F., 25, 51
Hitchins, D. K., 18, 47, 51, 72, 82, 84, 86, 87, 94, 95, 285, 332
Howard, R. A., 63, 94
Howe, R. J., 197, 210
Hubbard, D., 267, 278
Huitt, W., 10, 51
Huynh, T., 18, 51

I

Irwin, B., 44, 51

J

Jackson, M. C., 50, 79, 80, 94
Jain, R. K., 111, 113
Jonassen, D. H., 71, 95
Jones, M. R., 22, 51
Jordan, R. H., 121, 194
Jos, P. H., 356, 365
Juran, J. M., 1, 101, 106, 113

K

Kast, F. F., 25, 26, 51
Kenley, C. R., 274, 278
Keys, P., 79, 80, 94
Kezsbom, D. S., 102, 113
Kilmann, R. H., 341, 351
Kintsch, W., 79, 95
Kipling, J. R., 39, 51
Kuhn, T. S., 93, 95

L

Lano, R, 121, 193
Larson, E. W., 104, 113, 132, 151
Lawler III, E., 26, 51
Leifer, L. J., 90, 95
Levinson, J. C., 23, 51
Leviticus, 353, 362, 365
Lewicki, R. J., 338, 351
Lister, T., 200, 210

Litterer, J. A., 338, 351
Lukas, J. A., 299, 333
Lutz, S., 10, 51
Luzatto, M. C., 13, 52

M

MacEachron, A. E., 157, 193
McGregor, D., 23, 52
Machiavelli, N., 43, 52
Mackley, T., 14, 51
Mali, P., 48, 52, 104, 113, 242, 249, 329, 333
Malotaux, N., 105, 113, 145, 151, 193, 202, 210, 212, 227, 336, 351
Mankins, J. C., 273, 279
Manvel, A. D., 121, 194
Martin, D., 355, 356, 365
Maslow, A. H., 19, 24, 25, 52
Mellor, S. J., 285, 333
Merton, R. K., 26, 52, 66, 67, 95
Michaels, J. V., 233, 249
Miles Jnr, R. F., 8
Miller, G., 14, 30, 52, 119, 121, 125, 126, 138, 193, 270, 279
Mills, H. R., 27, 30, 52
Monostori, L., 79, 80, 94
Morgan, G., 13, 52

N

Needham, J., 82, 95
Needham, M., 253, 266

O

OAS, 236, 249
O'Keefe, R. D., 112, 113
O'Reilly, J., 150, 266
Osborn, A. F., 10, 52, 120, 194
Ouchi, W. G., 24, 52
Overbaugh, R. C., 156, 194

P

Palmer, K., 80, 95
Palmer, S. R., 300, 333
Paul, R., 11, 52
Peters, T., 10, 19, 52, 346, 351
Pinto, J. K., 43, 45, 52
PMI, 9, 47, 52, 102, 104, 113
Prince, H., 34, 52
Project Smart, 348, 351

Q

Quesada, J., 79, 95

Author Index

R

Ramo, S., 1, 8
Reardon, K., 44, 52
Recklies, D., 348, 353
Richmond, B., 13, 52
Rittel, H. W., 71, 95
Rodgers, T. J., 10, 19, 52
Rodkinson, M. L., 353, 365
Rogers, C. R., 41, 52
Rosenzweig, J. E., 25, 26, 51
Royce, W. W., 48, 52, 133, 159, 194.242, 249
Rumsfeld, D., 83, 96
Russo, J. E., 10, 52, 69, 73, 83, 96

S

Santayana, G., 242, 249
Savage, S. L., 59, 69, 96
Schermerhorn, R., 140, 193
Schilling, D. L., 102, 113
Schoemaker, P. H., 10, 52, 69, 73, 83, 96
Schön, D. A., 74, 96
Schultz, L., 156, 194
Schunk, D. H., 20, 52
Senge, P. M., 12, 13
Sen, S., 275, 279
Serrat, O., 40, 52
Shaw, G. B., 36, 52
Shell, R. G., 347, 351
Shenhar, A. J., 272, 279
Sherwood, D., 13, 53
Shiffrin, R., 10, 22, 50
Shoshany, S., 36, 51
Shoval, S., 233, 248
Sillitto, H., 80, 96
Simon, H. A., 71, 83, 91, 96
Skinner, B. F., 25, 53
Slocum M. S., 69, 94
SOH, 105, 113
Stauber, B. R., 121, 194
Steinert, M., 90, 95

T

TAF, 356, 365
Taylor, W., 10, 19, 52
Thomas, K. W., 341, 351
Tittle, P., 11, 53
Tomiyama, T., 79, 80, 94
Tompkins, M. E., 356, 365
Tran, X.-L., 252, 255, 266
Triandis, H. C., 111, 113
Tuckman, B. W., 197, 210

V

VOYAGES, 1, 8, 106, 113
Vroom, V., 26, 53

W

Ward, P. T., 285, 333
Waring, A., 55, 56, 96
Waterman, H. V., 10, 19, 52
Webber, M. M., 71, 95
Weinfeld, W., 121, 194
Weiss, M. P., 233, 248
Wilson, O. R., 23, 51
Wilson, T. D., 82, 96
Wolcott, S. K., 18, 22, 53
Wood, W. P., 233, 249
Woolfolk, A. E., 10, 20, 53
Wu, W., 10, 53

Y

Yen, D. H., 12, 53, 92, 96
Yoshikawa, H., 10, 53

Z

Zhao, Y.-Y., 56, 87, 74, 81, 85–87, 89, 90, 91, 93, 95, 96, 147, 193, 261, 266
Zonnenshain, A., 233, 248

Subject Index

A

AC, *see* Actual cost (AC)
Acceptable solution, 57, 70, 71, 82
Acceptance criteria, 127, 129, 144, 145, 189, 307
Active brainstorming, 42, 65, 69, 89
Actual cost (AC), 3, 6, 127, 230, 264, 265, 274, 281, 289
Actual cost of work performed (ACWP), 3, 289, 296
ACWP, *see* Actual cost of work performed (ACWP)
Analysis, 1, 3, 7, 14, 15, 19, 22, 42, 63, 66, 84, 89, 116, 133, 236, 238, 270, 271, 277, 282, 284, 286, 287, 289, 364, 375, 376, 378, 381, 382, 384, 386, 390
Analysis of alternatives (AoA), 3, 236–238
Analysis paralysis, 19
Anchor point, 13, 14
Annotated outline, 131, 143
AoA, *see* Analysis of alternatives (AoA)
As-is, 77
Attribute profile, 128, 270, 276, 302, 322

B

BAC, *see* Budget at completion (BAC)
Bar chart, 118, 154, 245, 246, 298, 338
BCWP, *see* Budgeted cost of work performed (BCWP)
BCWS, *see* Budgeted cost of work scheduled (BCWS)
Beyond systems thinking, 14, 15, 92
BPR, *see* Business process re-engineering (BPR)
Budget, 27, 99, 102, 104, 107, 115, 127, 131, 139, 151, 154, 212, 231, 238, 242, 256, 262, 269, 283, 286–291, 293–295, 299, 304, 323, 325, 327, 369, 380
Budget at completion (BAC), 3, 127, 256, 288–296, 298
Budget cut, 139
Budgeted cost of work performed (BCWP), 3, 289
Budgeted cost of work scheduled (BCWS), 3, 289, 296
Business process re-engineering (BPR), 3, 77

C

CAIV, *see* Cost as an independent variable (CAIV)
Cataract methodology, 136, 138–141
Categorized requirements in process (CRIP) 2, 3, 7, 116, 138, 146, 269, 276, 282, 283, 296, 299–322, 325, 328, 332, 362–364, 367, 371
Causal loop, 5, 12, 13, 47, 55, 62, 94
CCB, *see* Change control board (CCB)
CDR, *see* Critical design review (CDR)
CDTC, *see* Conceptual design to cost (CDTC)
Change control board (CCB), 3, 140, 141, 349
Checkland's soft systems methodology, 89
CM, *see* Configuration management (CM)
CMM, *see* Competency maturity model (CMM)
COBOL, *see* Common Business-Oriented Language (COBOL)
Cognitive filter, 10
Commercial-off-the-shelf (COTS), 3, 65, 236, 240, 243, 252, 298
Common Business-Oriented Language (COBOL), 3, 201
Competency maturity model (CMM), 3, 11, 19–22, 196, 202
Complex
  complex problem, 1, 5, 55, 71, 75, 79, 84, 85, 88, 91–94, 390
  complex system, 79, 80, 82–84, 87
Complexity, 5, 17, 55, 78–84, 90, 94, 130, 201, 232, 241, 300, 301, 367
  artificial complexity, 11, 80–82
  objective complexity, 78, 80, 81, 84, 120
  real-world complexity, 80
  subjective complexity, 80–82, 84, 120, 126
Compliance matrix, 192, 209, 227, 248, 278, 364, 372
Complicated, 11, 14, 35, 80, 84, 124
Compound line and bar chart, 245, 298
Concept map, 117, 122
Concept of operations (CONOPS), 3, 127–129, 140, 241
Conceptual design to cost (CDTC), 3, 58, 232–238
Configuration management (CM), 3, 297, 305, 377, 378, 385, 386
Conflict, 7, 43, 46, 104, 217, 335, 339–341, 346, 347, 351
CONOPS, *see* Concept of operations (CONOPS)
Control chart, 338
Cost as an independent variable (CAIV), 3, 236, 238
Cost performance index (CPI), 3, 289, 290
Cost variance (CV), 3, 289–291, 295, 298, 299
COTS, *see* Commercial-off-the-Shelf (COTS)

397

CPI, *see* Cost performance index (CPI)
CRIP, *see* Categorized requirements in process (CRIP)
Critical chain, 211, 219–221
Critical design review (CDR), 3, 124, 134–136, 154, 155, 258, 296–299, 307, 310, 314–318, 321, 350
Critical path, 5, 116, 121–123, 125, 127, 147, 211, 214–219, 227, 236, 253, 254, 256, 271, 369
Critical thinking, 11, 14, 16, 18, 20, 22, 340
CV, *see* Cost variance (CV)

D

Decision frame, 10
Decision-making process, 19, 48, 72, 84, 86, 198, 343
Deductive reasoning, 68
Defence Evaluation and Research Agency (DERA), 3, 139
Delivery readiness review (DRR), 3, 134–136, 153–155, 189, 207, 296, 307, 311, 312, 317–319, 363, 364
Department of Defense (DOD), 3, 37, 238, 270, 271
DERA, *see* Defence Evaluation and Research Agency (DERA)
Design to cost (DTC), 3, 233, 235
Diminishing manufacturing sources and material shortages (DMSMS), 3, 274
Discrepancy report (DR), 3, 140
DMSMS, *see* Diminishing manufacturing sources and material shortages (DMSMS)
DOD, *see* Department of Defense (DOD)
Domain, 2, 12, 17, 21, 22, 68, 73, 89, 100, 201
  implementation domain, 20, 66, 72, 98, 147
  problem domain, 20, 66, 71, 72, 147
  research domain, 69
  solution domain, 20, 22, 66, 72, 147
Don't care, 64, 65, 341, 342
Do statement, 153, 329, 330
DR, *see* Discrepancy report (DR)
DRR, *see* Delivery readiness review (DRR)
DTC, *see* Design to cost (DTC)
dTRL, *see* Dynamic TRL (dTRL)
Dynamic TRL (dTRL), 3, 275, 322

E

EAC, *see* Estimate at completion (EAC)
Earned value (EV), 3, 288–291, 295, 298, 299
Earned value analysis (EVA) 3, 7, 116, 127, 128, 138, 274, 282, 288–300, 304, 314, 327, 329, 332, 335, 351, 364
Employment and training administration (ETA), 3, 7, 22

Enhanced traffic light (ETL), 2, 3, 7, 116, 138, 146, 271, 282, 283, 289, 297–299, 304, 322–329, 332, 362, 371
Errors of commission, 66
Errors of omission, 66, 131
Estimate at completion (EAC), 3, 289–291, 295, 298, 299
ETA, *see* Employment and training administration (ETA)
ETL, *see* Enhanced traffic light (ETL)
EV, *see* Earned value (EV)
EVA, *see* Earned value analysis (EVA)

F

FCFDS, *see* Feasible conceptual future desirable solution (FCFDS)
Feasible conceptual future desirable solution (FCFDS), 3, 61, 67, 69, 70, 71, 74–78, 91, 92, 142, 158, 340
Feature driven development, 300
Federated aerospace, 232, 237, 306, 307, 367, 373–392

G

Gantt chart, 1, 48, 70, 102, 116, 118–121, 124, 125, 128, 133, 138, 154, 189, 192, 211, 212, 221, 222, 226, 227, 258, 259, 299, 327, 330
Generic template, 137, 147
Golden rule, 353, 354, 355
Graph, 232, 245, 274, 275
Graphics, 29, 30, 32, 34, 156, 323

H

Histogram, 276, 302, 322
Hitchins–Kasser–Massie Framework (HKMF), 3, 82, 134, 136
HKMF, *see* Hitchins–Kasser–Massie Framework (HKMF)
Holistic thinking, 14, 15, 150, 157
Holistic thinking perspective
  Big Picture, 15–17, 43, 45, 55, 79, 97, 101, 109, 239, 340, 375
  Continuum, 15–19, 24, 33, 35, 41, 44, 55–57, 62, 63, 67, 70, 72, 73, 78–80, 84, 92, 99, 100, 101, 103, 104, 111, 136, 157, 240, 242, 253, 340, 341, 342, 343, 349, 353
  Functional, 10, 15–17, 55, 72, 87, 99, 101, 110, 133, 234, 242, 260, 341, 353
  Generic, 15–17, 21, 46, 65, 67, 69, 81, 91–93, 136, 157, 158, 212, 219, 231, 232, 234, 236, 269, 274, 275, 299, 322, 325, 326, 328, 329, 338, 347, 349

Subject Index

Operational, 15, 16, 43, 55, 73, 98, 101, 110, 234, 242, 350, 353
Quantitative, 15–17, 55, 59, 101, 239, 290, 341
Scientific, 15–17, 24, 55, 76, 77, 78, 81, 93, 110, 136, 198, 274, 275, 342, 345, 349
Structural, 15, 16, 55, 61, 101, 110
Temporal, 15–17, 44, 78, 81, 92, 101, 109, 119, 133, 211, 231, 232, 274, 323, 324, 362

I

Idea storage template, 15
Implementation domain, 20, 66, 72, 73, 98, 142, 147
Integrated customer driven conceptual design method, 233
Integration readiness review (IRR), 3, 134–136, 153, 155, 296, 298, 299, 307, 311, 312, 316, 317, 319, 320
IRR, *see* Integration readiness review (IRR)

J

Just-in-time, 28, 139, 219, 254

K

Kipling questions, 39, 241
Knowledge readings, 155

L

Leadership, 46, 100, 104, 107, 197, 198, 335, 341–346, 351, 380, 385, 386, 388
Lessons learned, 7, 92, 132, 133, 138, 157, 167, 170, 177, 179, 187, 188, 192, 205, 206, 208, 209, 240, 242, 248, 261, 278, 284, 353, 357, 358, 362, 364, 372
Lisa, 203–208, 223–245, 261–264

M

Management
 applied management, 98–101
 general management, 5, 97, 101, 113, 168, 219
 pure management, 20, 99–101
Management by exception (MBE), 3, 7, 116, 282, 324, 327, 330–332
Management by objectives (MBO), 4, 7, 48, 104, 116, 124, 125, 242, 282, 329–330, 332
Management by walking around (MBWA), 4, 282, 346
MBE, *see* Management by exception (MBE)
MBO, *see* Management by objectives (MBO)
MBUM, *see* Micromanagement by upper management (MBUM)

MBWA, *see* Management by walking around (MBWA)
MCSSRP, *see* Multi-satellite Communications Switching System Replacement Project (MCSSRP)
Micromanagement by upper management (MBUM), 4, 305, 324, 327, 331
Miller's rule, 14, 30, 119, 121, 125, 126, 138
Multi-satellite Communications Switching System Replacement Project (MCSSRP), 4, 7, 185, 189, 192, 203, 209, 226, 232, 247, 265, 278, 332, 363, 367, 368

N

National Aeronautics and Space Administration (NASA), xxxi, xxxii, 4, 122, 273
N^2 chart, 121, 133
Negotiation, 7, 17, 43, 45, 46, 195, 335, 338, 341, 346–348, 351
Non-systems approach, 123, 129

O

Observe-Orient-Decide-Act (OODA), 4, 136
OCR, *see* Operations concept review (OCR)
OODA, *see* Observe-Orient-Decide-Act (OODA)
Operations concept review (OCR), 4, 134–136, 144, 145, 154
Organizing, 15, 17, 30, 98, 100, 101, 292, 375

P

PBS, *see* Product breakdown structure (PBS)
PDR, *see* Presentation design review (PDR)
Presentation design review (PDR), 3, 124, 134, 135, 160, 176, 185, 223, 224, 296–299, 307, 309, 310, 311, 314, 315, 318–322, 350
Perspectives perimeter, 37, 301
PERT, *see* Program evaluation review technique (PERT)
PID, *see* Project initiation document (PID)
PIR, *see* Presentation integration review (PIR)
Planning
 contingency planning, 218, 242, 270
 decision-making process, 19, 48, 72, 84, 86, 198, 343
 generic planning, 142–145
 planning process, 5, 115–194, 219, 220, 242, 271, 282, 330
 prevention planning, 271
 product-based planning, 5, 115, 192
 project planning, 46, 48, 100, 115–194, 231, 276, 277, 282, 288
 social policy planning, 71
 specific planning, 147, 201

Subject Index

Planned value (PV), 3, 289–291, 295, 296, 298, 299
Politics, 2, 9, 42–46, 97, 103, 322, 327, 362
PPMR, *see* Presentation Post-Mortem Review (PPMR)
Presentation delivery readiness review, 160, 176
Presentation integration review (PIR), 3, 4, 160, 185, 206, 224
Presentation post-mortem review (PPMR), 3, 160, 185, 190, 207
Presentation requirements review (PRR), 3, 160, 167, 176, 204, 223
Prevention, xxvii, 2, 5, 6, 9, 46, 49, 115, 155, 192, 230, 232, 239, 242, 267, 271, 277, 279, 283, 285
Prioritization, 234, 305, 335
Problem
　complex problem, 1, 5, 55, 71, 75, 79, 84, 85, 88, 91–94, 390
　components of a problem, 59
　ill-structured problem, 71, 82, 84, 89, 91, 92, 350
　intervention problem, 59, 62, 67, 69, 86, 87, 89
　open-ended problems, 18
　problem classification framework, 84
　problem domain, 20, 66, 71, 72, 147
　problem formulation template, 67, 69, 70, 76, 78, 116, 142, 158, 189, 192, 209, 226, 247, 248, 278, 364, 371
　problem identification, 20
　problem language, 285
　problem-solving, 1, 2, 3, 5, 10, 12, 14, 16, 17, 20, 42, 47, 48, 55–94, 97, 109, 113, 120, 129, 134–136, 157, 196, 340, 357, 378, 380, 381, 386, 391
　problem space, 58, 60, 74, 75, 86
　research problem, 59, 62, 67, 69, 86, 88, 92
　well-structured problem, 5, 56, 57, 70–71, 83, 84, 86, 87, 92, 349, 350
　wicked problem, 71, 82–84, 91, 93
Process
　change control process, 109, 242, 286, 287
　change management process, 142, 283
　change request process, 283, 286
　decision-making process, 19, 48, 72, 84, 86, 198, 343
　planning process, 5, 115, 120, 123–125, 142, 147–152, 155, 192, 201, 219, 220, 242, 271, 282
　problem-solving process, 5, 12, 14, 42, 47, 48, 55–60, 62, 67, 69, 70–78, 84–94, 120, 129, 134, 136, 340, 357
　process improvement, 338
　system development process, 14, 85, 100, 130, 234, 256, 271, 281
　transition process, 69, 70, 76

Product-activity-milestone (PAM) chart, 3, 116–118, 122, 143, 161, 330
Product breakdown structure (PBS), 127, 152
Program evaluation review technique (PERT), 4, 70, 104, 116, 117, 119, 121–127, 154, 168, 176, 185, 215, 216, 218, 222–224, 227, 330
Project
　project manager, 2, 4, 6, 18, 20, 22, 28, 45, 46, 72, 76, 98, 116, 137, 195, 229, 235, 281, 368
　project plan, 9, 46, 70, 115–117, 120, 125, 130, 131, 142, 148, 152–155, 158, 277, 282, 287, 315
Project initiation document (PID), 4, 130, 131
PRR, *see* Presentation requirements review (PRR)
PV, *see* Planned value (PV)

Q

Q, *see* Quality leader (Q), 4, 108
Quadruple constraints, xxvii, 106, 107, 146, 255, 286, 300
Quality Leader (Q), 4, 108

R

Request for proposal (RFP), 4, 306, 307
Requirements traceability matrix (RTM), 4, 161, 165, 167–169, 172, 177, 178, 182, 186, 188, 189, 204–208, 262, 375
Return on investment (ROI), 4, 5, 49, 102, 104, 229, 231, 233, 236, 248
RFP, *see* Request for proposal (RFP)
Risk
　risk analysis, 170, 171, 270, 277
　risk-indicator, 105, 106, 284–286, 299, 346
　risk management, 6, 49, 217, 219, 230, 240, 242, 267, 269–271, 277, 279, 297, 300, 322, 326, 327, 329, 358, 377
　risk mitigation, 128, 129, 147, 154, 271, 273, 277, 278
　risk prevention, 6, 230, 267, 277, 279
　risk profile, 267, 270, 276, 278
　risk rectangle, 6, 268–270, 278
ROI, *see* Return on investment (ROI)
RTM, *see* Requirements traceability matrix (RTM)

S

Satisfice, 281
Satisfy, 26, 37, 70, 93, 104, 108, 350, 354
Schedule performance index (SPI), 4, 289, 290
Schedule variance (SV), 4, 283, 289–291, 295, 297, 298

Subject Index

Scientific method, 67, 68, 88, 93
Sensitivity analysis, 286, 287
Session Presentation Planning Completed Review (SPPCR), 4, 160, 164, 204, 261
Simpson's paradox, 59, 69
Soft systems methodology (SSM), 89
Solution domain, 20, 22, 66, 72, 147
SPI, *see* Schedule performance index (SPI)
SPPCR, *see* Session Presentation planning completed review (SPPCR)
SRR, *see* Systems requirements review (SRR)
SSM, *see* Soft systems methodology (SSM)
Stakeholder, 46, 86, 104, 132, 348
    stakeholder consensus, 144
    stakeholder management, 43, 348, 349
Stay calm, Think, Ask questions, Listen, Listen (STALL), 4, 41, 340
STALL, *see* Stay calm, Think, Ask questions, Listen, Listen (STALL)
Subsystem test readiness review (TRR), 4, 135, 136, 153–155, 307, 310, 311, 313, 315, 316, 318, 319
Susan, 203–208, 224, 245, 255, 261–265, 387
SV, *see* Schedule variance (SV)
Synthesis, 14, 85, 386
Systematic thinking, 12, 14
Systemic thinking, 12, 13, 14
Systems approach to
    budget cuts, 256
    change management, 286
    design, 90
    problem-solving, 2
    project organization, 107
    project planning, 105, 125, 142, 148, 152–155, 159
    project management, 101, 121, 127, 129, 200, 220, 288, 299, 322, 330, 379, 381
    managing complexity, 5, 94
    risk management, 6, 269, 270, 271, 277
    scheduling, 212, 221
    staffing a team, 195, 200–203
Systems requirements review (SRR), 4, 134–136, 139, 140, 142, 144, 145, 154, 265, 278, 296–298, 301, 306–309, 313, 318–320, 329, 332
Systems thinking, xxvii, 2, 12–15, 20, 69, 72, 92, 124, 253, 340, 353

T

Task leader (TL), 4, 108, 109, 198, 199, 203, 364
TAWOO, *see* Technology Availability Window of Opportunity (TAWOO)
Teams, 197–203
    costing teams, 238
    cross-functional teams, 199, 377
    decision-making team, 73
    design teams, 72, 140, 234, 298
    effective teams, 198
    high-performance teams, 5, 195, 210
    lifecycle of teams, 197–198
    new product development teams, 90, 141, 233, 238, 387
    planning teams, 142
    project teams, 5, 18, 22, 43, 46, 98, 99, 104, 107, 132, 152, 195, 213, 234, 306, 307, 376, 377, 389
    proposal team, 306
    quality assurance team, 285
    test teams, 123, 141
The technology availability window of opportunity (TAWOO), 4, 112, 273–275, 278
Technology readiness level (TRL), 4, 88, 273–275
Template
    problem formulation template, 67, 69, 70, 76, 78, 116, 142, 158, 189, 192, 209, 226, 247, 248, 278, 364, 371
    template for a document, 168
    template for a management review presentation, 137, 192, 209, 226, 247, 278, 371
    template for a presentation, 167
    template for a project plan, 145
    template for a requirements traceability matrix, 167
    template for a student exercise presentation, 209, 226, 247, 265, 278, 332, 350, 363
    template for a system development process, 329
    template for a timeline, 143
    template for a work package, 143, 161, 368
    template for MBO, 242
Test and evaluation, 49, 87
Thank You, 26, 336, 353
Thinking
    systematic, 12, 14
    systemic, 12, 13, 14
    systems, xxvii, 2, 12–15, 20, 69, 72, 92, 124, 253, 340, 353
Three streams of activities, 48, 49, 85, 124, 125, 128, 129, 142, 143, 147, 153, 158, 200, 201
Timeline, 6, 48, 102, 104, 119–121, 123–125, 127, 143, 146, 155, 202, 229, 238, 258–260, 274, 281, 293, 376, 382
TL, *see* Task Leader (TL)
Top-down, 14
Traffic light chart, 122, 323, 324
Trend chart, 332
Triple constraints, xxvii, 106–107, 286
Theory of Inventive Problem-Solving (TRIZ), 4, 69

TRIZ, *see* Theory of Inventive Problem-Solving (TRIZ)
TRL, *see* Technology readiness level (TRL)
TRR, *see* Subsystem test readiness review (TRR)

V

Variance at completion (VAC), 4, 289–291, 295, 298, 299

W

Waterfall chart, 48, 59, 115, 133, 135, 136, 138, 275, 329, 330

WBS, *see* Work breakdown structure (WBS)
What if, 65, 125, 238
Widget, 257–261
Willing to pay, 233–235, 238, 244
Willingness to pay, 234, 244
Work breakdown structure (WBS), 4, 5, 115, 116, 127, 150–153, 161, 167, 168, 176, 185, 189, 192, 270, 299
Working backwards, 125, 143, 147, 282
Work package, 4, 5, 49, 108, 115, 198, 211, 232, 252, 271, 282, 335, 368, 371